FRACTURE MECHANICS CRITERIA AND APPLICATIONS

ENGINEERING APPLICATION OF FRACTURE MECHANICS

Editor-in-Chief: George C. Sih

VOLUME 10

The titles published in this series are listed at the end of this volume.

Fracture Mechanics Criteria and Applications

E.E. GDOUTOS

Division of Engineering Mechanics
Department of Civil Engineering
Democritus University of Thrace, Xanthi, Greece

KLUWER ACADEMIC PUBLISHERS
DORDRECHT / BOSTON / LONDON

Library of Congress Cataloging-in-Publication Data

```
Gdoutos, E. E., 1948-
   Fracture mechanics criteria and application / E.E. Gdoutos.
      p.   cm. -- (Engineering application of fracture mechanics ;
   10)
   ISBN 0-7923-0605-8 (U.S. : alk. paper)
   1. Fracture mechanics.   I. Title.   II. Series.
TA409.G37   1990
620.1'126--dc20                                           90-4214
```

ISBN 0-7923-0605-8

Published by Kluwer Academic Publishers,
P.O. Box 17, 3300 AA Dordrecht, The Netherlands

Kluwer Academic Publishers incorporates the publishing programmes of
D. Reidel, Martinus Nijhoff, Dr W. Junk, and MTP Press.

Sold and distributed in the U.S.A. and Canada
by Kluwer Academic Publishers,
101 Philip Drive, Norwell, MA 02061, U.S.A.

In all other countries, sold and distributed
by Kluwer Academic Publishers Group,
P.O. Box 322, 3300 AH Dordrecht, The Netherlands

printed on acid free paper

All Rights Reserved
© 1990 Kluwer Academic Publishers
No part of the material protected by this copyright notice may be reproduced or
utilized in any form or by any means, electronic or mechanical,
including photocopying, recording or by any information storage and
retrieval system, without written permission from the copyright owner.

Printed in The Netherlands

Contents

Editor's Preface		ix
Preface		xi
1. Introductory chapter		1
1.1.	Conventional failure criteria	1
1.2.	Characteristic brittle failures	3
1.3.	Griffith's work	5
1.4.	Fracture mechanics	10
	References	13
2. Linear elastic stress field in cracked bodies		15
2.1.	Introduction	15
2.2.	Crack deformation modes and basic concepts	16
2.3.	Eigenfunction expansion method for a semi-infinite crack	18
2.4.	Westergaard method	23
2.5.	Singular stress and displacement fields	29
2.6.	Method of complex potentials	36
2.7.	Numerical methods	41
2.8.	Experimental methods	55
2.9.	Three-dimensional crack problems	61
2.10.	Cracks in bending plates and shells	66
	References	71
3. Elastic–plastic stress field in cracked bodies		76
3.1.	Introduction	76
3.2.	Approximate determination of the crack-tip plastic zone	78
3.3.	Small-scale yielding solution for antiplane mode	84
3.4.	Complete solution for antiplane mode	92
3.5.	Irwin's model	93
3.6.	Dugdale's model	96
3.7.	Singular solution for a work-hardening material	100

v

3.8.	Numerical solutions	105
References		109

4. Crack growth based on energy balance — 112
4.1.	Introduction	112
4.2.	Energy balance during crack growth	113
4.3.	Griffith theory	116
4.4.	Graphical representation of the energy balance equation	122
4.5.	Equivalence between strain energy release rate and stress intensity factor	127
4.6.	Compliance	129
4.7.	Critical stress intensity factor fracture criterion	132
4.8.	Experimental determination of K_{Ic}	137
4.9.	Crack stability	142
4.10.	Crack growth resistance curve (R-curve) method	147
4.11.	Mixed-mode crack propagation	154
References		159

5. J-Integral and crack opening displacement fracture criteria — 162
5.1.	Introduction	162
5.2.	Path-independent integrals	163
5.3.	J-integral	164
5.4.	Relationship between the J-integral and potential energy	168
5.5.	J-integral fracture criterion	170
5.6.	Experimental determination of the J-integral	173
5.7.	Stable crack growth studied by the J-integral	180
5.8.	Mixed-mode crack growth	183
5.9.	Crack opening displacement (COD) fracture criterion	185
References		191

6. Strain energy density failure criterion — 195
6.1.	Introduction	195
6.2.	Volume strain energy density	197
6.3.	Basic hypotheses	201
6.4.	Two-dimensional linear elastic crack problems	203
6.5.	Uniaxial extension of an inclined crack	205
6.6.	Three-dimensional linear elastic crack problems	212
6.7.	Bending of cracked plates	216
6.8.	Ductile fracture	219
6.9.	Failure initiation in bodies without pre-existing cracks	223
6.10.	Other criteria based on energy density	225
References		226

7. Dynamic fracture — 230
7.1.	Introduction	230
7.2.	Mott's model	231

7.3.	Stress field around a rapidly propagating crack	234
7.4.	Strain energy release rate	239
7.5.	Transient response of cracks to impact loads	241
7.6.	Standing plane waves interacting with a crack	244
7.7.	Crack branching	247
7.8.	Crack arrest	249
7.9.	Experimental determination of crack velocity and dynamic stress intensity factor	250
References		252

8. Fatigue and environment-assisted fracture — 255

8.1.	Introduction	255
8.2.	Fatigue crack propagation laws	257
8.3.	Fatigue life calculations	261
8.4.	Variable amplitude loading	262
8.5.	Mixed-mode fatigue crack propagation	265
8.6.	Nonlinear fatigue analysis based on the strain energy density theory	270
8.7.	Environment-assisted fracture	272
References		275

9. Engineering applications — 278

9.1.	Introduction	278
9.2.	Fracture mechanics design philosophy	279
9.3.	Design example problems	281
9.4.	Fiber-reinforced composites	290
9.5.	Concrete	295
9.6.	Crack detection methods	301
References		303

Author Index — 307

Subject Index — 311

Editor's preface

It is difficult to do justice to fracture mechanics in a textbook, for the subject encompasses so many disciplines. A general survey of the field would serve no purpose other than give a collection of references. The present book by Professor E. E. Gdoutos is refreshing because it does not fall into the esoteric tradition of outlining equations and results. Basic ideas and underlying principles are clearly explained as to how they are used in application. The presentations are concise and each topic can be understood by advanced undergraduates in material science and continuum mechanics. The book is highly recommended not only as a text in fracture mechanics but also as a reference to those interested in the general aspects of failure analysis.

In addition to providing an in-depth review of the analytical methods for evaluating the fundamental quantities used in linear elastic fracture mechanics, various criteria are discussed reflecting their limitations and applications. Particular emphases are given to predicting crack initiation, subcritical growth and the onset of rapid fracture from a single criterion. Those models in which it is assumed that the crack extends from tip to tip rely on the *specific* surface energy concept. The differences in the global and energy states before and after crack extension were associated with the energy required to create a unit area of crack surface. Applications were limited by the requirement of self-similar crack growth. Quantities such as energy release rate, stress intensity factor, path independent integral, etc. fall into this category. An alternative view based on physical observation is that crack growth is not continuous but is a discrete process of failure initiation, repeated many times, at a finite distance from the crack tip. Attention would then be focused on the energy stored in a unit volume of material reaching a critical state. Crack growth is considered as the loci of fracture initiation sites. Consistency can thus be achieved not only in assessing initiation, slow growth and rapid fracture but also in the simultaneous description of yielding and fracture. Macroplasticity off to the side of a macrocrack extension is distinguished from the microplasticity that may prevail ahead of a growing macrocrack. By the same token, microcracks may occur in regions of macroplasticity. The time and location at which energy is dissipated to deform and fracture at a specified size scale of material need to be identified. The book gives a comprehensive explanation of how the stationary values of the strain energy density function can be applied to locate the sites of yield and fracture initiation. Yielding and fracture at the macroscopic level are shown to occur at different locations. The former always precedes the latter. Many illustrative examples are presented and critiqued against the other criteria. Fatigue crack growth can be treated in the same way by considering the cumulation of the strain on volume energy density. When the disturbance becomes time dependent, inertia effects of material elements around the crack come into play. This can occur regardless of whether the crack is stationary or moving. Further material on dynamic fracture can be found in the references.

The important contribution on applying fracture mechanics principles is not the number of examples but the methodology for resolving the loading rate and specimen size effect. Straight line relations obtained from the constant rate change of strain energy density factor with crack growth permit linear interpolation of the results. Data on crack growth can thus be obtained with a limited number of experiments and can be used to predict situations other than those tested. A realistic application of fracture mechanics could not be made without a sound understanding of the fundamentals. To this end, the book has met the objective.

Bethlehem, Pennsylvania G. C. SIH
December, 1989

Preface

The objective of engineering design is the determination of the geometry and dimensions of machine or structural elements and the selection of material in such a way that the elements perform their operating function in an efficient, safe and economic manner. For this reason the results of stress and displacement analysis are coupled with an appropriate failure criterion, which is basically a postulate predicting the event of failure itself. Traditional failure criteria cannot adequately explain the number of structural failures that occur at stress levels considerably lower than the ultimate strength of the material. Example problems include bridges, tanks, pipes, weapons, ships, railways and aerospace structures. On the other hand, experiments performed by Griffith in 1921 on glass fibers led to the conclusion that the strength of real materials is much smaller, typically by two orders of magnitude, than their theoretical strength. In an effort to explain these phenomena the discipline of fracture mechanics has been created. It is based on the realistic assumption that all materials contain crack-like defects which constitute the nuclei of failure initiation.

Defects can appear in a structure in three major different ways: first, they can exist in a material due to its composition, as second-phase particles, debonds in composites, etc.; second, they can be introduced into a structure during fabrication, as in welds; and, third, they can be created during the service life of a component, like fatigue, environment-assisted or creep cracks. Fracture mechanics studies the load-bearing capacity of structures in the presence of initial defects, where a dominant crack is usually assumed to exist.

A new design philosophy is therefore introduced by fracture mechanics as opposed to the use of the conventional failure criteria. As catastrophic fracture is a consequence of the unstable propagation of a crack from a pre-existing defect, we are faced with the question: 'Can fracture be prevented by constructing structures that have no defects?' The answer is 'no', on the grounds of practicality. Then, the safe design of structures should proceed along two lines: either the *safe operating load* should be determined when a crack of a prescribed size is assumed to exist in the structure; or, given the operating load, the *size of the crack that is created in the structure* should be determined.

Fracture mechanics is searching for parameters which characterize the propensity of a crack to extend. Such a parameter should be able to relate laboratory test results to structural performance, so that the response of a structure with cracks can be predicted from laboratory test data. This is determined as a function of material behavior, crack size, structural geometry and loading conditions. On the other hand, the critical value of this parameter – known as *fracture toughness*, a property of the material – is determined from laboratory tests. Fracture toughness expresses the ability of the material to resist fracture in the presence of cracks. By equating this parameter to its critical value a relation is obtained between applied load, crack and structure geometry which gives the necessary information for structural design. Fracture toughness is used to rank a material's ability to resist fracture within the framework of fracture mechanics, in the same way that yield or ultimate strength is used to rank a material's resistance to yield or fracture in the conventional design criteria. In selecting materials for structural applications a choice has to be made between materials with a high yield strength but comparatively low fracture toughness or a lower yield strength but higher fracture toughness.

The phenomenon of fracture of a solid is complicated and depends on a wide variety of factors, including the macroscopic effects, the microscopic phenomena which take place at the locations where the fracture nucleates or grows, and the composition of the material. The study of the fracture process depends on the scale level at which it is considered. At one extreme is the rupture of cohesive bonds in the solid, and the associated phenomena take place within distances of the order of 10^{-7} cm. For such studies the principles of quantum mechanics should be used. At the other extreme the material is considered as a homogeneous continuum and the phenomenon of fracture is studied within the framework of continuum mechanics and classical thermodynamics. Fracture studies that take place at scale levels between these two extremes concern movement of dislocations, formation of subgrain boundary precipitates and slip bands, grain inclusions and voids. Thus, the understanding of the phenomenon of fracture depends to a large extent on the successful integration of continuum mechanics with materials science, metallurgy, physics and chemistry. Due to the insurmountable difficulties encountered in an interdisciplinary approach the phenomenon of fracture is usually studied within only one of the three scale levels: namely, the atomic, the microscopic and the continuum. Attempts are under way to find a unified, interdisciplinary approach to the phenomenon of the failure of solids.

The purpose of this book is to present a clear, straightforward and unified interpretation of the basic problems of fracture mechanics with particular emphasis given to fracture mechanics criteria and their application in engineering design. The book is divided into nine chapters.

The first, introductory, chapter gives a brief account of some characteristic failures that could not be explained by the traditional failure criteria, and of Griffith's experiments which gave impetus to the development of a new philosophy in engineering design based on the discipline of fracture mechanics. The next two chapters deal with the determination of the stress and deformation fields in cracked bodies and provide the necessary prerequisite for the development of the criteria of fracture mechanics. More specifically, Chapter 2 covers the basic analytical, numerical and experimental methods for determining the linear elastic stress field in cracked bodies, with particular emphasis on the local behavior around the crack tip, and Chapter 3 is devoted to the determination of the elastic–plastic stress and displacement distribution around cracks for time-independent plasticity. Addressed in the fourth chapter is the theory of crack growth, based on the global energy balance of the entire system. The fifth chapter deals with the theoretical foundation of the path-independent J-integral and its use as a fracture criterion. Furthermore, a brief presentation of the crack opening displacement fracture criterion is given. Chapter 6 studies the underlying principles of the strain energy density theory and demonstrates its usefulness and versatility in solving a host of two- and three-dimensional problems of mixed-mode crack growth in brittle and ductile fracture. Chapter 7 presents in a concise form the basic concepts and the salient points of dynamic fracture mechanics. Addressed in Chapter 8 is the phenomenon of fatigue and environment-assisted crack growth which takes place within the framework of the macroscopic scale level. Finally, Chapter 9 presents the basic principles of engineering design based on the discipline of fracture mechanics and gives a number of example problems. The applicability of fracture mechanics to composites and concrete is also discussed. The chapter concludes with a brief description of the more widely used nondestructive testing methods for defect detection.

Particular care was taken throughout the book to give a clear, consistent, simple and straightforward presentation of the basic concepts of the discipline of fracture mechanics from a continuum mechanics viewpoint. The book is self-contained and can be used as a textbook in undergraduate and postgraduate courses and as a reference book by all those who are interested in developing design methodologies and promoting research to include the influence of initial defects or cracks.

The author wishes to express his gratitude to Professor G. C. Sih of Lehigh University for his pioneering work on fracture mechanics on which parts of the book were based, his very stimulating discussions and his comments and suggestions during the writing of the book. Thanks are also due to my secretary, Mrs L. Adamidou, for typing the manuscript. Finally, I wish to express my gratitude to my wife, Maria, not only for proofreading the manuscript, but for her understanding and patience during the writing of the book.

Xanthi, Greece, 1989 EMMANUEL E. GDOUTOS

Introductory chapter

1.1. Conventional failure criteria

The mechanical design of engineering structures usually involves an analysis of the stress and displacement fields in conjunction with a postulate predicting the event of failure itself. Sophisticated methods for determining stress distributions in loaded structures are available today. Detailed theoretical analyses based on simplifying assumptions regarding material behavior and structural geometry are undertaken to obtain an accurate knowledge of the stress state. For complicated structure or loading situations experimental or numerical methods are preferable. Having performed the stress analysis, a suitable failure criterion is selected for an assessment of the strength and integrity of the structural component.

Conventional failure criteria have been developed to explain strength failures of load-bearing structures which can be classified roughly as ductile at one extreme and brittle at another. In the first case, breakage of a structure is preceded by large deformation which occurs over a relatively long time period and may be associated with yielding or plastic flow. The brittle failure, on the other hand, is preceded by small deformation and is usually sudden. Defects play a major role in the mechanism of both these types of failure and those associated with ductile failure differ significantly from those influencing brittle fracture. For ductile failures, which are dominated by yielding before breakage, the important defects (dislocations, grain boundary spacings, interstitial and out-of-size substitutional atoms, precipitates) tend to distort and warp the crystal lattice planes. Brittle fracture, however, which takes place before any appreciable plastic flow occurs, initiates at larger defects such as inclusions, sharp notches, surface scratches or cracks.

Materials that fail in a ductile manner undergo yielding before they ultimately fracture. Postulates for determining those macroscopic stress combinations that result in initial yielding of ductile materials have been developed and are known as yield criteria. At this point it should become clear that a material may behave in a ductile or brittle manner, depending on the temperature, rate of loading and other variables present. Thus, when we speak about ductile or brittle materials we actually mean the ductile or brittle *state* of materials. Although the onset of yielding is influenced by factors such as temperature, time, size effects, there is a wide range of circumstances where yielding is mainly determined by the stress state itself. Under such conditions for isotropic materials there is extensive evidence that yielding is a result of distortion and is mainly influenced by shear stresses. Hydrostatic stress states, however, play a minor role in the initial yielding of metals. Following these reasonings the Tresca and von Mises yield criteria have been developed [1.1].

The Tresca criterion states that a material element under a multiaxial stress state enters a state of yielding when the maximum shear stress becomes equal to the critical shear stress in a pure shear test at the point of yielding. The latter is a material parameter. Mathematically speaking, this criterion is expressed by [1.2, 1.3]

$$\frac{|\sigma_1 - \sigma_3|}{2} = k, \quad \sigma_1 > \sigma_2 > \sigma_3, \tag{1.1}$$

where σ_1, σ_2, σ_3 are the principal stresses and k is the yield stress in a pure shear test.

The von Mises criterion is based on the distortional energy and states that a material element initially yields when it absorbs a critical amount of distortional strain energy which is equal to the distortional energy in uniaxial tension at the point of yield. The yield condition is written in the form [1.4, 1.5]

$$(\sigma_1 - \sigma_2)^2 + (\sigma_2 - \sigma_3)^2 + (\sigma_3 - \sigma_1)^2 = 2\sigma_y^2 \tag{1.2}$$

where σ_y is the yield stress in uniaxial tension.

However, for porous or granular materials as well as for some glassy polymers it was established that the yield condition is sensitive to hydrostatic stress states. For such materials the yield stress in simple tension is not equal in general to the yield stress in simple compression. A number of pressure-dependent yield criteria have been proposed in the literature.

On the other hand, brittle materials – or, more strictly, materials in the brittle state – fracture without appreciable plastic deformation. For such cases the maximum tensile stress and the Coulomb–Mohr [1.6, 1.7] criterion gained popularity. The latter criterion was mainly employed in rock and soil mechanics. The maximum tensile stress criterion assumes that rupture of a material occurs when the maximum tensile stress exceeds a specific stress which is a material parameter. The Coulomb–Mohr criterion states that fracture occurs when the shear stress τ on a given plane becomes equal to a critical value which depends on the normal stress σ on that plane. The fracture condition can be written as [1.6, 1.7]

$$|\tau| = F(\sigma), \tag{1.3}$$

where the curve $\tau = F(\sigma)$ on the σ–τ plane is determined experimentally and is considered as a material parameter.

The simplest form of the curve $\tau = F(\sigma)$ is the straight line, which is expressed by

$$\tau = c - \mu\sigma. \tag{1.4}$$

Under such conditions the Coulomb–Mohr fracture criterion is expressed by

$$\left(\frac{1+\sin\omega}{2c\cos\omega}\right)\sigma_1 - \left(\frac{1-\sin\omega}{2c\cos\omega}\right)\sigma_3 = 1, \tag{1.5}$$

where $\tan\omega = \mu$ and $\sigma_1 > \sigma_2 > \sigma_3$.

Equation (1.5) suggests that fracture is independent of the intermediate principal stress σ_2. Modifications to the Coulomb–Mohr criterion have been introduced to account for the influence of the intermediate principal stress on the fracture of pressure-dependent materials.

The above briefly outlined macroscopic failure criteria for describing the onset of yield in materials with ductile behavior or fracture in materials with brittle behavior have been used extensively in the design of engineering structures. In order to take into account uncertainties in the analysis of service loads, material or fabrication defects and high local or residual stresses, a safety factor is employed to limit the calculated critical equivalent yield or fracture stress to a portion of the nominal yield or fracture stress of the material. The latter quantities are determined experimentally. This procedure of design has been succesful for the majority of structures for many years.

However, it was early realized that there is a broad class of structures, especially those made of high-strength materials, whose failure could not be adequately explained by the conventional design criteria. On the other hand, Griffith [1.8, 1.9], from a series of experiments run on glass fibers, came to the conclusion that the strength of real materials is much smaller, typically by two orders of magnitude, than their theoretical strength. The theoretical strength is determined by the properties of a material's internal structure, and is defined as the highest stress level that the material can sustain. In the following two sections we shall give a brief account of some characteristic failures which could not be explained by the traditional failure criteria, and some of Griffith's experiments will be detailed. These were the major events that gave impetus to the development of a new philosophy in structural design based on the discipline of fracture mechanics.

1.2. Characteristic brittle failures

The phenomenon of brittle fracture is frequently encountered in many aspects of everyday life. It is involved, for example, in splitting logs with wedges, in the art of sculpture, in cleaving layers of mica, in machining materials and in many manufacturing and constructional processes. On the other hand, many catastrophic structural failures involving loss of life have occurred as a result of sudden, unexpected brittle fracture. The history of technology is full of such incidents. It is not the intent here to overwhelm the reader with the vast number of disasters involving failures of bridges, tanks, pipes, weapons, ships, railways and aerospace structures, but rather to present a few characteristic cases which substantially influenced the development of fracture mechanics.

Although brittle fractures have occurred in many structures over the centuries, the problem arose in acute form with the introduction of all-welded designs. In riveted structures, for example, fractures usually stopped at the riveted joints and did not propagate into adjoining plates. A welded structure, however, appears to be continuous and a crack growth may propagate from one plate to the next through the welds, resulting to global structural failure. Furthermore, welds may have defects of various kinds, including cracks, and usually introduce high-tensile residual stresses.

The most extensive and widely known massive failures are those that occurred in tankers and cargo ships that were built, mainly in the U.S.A., under the emergency shipbuilding programs of the Second World War [1.10–1.14]. Shortly after these ships were commissioned several serious fractures appeared in some of them. The fractures were usually sudden and were accompanied by a loud noise. Of approximately 5000 merchant ships built in U.S.A., more than one-fifth developed cracks before April 1946. Most of the ships were less than three years old. In the period between November 1942 and December 1952 more than 200 ships experienced serious failures. Ten tankers and three Liberty ships broke completely in two, while about 25 ships suffered complete fractures of the deck and bottom plating. The ships experienced more failures in heavy seas than in calm seas and a number of failures took place at stresses that were well below the yield stress of the material. A characteristic brittle fracture concerns the tanker *Schenectady*, which suddenly broke in two while in the harbor in cool weather after she had completed successful sea trials. The fracture occurred without warning, extended across the deck just aft of the bridge about midship, down both sides and around the bilges. It did not cross the bottom plating [1.15].

Extensive brittle fractures have also occurred in a variety of large steel structures. Shank [1.16], in a report published in 1954, covers over 60 major structural failures including bridges, pressure vessels, tanks and pipelines. Following Shank the earliest structural brittle failure on record is a riveted standpipe 250 ft high in Long Island that failed in 1886 during a hydrostatic acceptance test. After pumping water to a height of 227 ft, a 20 ft long vertical crack appeared in the bottom, accompanied by a sharp rending sound, and the tower collapsed. In 1938 a welded bridge of the Vierendeel truss type built across the Albert Canal in Belgium with a span of 245 ft collapsed into the canal in quite cold weather. Failure was accompanied by a sound like a shot and a crack appeared in the lower cord. The bridge was only about one year old. In 1940 two similar bridges over the Albert Canal suffered major structural failures. In 1962 the one-year-old King's Bridge in Melbourne, Australia, fractured after a span collapsed as a result of cracks that developed in a welded girder [1.17]. A spherical hydrogen welded tank of 38.5 ft diameter and 0.66 in. thickness in Schenectady, New York, failed in 1943 under an internal pressure of about 50 lb/in^2 and at ambient temperature of 10°F [1.16]. The tank burst catastrophically into 20 fragments with a total of 650 ft of herringboned brittle tears. Concerning early aircraft failures, two British de Havilland jet-propelled airliners known as Comets (the first jet airplane designed for commercial service) crashed near Elba and Naples in the Mediterranean in 1954 [1.18]. After these accidents, the entire fleet of these passenger aircraft was grounded. In order to shed light into the cause of

Introductory chapter 5

the accident a water tank was built at Farnborough into which was placed a complete Comet aircraft. The fuselage was subjected to a cyclic pressurization, and the wings to air loads that simulated the corresponding loads during flight. The plane tested had already flown for 3500 hours. After tests giving a total lifetime equivalent to about 2.25 times the former flying time, the fuselage burst in a catastrophic manner after a fatigue crack appeared at a rivet hole attaching reinforcement around the forward escape hatch. For a survey and analysis of extensive brittle failures the interested reader is referred to reference [1.19] for large rotating machinery, to [1.20] for pressure vessels and piping, to [1.21] for ordnance structures and to [1.22] for airflight vehicles.

From a comprehensive investigation and analysis of the above structural failures , the following general remarks can be drawn.

(a) Most fractures were mainly brittle in the sense that they were accompanied by very little plastic deformation, although they were made of materials with ductile behavior at ambient temperatures.

(b) Most brittle failures occurred in low temperatures.

(c) Usually, the nominal stress in the structure was well below the yield stress of the material at the moment of failure.

(d) Most failures originated from structural discontinuities including holes, notches, re-entrant corners, etc.

(e) The origin of most failures, excluding those due to poor design, was pre-existing defects and flaws, such as cracks, accidentally introduced into the structure. In many cases the flaws that triggered fracture were clearly identified.

(f) The structures that were susceptible to brittle fracture were mostly made of high-strength materials which have low notch or crack toughness (ability of the material to resist loads in the presence of notches or cracks).

(g) Fractures usually propagated at high speeds which, for steel structures, were in the order of 1000 m/s. The observed crack speeds were a fraction of the longitudinal sound waves in the medium.

These findings were very essential for the development of a new philosophy in structural design based on the discipline of fracture mechanics.

1.3. Griffith's work

Long before 1921, when Griffith published his monumental theory on the rupture of solids, a number of pioneering results had appeared which gave evidence of the existence of a size effect on the strength of solids. These findings, which could be considered as a prelude to the Griffith theory, will now be briefly described. Leonardo da Vinci (1452–1519) ran tests to determine the strength of iron wires [1.23]. He found an inverse relationship between the strength and the length for wires of constant diameter. We quote from an authoritative translation of da Vinci's sketch book, according to reference [1.24], the passage:

> Observe what the weight was that broke the wire, and in what part the wire

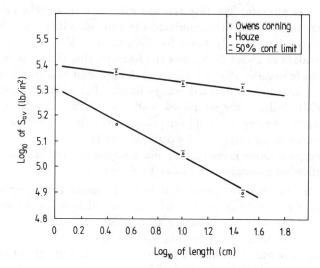

Figure 1.1. Logarithm of average tensile strength versus logarithm of specimen length for carefully protected glass fibers (x) and fibers damaged by rough handling (o) [1.29].

broke ... Then shorten this wire, at first by half, and see how much more weight it supports; and then make it one quarter of its original length, and so on, making various lengths and noting the weight that breaks each one and the place in which it breaks.

Todhunter and Pearson [1.25] refer to two experimental results analogous to those of da Vinci. According to [1.25], Lloyd (about 1830) found that the average strength of short iron bars is higher than that of long iron bars and Le Blanc (1839) established long iron wires to be weaker than short wires of the same diameter. Stanton and Batson [1.26] reported the results of tests conducted on notched-bar specimens at the National Physical Laboratory, Teddington, after the First World War. From a series of tests it was obtained that the work at fracture per unit volume was decreased as the specimen dimensions were increased. Analogous results were obtained by Docherty [1.27, 1.28] who found that the increase of the plastic work at fracture with the specimen size was smaller than that obtained from geometrical similarity of the strain patterns. This means that the specimens behaved in a more brittle fracture as their size was increased.

All these early results gave indication of the so-called size effect of the strength of solids, which is expressed by an increase in strength as the dimensions of the testpiece decrease. Newly derived results at the U.S. Naval Research Laboratory on the strength of glass fibers [1.29] corroborated the early findings of Leonardo da Vinci. Figure 1.1, taken from reference [1.29], shows a decrease of the logarithm of the average strength of glass as a function of the logarithm of the specimen length. The upper line refers to fibers for which precautions have been taken to prevent damage in handling. The lower line was obtained for fibers that came in a loose skein and had a number of flaws.

A plausible explanation of these results can be attributed to the fact that all structural materials contain flaws which have a deteriorating effect on the strength of the material. The larger the volume of the material tested, the higher the possibility that large cracks exist which, as will be seen, reduce the material strength in a square root relation to their dimensions. However, the first systematic study of the size effect was made by Griffith [1.8, 1.9] who, with his keystone ideas about the strength of solids, laid down the foundation of the contemporary theory of fracture.

Griffith was motivated in his work by the study of the effect of surface treatment on the strength of metallic machine parts. Early results by Kommers [1.30] indicated that the strength of polished specimens was about 45–50 percent higher than the strength of turned specimens. Other results indicated a decrease of the order of 20 percent. Furthermore, the strength was increased by decreasing the size of the scratches. Following the Inglis solution [1.31] of the stress field in an infinite plate weakened by an elliptic hole, Griffith observed that tensile stresses – higher than the applied stress by almost an order of magnitude – appeared near the holes according to their shape. Furthermore, he noticed that these maximum stresses were independent of the absolute size of the hole and depended only on the ratio of the axes of the elliptic hole. Indeed, according to [1.31] the maximum stress in the plate, $\sigma_{\max}$, occurs at the end point of the major axis of the ellipse and is given by

$$\sigma_{\max} = \sigma\left(1 + \frac{2a}{b}\right) \simeq 2\sigma\sqrt{\frac{a}{\rho}}, \qquad (1.6)$$

where σ is the applied stress at infinity in a direction normal to the major axis of the hole, $2a$ and $2b$ are the lengths of the major and minor axes of the ellipse and ρ is the radius of curvature at the ends of the major axis of the ellipse.

These results were in conflict with experiments. Indeed, first, the strength of scratched plates depends on the size and not only on the shape of the scratch, and, second, higher stresses could be sustained by a scratched plate than those observed in an ordinary tensile test. In experiments performed on cracked circular tubes made of glass, Griffith observed that the maximum tensile stress in the tube was of the magnitude of 344 kip*/in^2, while the tensile strength of glass was 24.9 kip/in^2. These results led him to raise the following questions (we quote from reference [1.8]):

> If the strength of this glass, as ordinarily interpreted, is not constant, on what does it depend? What is the greatest possible strength, and can this strength be made available for technical purposes by appropriate treatment of the material?

In order to explain these discrepancies Griffith attacked the problem of rupture of elastic solids from a different standpoint. He extended the theorem of minimum potential energy to enable it to be applied to the critical moment at which

---

* kip = 1000 lb.

rupture of the solid occurs. Thus, he considered the rupture position of the solid to be an equilibrium position. And, in applying the theorem he took into account the increase of potential energy due to the formation of new material surfaces in the interior of solids. Using the Inglis solution Griffith obtained the critical breaking stress of a cracked plate, which was found to be inversely proportional to the square root of the length of the crack. Thus, he resolved the paradox arising from the Inglis solution that the strength of the plate is independent of the size of the crack. Griffith corroborated his theoretical predictions by experiments performed on cracked spherical bulbs and circular tubes made of glass. The Griffith theory, and his accompanying experiments on cracked specimens, will be presented in detail in chapter four. Here, in respect of the size effect that was mentioned at the beginning of this section, some further experiments performed by Griffith on the strength of thin glass will be briefly described.

Glass fibers of various diameters were prepared and tested in tension until they broke. The fibers were drawn by hand as quickly as possible from glass bead heated to about 1400–1500°C. For a few seconds after preparation the strength of the fibers was found to be very high. Values of tensile strength in the range 220–900 kip/in^2 for fibers of about 0.02 in. diameter were observed. These values were obtained by bending the fibers to fracture and measuring the critical radius of curvature. It was found that the fibers remained almost perfectly elastic until breakage. The strength of the fibers decreased with time and after a few hours a steady state was reached where the strength of the fibers depended upon the diameter only. These fibers were then tested in order to obtain a relation between the strength and the diameter. The fiber diameter ranged from 0.13×10^{-3} to 4.2×10^{-3} in. and the fibers were left for about 40 hours before being tested. The specimens had a constant length of about 0.05 in. and were obtained after breaking the long fibers several times. Thus, the probability of material defects along the entire specimen length was low, and this was the same for all specimens. The results of the tests are shown in Table 1.1, taken from reference [1.8].

Note that the strength increases as the fiber diameter decreases. The strength tends to that of bulk glass for large thicknesses. The limit as the diameter decreases was obtained by Griffith by plotting the reciprocals of the strength and extrapolating to zero diameter. The maximum strength of glass was found to be 1600 kip/in^2 and this value agreed with that obtained from experiments on cracked plates in conjunction with the Griffith theory.

Analogous results on the maximum strength of other materials had been obtained long before Griffith's results. Based on the molecular theory of matter, it had been established that the tensile strength of an isotropic solid or liquid is of the same order as, and always less than, its intrinsic pressure. The latter quantity can be determined using the Van der Waals equation or by measuring the heat that is required to vaporize the substance. According to Griffith [1.8], Traube [1.32] gives values of the intrinsic pressures of various metals including nickel, iron, copper, silver, antimony, zinc, tin and lead which are from 20 to 100 times the tensile strength of the metals. Based on these results the conclusion was reached that the actual strength is always a small fraction of that estimated

Table 1.1 Strength of glass fibers according to Griffith's experiments

Diameter (10^{-3}in.)	Breaking stress (lb/in^2)	Diameter (10^{-3}in.)	Breaking stress (lb/in^2)
40.00	24 900	0.95	117 000
4.20	42 300	0.75	134 000
2.78	50 800	0.70	164 000
2.25	64 100	0.60	185 000
2.00	79 600	0.56	154 000
1.85	88 500	0.50	195 000
1.75	82 600	0.38	232 000
1.40	85 200	0.26	332 000
1.32	99 500	0.165	498 000
1.15	88 700	0.130	491 000

by molecular theory.

Long before Griffith established the dependence of the strength of glass fibers on the fiber diameter, Karmarsch [1.33] in 1858 gave the following expresion for the tensile strength of metal wires,

$$\sigma_{\max} = A + \frac{B}{d} \tag{1.7}$$

where d is the diameter of the wire and A and B are constants. Griffith's results of Table 1.1 can be represented by the expression

$$\sigma_{\max} = 22\,400 \frac{4.4 + d}{0.06 + d} \tag{1.8}$$

where $\sigma_{\max}$ is in lb/in^2 and d in thousandths of an inch. For the range of diameters available to Karmarsch, Equation (1.8) differs little from

$$\sigma_{\max} = 22\,400 + \frac{98600}{d} \tag{1.9}$$

which is of the same form as Equation (1.7).

Griffith's experiments on glass fibers established the 'size effect' in solids and gave an explanation of the paradox arising from Griffith that 'the maximum tensile stress in the corners of the crack is more than ten times as great as the tensile strength of the material, as measured in an ordinary test' [1.8]. The maximum tensile stress in a cracked plate was estimated from the Inglis solution by measuring the radius of curvature ρ at the ends of the crack. The latter quantity was measured by Griffith by inspection of the interference colors there. It was inferred that the width of the crack at its end is about one-quarter of the shortest wavelength of visible light. It was found that $\rho = 2 \times 10^{-6}$ in. Then Equation (1.6) gives $\sigma_{\max} = 350$ kip/in^2 which is almost one-fifth of the theoretical strength of glass. Thus, near very small distances from the crack ends the stresses could approach the theoretical strength of the material. For such small distances, however, Griffith raised the question of appropriateness of the continuum theory. We quote from reference [1.8]: 'The theory of isotropic homogeneous solids may break down if applied to metals in cases where the smallest linear dimension involved is not many times the length of a crystal.' The consequences of this observation will be discussed later.

1.4. Fracture mechanics

Griffith attributed the observed low strength of glass tension test specimens of the order of 24.9 kip/in^2 – as compared to the maximum stress observed in cracked bodies of the order of 344 kip/in^2 and to the theoretical strength of glass of the order of 1600 kip/in^2 – to the presence of discontinuities or flaws. For the tension specimen he calculated that flaws of length 2×10^{-4} in. should exist.

By his flaw hypothesis Griffith gave a solid explanation of the size effect and laid down the foundations of a new theory of fracture of solids. This theory received no further consideration until almost after the Second World War due to the massive failures of tankers and cargo ships and other catastrophic fractures reported in Section 1.2. These failures could not have been explained by the conventional design criteria which existed at that time for engineering design. Attempts then were made for the use of Griffith's ideas in the formulation of a new philosophy for structural design. These efforts led to the development of a new discipline, which is well known as *fracture mechanics*.

Before further discussing the basic concepts of the discipline of fracture mechanics, it would be appropriate at this point to pay attention to the phenomenon of the fracture of solids. During the fracture process of solids new material surfaces are formed in the medium in a thermodynamically irreversible manner. The fracture may roughly be classified from the macroscopic point of view as brittle and ductile. Brittle fracture is associated with low energy, and for unstable loading conditions it usually takes place under high fracture velocities. On the other hand, ductile fracture is associated with large deformations, high energy dissipation rates and slow fracture velocities. The phenomenon of the fracture of solids is very complicated and depends on a wide variety of factors, including the macroscopic effects, the microscopic phenomena which take place at the locations where the fracture nucleates or grows, and the composition of the material. The study of the fracture process depends on the scale level at which it is considered. At one extreme there is a rupture of cohesive bonds in the solid and the associated phenomena take place within distances of the order of 10^{-7} cm. For such studies the principles of quantum mechanics should be used. At the other extreme the material is considered as a homogeneous continuum and the phenomenon of fracture is studied within the framework of continuum mechanics and classical thermodynamics. Fracture studies which take place at scale levels between these two extremes concern movement of dislocations, formation of subgrain boundary precipitates and slip band, grain inclusions and voids. The size range of significant events involved in the process of crack extension is shown in Figure 1.2. This has been discussed in connection with the size of the crack tip radius of curvature relative to the characteristic dimension of the material microstructure [1.34]. Thus, an understanding of the phenomenon of fracture depends to a large extent on the successful integration of continuum mechanics with materials science, metallurgy, physics and chemistry. Due to the insurmountable difficulties encountered in an interdisciplinary approach the phenomenon of fracture is usually studied within only one of the three scale levels:

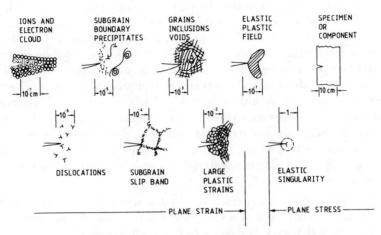

Figure 1.2. Fracture mechanisms at different scale levels.
(After McClintock and Irwin, in *Fracture Toughness Testing and its Applications*, ASTM STP 381, p.84, 1962, with permission.)

namely, the atomic, the microscopic and the continuum. Attempts have been made to bridge the gap between these three approaches.

The demonstration, first by Griffith and later by other investigators, of the size effect – which is expressed by the fact that the strength of a material measured from a laboratory specimen is many times lower than that predicted from theoretical calculations – and of the inadequacy of the traditional failure criteria to explain failures which occur at a nominal stress level considerably lower than the ultimate strength of the material gave impetus to the development of fracture mechanics. This discipline is based on the principle that all materials contain initial defects in the form of cracks, voids or inclusions which can affect the load carrying capacity of engineering structures. This is revealed experimentally. Near the defects, high stresses prevail that are often responsible for lowering the strength of the material. One of the objectives of fracture mechanics as applied to engineering design is the determination of the critical load of a structure by accounting for the size and location of initial defects. Thus, the problems of initiation, growth and arrest of cracks play a major role in the understanding of the mechanism of failure of structural components.

There are three major different ways in which defects can appear in a structure: first, they can exist in a material due to its composition, as second-phase particles, debonds in composites, etc.; second, they can be introduced in a structure during fabrication, as in welds; and third, they can be created during the service life of a component like fatigue, environment assisted or creep cracks. Fracture mechanics studies the load-bearing capacity of structures in the presence of initial defects. For engineering applications the nature of the initial defects is of no major significance. Thus, defects, basically in the form of cracks, are hypothesized to exist in structures and fracture mechanics studies the conditions of initiation, growth and arrest of cracks. Usually a dominant crack is assumed to exist in the structural component under study.

A new design philosophy is therefore introduced by fracture mechanics as opposed to the use of the conventional failure criteria. As catastrophic fracture is the consequence of the unstable propagation of a crack from a pre-existing defect, we are faced with the question: 'Can fracture be prevented by constructing structures that have no defects?' The answer is 'no', on the grounds of practicality. Then, the safe design of structures should proceed along two lines: either the safe operating load should be determined when a crack of a prescribed size is assumed to exist in the structure; or, given the operating load, the size of the crack that is created in the structure should be determined. In this case the structure should be inspected periodically to ensure that the actual crack size is smaller than the crack size the material can sustain safely. Then the following questions arise:

(a) What is the maximum crack size that a material can sustain safely?
(b) What is the strength of a structure as a function of crack size?
(c) How does the crack size relate to the applied loads?
(d) What is the critical load required to extend a crack of known size, and is the crack extension stable or unstable?
(e) How does the crack size increase as function of time?

In answering these questions fracture mechanics is searching for parameters which characterize the propensity of a crack to extend. Such a parameter should be able to relate laboratory test results to structural performance, so that the response of a structure with cracks can be predicted from laboratory test data. If we call such a parameter the *crack driving force* we should be able to determine that force as a function of material behavior, crack size, structural geometry and loading conditions. On the other hand, the critical value of this parameter, which is taken as a property of the material, should be determined from laboratory tests. The critical value of the crack driving force, known as the *fracture toughness* [1.35], expresses the ability of the material to resist fracture in the presence of cracks. By equating the crack driving force to its critical value, a relation is obtained between applied load, crack size and structure geometry which gives the necessary information for structural design.

An additional material parameter, the fracture toughness, is therefore introduced into structural design by the methodology of fracture mechanics. This parameter is used to rank a material's ability to resist fracture within the framework of fracture mechanics in the same way that yield or ultimate strength ranks a material's resistance to yield or fracture in the conventional design criteria. In selecting materials for structural applications a choice has to be made between materials with a high yield strength but comparatively low fracture toughness or a lower yield strength but higher fracture toughness. As Griffith discovered, the fracture strength is inversely proportional to the square root of the crack size for brittle fracture behavior. Failure by general yield, however, intervenes at some point. Figure 1.3 presents the variation of the strength of a structure versus crack size for two materials A and B differing in yield strength and fracture toughness. Material A has higher yield strength but lower fracture toughness than material

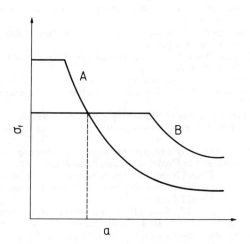

Figure 1.3. Failure strength versus crack size for two different materials A and B.

B. The two horizontal lines in the figure represent the failure strength governed by the general yield, while the two downward sloping curves depict the failure strength according to linear elastic fracture mechanics. It is observed that for crack sizes smaller than the crack size corresponding to the intersection of the curves, the strength of the structure is higher for the lower toughness material. Thus, for a structural design in situations where small cracks are anticipated to exist, a material with a higher yield strength should be used, whereas for larger crack sizes a material with a higher fracture toughness would be preferable.

References

1.1. Nadai, A., *Theory of Flow and Fracture of Solids*, McGraw-Hill, New York (1950).
1.2. Tresca, M.H., Mémoire sur l'écoulement des corps solides soumis à de fortes pressions, *Compt. Rend.* **59**, 754–758 (1864).
1.3. Tresca, M. H., Mémoire sur le poinconnage des métaux et des matières plastiques, *Compt. Rend.* **70**, 27–31 (1870).
1.4. Mises, R. von, Mechanik der festen Koerper im plastischen deformable Zustand, in *Nachrichten der Gesellschaft der Wissenschaften Goettingen, Mathematisch-Physisch Klasse*, Goettingen (1913).
1.5. Mises, R. von, Mechanik der plastischen Formänderung von Kristallen, *Zeitschrift für Angewandte Mathematik und Mechanik* **8**, 161–185 (1928).
1.6. Coulomb, C. A., Mémoires de Mathématique et de Physique, *Academie Royal des Sciences par Divers Savans*, Vol. 7, pp. 343, 382 (1773).
1.7. Mohr, O., Welche Umstände bedingen die Elastizitätsgrenze und den Bruch eines Materiales?, *Zeitschrift des Vereines Deutscher Ingenieure* **44**, 1524–1530 (1900).
1.8. Griffith, A. A., The phenomena of rupture and flow in solids, *Philosophical Transactions of the Royal Society of London* **A221**, 163–198 (1921).
1.9. Griffith, A. A., The theory of rupture, *Proceedings of First International Congress of Applied Mechanics*, Delft, pp. 55–63 (1924).

1.10. Final report of the board to investigate 'The design and methods of construction of welded steel merchant vessels', July 15 (1946), Government Printing Office, Washington (1947); reprinted in part in *Welding Journal* **20**, 569–619 (1947).
1.11. Technical progress report of the ship structure committee, *Welding Journal* **27**, 337s (1948).
1.12. Second technical report of the ship structure committee, July 1 (1950); reprinted in *Welding Journal* **30**, 169s–181s (1951).
1.13. Williams, M. L. and Ellinger, G. A., Investigation of structural failures of welded ships, *Welding Journal* **32**, 498s–527s (1953).
1.14. Boyd, G. M., Fracture design practices for ship structures, in *Fracture–An Advance Treatise*, Vol. V, *Fracture Design of Structures* (ed. H. Liebowitz), Pergamon Press, pp. 383–470 (1969).
1.15. Parker, E. R., *Brittle Behavior of Engineering Structures*, Wiley, New York (1957).
1.16. Shank, M. E., Brittle failure of steel structures–a brief history, *Metal Progress* **66**, 83–88 (1954).
1.17. Fractured girders of the King's Bridge, Melbourne, *Engineering* **217**, 520–522 (1964).
1.18. Bishop, T., Fatigue and the Comet disasters, *Metal Progress* **67**, 79–85 (1955).
1.19. Yukawa, S., Timo, D. P. and Rubio, A., Fracture design practices for rotating equipment, in *Fracture–An Advanced Treatise*, Vol. V, *Fracture Design of Structures* (ed. H. Liebowitz), Pergamon Press, pp. 65–157 (1969).
1.20. Duffy, A. R., McClure, G. M., Eiber, R. J. and Maxey, W. A., Fracture design practices for pressure piping, in *Fracture–An Advanced Treatise*, Vol. V, *Fracture Design of Structures* (ed. H. Liebowitz), Pergamon Press, pp. 159-232 (1969).
1.21. Adachi, J., Fracture design practices for ordnance structures, in *Fracture–An Advanced Treatise*, Vol. V, *Fracture Design of Structures* (ed. H. Liebowitz), Pergamon Press, pp. 285–381 (1969).
1.22. Kuhn, P., Fracture design analysis for airflight vehicles, in *Fracture–An Advanced Treatise*, Vol. V, *Fracture Design of Structures* (ed. H. Liebowitz), Pergamon Press, pp. 471–500 (1969).
1.23. Timoshenko, S. P., *History of the Strength of Materials*, McGraw-Hill, New York (1953).
1.24. Irwin, G. R. and Wells, A. A., A continuum-mechanics view of crack propagation, *Metallurgical Reviews* **10**, 223–270 (1965).
1.25. Todhunter, I. and Pearson, K., *History of the Theory of Elasticity and of the Strength of Materials*, Sections 1503 and 936, Cambridge Univ. Press (1986).
1.26. Stanton, T. E. and Batson, R. G. C., *Proceedings of the Institute of Civil Engineers* **211**, 67–100 (1921).
1.27. Docherty, J. G., Bending tests on geometrically similar notched bar specimens, *Engineering* **133**, 645–647 (1932).
1.28. Docherty, J. G., Slow bending tests on large notched bars, *Engineering* **139**, 211–213 (1935).
1.29. Irwin, G. R., Structural aspects of brittle fracture, *Applied Materials Research* **3**, 65–81 (1964).
1.30. Kommers, J. B., *International Association for Testing Materials* **4A**, **4B** (1912).
1.31. Inglis, C. E., Stresses in a plate due to the presence of cracks and sharp corners, *Transactions of the Institute of Naval Architects* **55**, 219–241 (1913).
1.32. Traube, I., Die physikalischen Eigenschaften der Elemente vom Standpunkte der Zustandsgleichung von van der Waals, *Zeitschrift für Anorganische Chemie* **XXXIV**, 413–426 (1903).
1.33. Karmarsch, I., *Mitteilungen des gew. Ver für Hannover*, pp. 138–155 (1858).
1.34. Sih G. C., The state of affairs near the crack tip, *Modeling Problems in Crack Tip Mechanics* (ed. J.T. Pindera), Martinus Nijhoff Publ., The Netherlands, pp. 65–90 (1984).
1.35. Sih G. C., *Fracture Toughness Concept*, ASTM STP 605, American Society of Testing Materials, pp. 3–15 (1976).

2

Linear elastic stress field in cracked bodies

2.1. Introduction

Fracture mechanics methodology is based on the assumption that all engineering materials contain cracks from which failure starts. The estimation of the remaining life of machine or structural components requires knowledge of the redistribution of stresses caused by the introduction of cracks in conjunction with a crack growth condition. Cracks result in high stress elevation in the neighborhood of the crack tip, which should receive particular attention since it is at that point that further crack growth takes place. It is the objective of this chapter to present a brief and comprehensive analysis of the methods used to determine the stress field in cracked bodies, particularly in the vicinity of the crack tip. Loading of a cracked body is usually accompanied by inelastic deformation and other nonlinear effects in the neighborhood of the crack tip, except for the case of ideally brittle materials. There are, however, situations where the extent of inelastic deformation and the nonlinear effects are very small compared to the crack size and any other characteristic length of the body. In such cases the linear theory is adequately justified to address the problem of stress distribution in the cracked body. Situations where the extent of inelastic deformation is pronounced necessitate the use of nonlinear theories and will be dealt with in the next chapter.

The first mathematical solution of a stress field in a linear elastic infinite flat plate subjected to uniform tension and weakened by an elliptical hole which could be degenerated into a crack was provided by Inglis [2.1]. Westergaard [2.2, 2.3] developed a semi-inverse method based on a complex representation of the Airy stress function suitable for a class of two-dimensional problems including the case of cracks. Sneddon [2.4], using the Westergaard stress function, was the first to give the stress field in the vicinity of the crack tip. Further, using the theory of Hankel transforms, he gave the stress and displacement fields around a penny-shaped crack which was first introduced by Sack for the application of the Griffith theory. The singular symmetric and antisymmetric crack tip stress field was obtained by Williams [2.5] using the eigenfunction expansion method. However, Irwin [2.6, 2.7] was the first to recognize the general applicability of the singular stress field and introduced the concept of the stress intensity factor

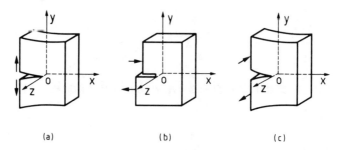

Figure 2.1. The three basic modes of crack extension. (a) Opening mode, I, (b) sliding mode, II, and (c) tearing (or antiplane) mode, III.

that measures the strength of the singular stress field. Sih and coworkers [2.8–2.11] used the method of complex potentials of Muskhelishvili [2.12] to obtain the stress field and the stress intensity factors in a variety of crack problems. Following these pioneering works, a vast number of publications have appeared in the literature concerning solutions of crack problems with emphasis on the stress intensity factors. For a comprehensive presentation of these solutions the reader is referred to [2.13–2.15].

The present chapter covers the basic methods for determining the linear elastic stress field in cracked bodies with emphasis on the problem of a single crack in an infinite plate. The eigenfunction expansion method, the semi-inverse method of Westergaard and the Hilbert–Riemann formulation for the solution of two-dimensional crack problems are briefly presented. Solutions for three-dimensional surface or embedded cracks in bodies and cracks in plates and shells are also given. Finally, the basic analytical, numerical and experimental methods for determining stress intensity factors are briefly outlined.

2.2. Crack deformation modes and basic concepts

Consider a plane crack extending through the thickness of a flat plate and let the crack plane occupy the plane xz and the crack front be parallel to the z-axis. Place the origin of the system $Oxyz$ at the midpoint of the crack front. It was first pointed out by Irwin [2.7] that there are three independent kinematic movements of the upper and lower crack surfaces with respect to each other. These three basic modes of deformation are illustrated in Figure 2.1, which presents the displacements of the crack surfaces of a local element containing the crack front. Any deformation of the crack surfaces can result from a superposition of these basic deformation modes, which are defined as follows:

(a) *Opening mode, I.* The crack surfaces separate symmetrically with respect to the planes xy and xz.

(b) *Sliding mode, II.* The crack surfaces slide relative to each other symmetrically with respect to the plane xy and skew-symmetrically with respect to the plane xz.

(c) *Tearing mode, III.* The crack surfaces slide relative to each other skew-symmetrically with respect to both planes xy and xz.

The stress and deformation fields associated with each of these three deformation modes will be determined in the sequel for the cases of plane strain and generalized plane stress. A body is said to be in a state of plane strain parallel to the plane xy if

$$u = u(x,y), \quad v = v(x,y), \quad w = 0 \tag{2.1}$$

where u, v and w denote the displacement components along the axes x, y and z. Then, the strains and stresses depend only on the variables x and y. Plane strain conditions are realized in long cylindrical bodies which are subjected to loads normal to the cylinder axis and uniform in the z-direction. In crack problems plane strain conditions are approximated in plates with large thickness relative to the crack length.

A generalized plane stress state parallel to the xy plane is defined by

$$\begin{aligned}&\sigma_z = \tau_{zx} = \tau_{zy} = 0\\ &\sigma_x = \sigma_x(x,y), \quad \sigma_y = \sigma_y(x,y), \quad \tau_{xy} = \tau_{xy}(x,y)\end{aligned} \tag{2.2}$$

where $\sigma_x, \sigma_y, \sigma_z$ and $\tau_{xy}, \tau_{zx}, \tau_{zy}$ denote the normal and shear stresses associated with the system xyz. The state of generalized plane stress is approximate, since equations (2.2) violate the compatibility equations of elasticity. Generalized plane stress conditions are realized in thin flat plates with traction-free surfaces. In crack problems, the generalized plane stress conditions are approximated in plates with crack lengths that are large in relation to the plate thickness.

It is recalled from the theory of elasticity that a plane strain problem may be solved as a generalized plane stress problem by replacing the value of Poisson's ratio ν by the value $\nu/(1+\nu)$. In the sequel, plane crack problems in flat plates which incorporate the plane strain and the generalized plane stress state will first be considered.

Following the method of complex potentials, the displacements u and v and the stress components $\sigma_x, \sigma_y, \tau_{xy}$ in plane elasticity problems can be expressed by the following equations [2.12]:

$$\begin{aligned}2\mu(u+iv) &= \kappa\phi(z) - z\overline{\phi'(z)} - \overline{\psi(z)}\\ \sigma_x + \sigma_y &= 2[\phi'(z) + \overline{\phi'(z)}]\\ \sigma_y - \sigma_x + 2i\tau_{xy} &= 2[\bar{z}\phi''(z) + \chi''(z)]\end{aligned} \tag{2.3}$$

with

$$\psi(z) = \chi'(z).$$

In these equations $\phi(z)$, $\psi(z)$ and $\chi(z)$ are analytic functions of the complex variable z, μ is the shear modulus and $\kappa = 3-4\nu$ or $\kappa = (3-\nu)/(1+\nu)$ for plane strain or generalized plane stress, with ν being Poisson's ratio. The prime in a function means differentiation with respect to z and the bar denotes a complex conjugate quantity.

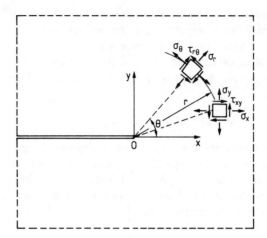

Figure 2.2. Rectangular and polar stress components around a crack tip in a plate subjected to in-plane loads.

The polar stress components $\sigma_r, \sigma_\theta, \tau_{r\theta}$ and the polar displacements u_r, u_θ are given by $(z = re^{i\theta})$

$$2\mu(u_r + iu_\theta) = e^{-i\theta}[\kappa\phi(z) - z\overline{\phi'(z)} - \overline{\chi'(z)}]$$
$$\sigma_r + \sigma_\theta = 2[\phi'(z) + \overline{\phi'(z)}] \qquad (2.4)$$
$$\sigma_\theta - \sigma_r + 2i\tau_{r\theta} = 2e^{2i\theta}[\bar{z}\phi''(z) + \chi''(z)]$$

2.3. Eigenfunction expansion method for a semi-infinite crack

The method of eigenfunction expansion introduced by Williams [2.5] is perhaps the most direct way of finding the structure of the stress field in the neighborhood of the crack tip. For crack problems the complex potentials $\phi(z)$ and $\chi(z)$ in Equations (2.3) and (2.4) can be put in the form

$$\phi(z) = \sum_0^\infty A_n z^{\lambda_n}, \qquad \chi(z) = \sum_0^\infty B_n z^{\lambda_n+1} \qquad (2.5)$$

where the eigenvalues λ_n are real.

Consider a semi-infinite crack in a plate subjected to general in-plane loads and let the crack be placed along the negative x-axis (Figure 2.2). If the upper and lower crack faces are free of traction we have

$$\sigma_\theta + i\tau_{r\theta} = 0, \qquad \theta = \pm\pi. \qquad (2.6)$$

Equation (2.6), in conjunction with Equations (2.4) and (2.5), renders

$$\sum_0^\infty \lambda_n r^{\lambda_n-1}[\lambda_n e^{i(\lambda_n-1)\theta}A_n + e^{-i(\lambda_n-1)\theta}\overline{A_n} + (\lambda_n+1)e^{i(\lambda_n+1)\theta}B_n] = 0 \quad (2.7)$$

which, for $\theta = \pm\pi$, gives the characteristic equation

$$\sin 2\pi\lambda_n = 0. \tag{2.8}$$

Equation (2.8) gives the eigenvalues

$$\lambda_n = \tfrac{n}{2}, \quad n = 1, 2, \cdots \tag{2.9}$$

Negative values of n are ignored because they produce infinite displacements at the crack tip ($r = 0$). For $\lambda = 0$, Equations (2.4) and (2.5) give stresses and strains of the form

$$\sigma = \frac{f(\theta)}{r}, \quad \epsilon = \frac{g(\theta)}{r} \tag{2.10}$$

where σ stands for the stresses $\sigma_r, \sigma_\theta, \tau_{r\theta}$, ϵ for the strains $\epsilon_r, \epsilon_\theta, \gamma_{r\theta}$ and $f(\theta)$ and $g(\theta)$ are functions of θ.

The strain energy density function dW/dV for plane elasticity problems is given by

$$\frac{dW}{dV} = \frac{1}{4\mu}\left[\frac{\kappa+1}{4}(\sigma_x + \sigma_y)^2 - 2(\sigma_x\sigma_y - \tau_{xy}^2)\right] \tag{2.11}$$

and from Equation (2.10) it should have the form

$$\frac{dW}{dV} = \frac{h(\theta)}{r^2} \tag{2.12}$$

where $h(\theta)$ is a function of θ.

Thus, the total strain energy of any circular region of radius $r < R$ surrounding the crack front is given by

$$W = \int_0^{2\pi}\int_{r_0}^{R} \frac{h(\theta)}{r^2} r\, dr\, d\theta \tag{2.13}$$

and becomes unbounded for $r_0 \to 0$ which is physically unacceptable. Thus, the eigenvalue $\lambda = 0$ should be excluded from the solution.

Equation (2.7) renders

$$\lambda_n A_n + (-1)^n \overline{A}_n + (\lambda_n + 1)B_n = 0, \quad n = 1, 2, \cdots \tag{2.14}$$

The smallest eigenvalue $\lambda_1 = \tfrac{1}{2}$ ($n = 1$) gives for the complex potentials $\phi(z)$ and $\chi(z)$ (Equations (2.5)) the following expressions

$$\phi(z) = A_1 z^{1/2}, \quad \chi(z) = B_1 z^{3/2} \tag{2.15}$$

while Equation (2.14) renders

$$3B_1 = 2\overline{A}_1 - A_1. \tag{2.16}$$

Introducing the values of $\phi(z)$ and $\chi(z)$ from Equations (2.15) and (2.16) into Equations (2.4), the following expressions for the stresses and displacements in the neighborhood of the crack tip are obtained:

$$\sigma_r = \frac{1}{4\sqrt{r}}\left[a_1\left(5\cos\frac{\theta}{2} - \cos\frac{3\theta}{2}\right) + a_2\left(-5\sin\frac{\theta}{2} + 3\sin\frac{3\theta}{2}\right)\right]$$

$$\sigma_\theta = \frac{1}{4\sqrt{r}}\left[a_1\left(3\cos\frac{\theta}{2} + \cos\frac{3\theta}{2}\right) + a_2\left(-3\sin\frac{\theta}{2} - 3\sin\frac{3\theta}{2}\right)\right] \quad (2.17)$$

$$\tau_{r\theta} = \frac{1}{4\sqrt{r}}\left[a_1\left(\sin\frac{\theta}{2} + \sin\frac{3\theta}{2}\right) + a_2\left(\cos\frac{\theta}{2} + 3\cos\frac{3\theta}{2}\right)\right]$$

and

$$u_r = \frac{\sqrt{r}}{4\mu}\left[a_1\left[(2\kappa-1)\cos\frac{\theta}{2} - \cos\frac{3\theta}{2}\right] + a_2\left[-(2\kappa-1)\sin\frac{\theta}{2} + 3\sin\frac{3\theta}{2}\right]\right]$$

$$u_\theta = \frac{\sqrt{r}}{4\mu}\left[a_1\left[-(2\kappa+1)\sin\frac{\theta}{2} + \sin\frac{3\theta}{2}\right] + a_2\left[-(2\kappa+1)\cos\frac{\theta}{2} + 3\cos\frac{3\theta}{2}\right]\right] \quad (2.18)$$

where

$$\kappa = 3 - 4\nu \quad \text{for plane strain}$$

$$\kappa = \frac{3-\nu}{1+\nu} \quad \text{for generalized plane stress.}$$

The constants a_1 and a_2 in the above equations are defined by

$$A_1 = a_1 - ia_2. \quad (2.19)$$

If now we separate the symmetric and the skew-symmetric terms with respect to the x-axis in Equations (2.17) and (2.18), the following equations for the polar components of stress and displacement for the opening and sliding modes of deformation are obtained.

Opening mode

$$\sigma_r = \frac{K_I}{\sqrt{2\pi r}}\left(\tfrac{5}{4}\cos\frac{\theta}{2} - \tfrac{1}{4}\cos\frac{3\theta}{2}\right)$$

$$\sigma_\theta = \frac{K_I}{\sqrt{2\pi r}}\left(\tfrac{3}{4}\cos\frac{\theta}{2} + \tfrac{1}{4}\cos\frac{3\theta}{2}\right) \quad (2.20)$$

$$\tau_{r\theta} = \frac{K_I}{\sqrt{2\pi r}}\left(\tfrac{1}{4}\sin\frac{\theta}{2} + \tfrac{1}{4}\sin\frac{3\theta}{2}\right)$$

and

$$u_r = \frac{K_I}{4\mu}\sqrt{\frac{r}{2\pi}}\left[(2\kappa-1)\cos\frac{\theta}{2} - \cos\frac{3\theta}{2}\right]$$

$$u_\theta = \frac{K_I}{4\mu}\sqrt{\frac{r}{2\pi}}\left[-(2\kappa+1)\sin\frac{\theta}{2} + \sin\frac{3\theta}{2}\right] \quad (2.21)$$

Sliding mode

$$\sigma_r = \frac{K_{II}}{\sqrt{2\pi r}}\left(-\tfrac{5}{4}\sin\frac{\theta}{2} + \tfrac{3}{4}\sin\frac{3\theta}{2}\right)$$
$$\sigma_\theta = \frac{K_{II}}{\sqrt{2\pi r}}\left(-\tfrac{3}{4}\sin\frac{\theta}{2} - \tfrac{3}{4}\sin\frac{3\theta}{2}\right) \quad (2.22)$$
$$\tau_{r\theta} = \frac{K_{II}}{\sqrt{2\pi r}}\left(\tfrac{1}{4}\cos\frac{\theta}{2} + \tfrac{3}{4}\cos\frac{3\theta}{2}\right)$$

and

$$u_r = \frac{K_{II}}{4\mu}\sqrt{\frac{r}{2\pi}}\left[-(2\kappa - 1)\sin\frac{\theta}{2} + 3\sin\frac{3\theta}{2}\right]$$
$$u_\theta = \frac{K_{II}}{4\mu}\sqrt{\frac{r}{2\pi}}\left[-(2\kappa + 1)\cos\frac{\theta}{2} + 3\cos\frac{3\theta}{2}\right] \quad (2.23)$$

where

$$K_I - iK_{II} = \frac{1}{\sqrt{2\pi}}(a_1 - ia_2). \quad (2.24)$$

Equations (2.20)–(2.23) give the stress and displacement components which correspond to the value $\lambda_1 = \tfrac{1}{2}$ of the eigenfunction expansion of the complex potentials $\phi(z)$ and $\chi(z)$ expressed by Equations (2.5). It is observed that the stresses have an inverse square root singularity at the crack tip. The stresses and displacements associated with the other values of $\lambda_n = 1, 3/2, 2, \cdots$ can be obtained in an analogous manner. It is seen that these stresses are finite at the crack tip. Thus, the stresses and displacements given by Equations (2.20)–(2.23) represent the asymptotic forms of the linear elastic stress and displacement fields, and dominate in the vicinity of the crack tip.

The quantities K_I and K_{II} defined by Equation (2.24) are called the opening mode or mode-I and sliding mode or mode-II stress intensity factors. They depend on the far field boundary conditions and the geometry of the cracked plate and express the strength of stress and displacement fields near to the crack tip. A more natural definition of the stress intensity factors and the way they are determined will be given later in this book.

In a similar manner, the singular stress and displacement fields for the tearing (or antiplane) mode of crack deformation can be obtained (Figure 2.3). For this situation the in-plane displacements u and v are zero, while the displacement w is a function of the in-plane coordinates x and y, that is

$$u = v = 0, \quad w = w(x, y). \quad (2.25)$$

Equations (2.25) suggest that the movement of the crack surfaces can be related to the warping action of noncircular cylinders subjected to torsion. Equations (2.25) render

$$\epsilon_r = \epsilon_\theta = \epsilon_z = \gamma_{r\theta} = 0$$
$$\gamma_{rz} = \frac{\partial w}{\partial r}, \quad \gamma_{\theta z} = \frac{1}{r}\frac{\partial w}{\partial \theta} \quad (2.26)$$

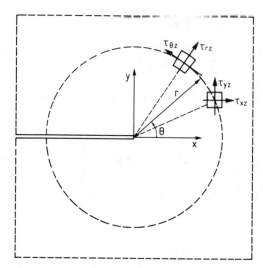

Figure 2.3. Rectangular and polar shear stress components around a crack tip in a plate subjected to antiplane mode of deformation.

and from Hooke's law we have

$$\sigma_r = \sigma_\theta = \sigma_z = \tau_{r\theta} = 0$$
$$\tau_{rz} = \mu \frac{\partial w}{\partial r}, \quad \tau_{\theta z} = \frac{\mu}{r} \frac{\partial w}{\partial \theta} \tag{2.27}$$

Substituting Equations (2.27) into the non-self-satisfied equilibrium equation

$$\frac{\partial \tau_{rz}}{\partial r} + \frac{1}{r} \frac{\partial \tau_{\theta z}}{\partial \theta} + \frac{\tau_{rz}}{r} = 0 \tag{2.28}$$

we obtain for w

$$\frac{\partial^2 w}{\partial r^2} + \frac{1}{r} \frac{\partial w}{\partial r} + \frac{1}{r^2} \frac{\partial^2 w}{\partial \theta^2} = 0. \tag{2.29}$$

Since w satisfies the Laplace equation it can be represented as the real part of a complex function $f(z)$, i.e.

$$w = \text{Re}[f(z)]. \tag{2.30}$$

Then Equation (2.27) gives

$$\tau_{rz} - i\tau_{\theta z} = \mu e^{i\theta} f'(z). \tag{2.31}$$

The complex function $f(z)$ can be put in the form

$$f(z) = \sum_{0}^{\infty} C_n z^{\lambda_n} \tag{2.32}$$

where the eigenvalues λ_n are assumed to be real.

Let the crack occupy the negative x-axis and be free of stress. Then

$$\tau_{\theta z} = 0, \qquad \theta = \pm \pi. \tag{2.33}$$

Equation (2.33), in conjunction with Equations (2.31) and (2.32), renders

$$\tau_{\theta z} = \frac{i\mu}{2} \sum_{0}^{\infty} \lambda_n r^{\lambda_n} \left(e^{i\lambda_n \theta} C_n - e^{-i\lambda_n \theta} \overline{C}_n \right) = 0, \qquad \theta = \pm \pi. \tag{2.34}$$

Equation (2.34) gives the same equations (Equations (2.8) and (2.9)) for the eigenvalues λ_n. It is also obtained from Equation (2.34)

$$(-1)^n C_n - \overline{C}_n = 0. \tag{2.35}$$

As in the previous case, the non-positive values of $\lambda(0, -\frac{1}{2}, -1, \cdots)$ should be excluded from the solution. The smallest eigenvalue $\lambda_1 = \frac{1}{2}$ $(n = 1)$ gives, for the function $f(z)$,

$$f(z) = C_1 z^{1/2} \tag{2.36}$$

where, as Equation (2.35) shows, the constant C_1 must be imaginary. Putting $C_1 = -(i\sqrt{2/\pi}/\mu) K_{\text{III}}$, Equations (2.27) and (2.30) give the crack-tip stresses τ_{rz} and $\tau_{\theta z}$:

$$\tau_{rz} = \frac{K_{\text{III}}}{\sqrt{2\pi r}} \sin \frac{\theta}{2}, \qquad \tau_{\theta z} = \frac{K_{\text{III}}}{\sqrt{2\pi r}} \cos \frac{\theta}{2}, \tag{2.37}$$

and the displacement

$$w = \frac{2 K_{\text{III}}}{\mu} \sqrt{\frac{r}{2\pi}} \sin \frac{\theta}{2}. \tag{2.38}$$

Equations (2.37) and (2.38) give the asymptotic forms of the stress and displacement fields which dominate in the vicinity of the crack tip for the tearing mode of deformation. They have the same structure as the corresponding stresses and displacements for the opening and sliding modes of deformation obtained previously. The coefficient K_{III}, which reflects the strength of the stress and displacement fields near to the crack tip, is called the tearing mode or mode-III stress intensity factor.

2.4. Westergaard method

(a) Description of the method

The Westergaard method [2.2, 2.3] constitutes a simple and versatile tool for solving a certain class of plane elasticity problems. Following the Airy stress function representation, the solution of a plane elasticity problem is reduced to finding a function U which satisfies the biharmonic equation

$$\nabla^2 \nabla^2 U = \frac{\partial^4 U}{\partial x^4} + 2 \frac{\partial^4 U}{\partial x^2 \partial y^2} + \frac{\partial^4 U}{\partial y^4} = 0 \tag{2.39}$$

where the stress components are given by

$$\sigma_x = \frac{\partial^2 U}{\partial y^2}, \quad \sigma_y = \frac{\partial^2 U}{\partial x^2}, \quad \tau_{xy} = -\frac{\partial^2 U}{\partial x \partial y} \tag{2.40}$$

and the appropriate boundary conditions.

Choosing the function U in the form

$$U = \Psi_1 + x\Psi_2 + y\Psi_3 \tag{2.41}$$

where the functions Ψ_i ($i = 1, 2, 3$) are harmonic, that is,

$$\nabla^2 \Psi_i = 0, \tag{2.42}$$

U will automatically satisfy Equation (2.39).

Following the Cauchy–Riemann conditions, the functions Ψ_i can be considered as the real or imaginary part of an analytic function of the complex variable z. Introducing the notation

$$\tilde{Z} = \frac{d\tilde{\tilde{Z}}}{dz}, \quad Z = \frac{d\tilde{Z}}{dz}, \quad Z' = \frac{dZ}{dz} \tag{2.43}$$

Westergaard defined an Airy function U_I for symmetric problems by

$$U_I = \operatorname{Re} \tilde{Z}_I + y \operatorname{Im} \tilde{Z}_I. \tag{2.44}$$

Using Equations (2.40) the stresses following from U_I are

$$\begin{aligned}\sigma_x &= \operatorname{Re} Z_I - y \operatorname{Im} Z'_I \\ \sigma_y &= \operatorname{Re} Z_I + y \operatorname{Im} Z'_I \\ \tau_{xy} &= -y \operatorname{Re} Z'_I.\end{aligned} \tag{2.45}$$

The limitation of this type of solution, which requires that

$$\sigma_x = \sigma_y, \quad \tau_{xy} = 0 \quad \text{for } y = 0, \tag{2.46}$$

is evident.

In solving crack problems, condition (2.46) is not always satisfied. Thus, Sih [2.16] extended the Westergaard method for situations where $\sigma_x \neq \sigma_y$ for $y = 0$. When the external loads are situated symmetrically with respect to the x-axis, it is obtained from the third of Equations (2.3):

$$\operatorname{Im}[\bar{z}\phi''(z) + \psi'(z)] = 0 \quad \text{for } y = 0. \tag{2.47}$$

Equation (2.47) yields

$$\bar{z}\phi''(z) + \psi'(z) + A = 0 \tag{2.48}$$

where A is a real constant.

Introducing the value of $\psi'(z)$ from Equation (2.48) into Equations (2.3) the following equations for the stresses and the displacements in terms of a single function are obtained:

$$\begin{aligned}\sigma_x &= \operatorname{Re} Z_I - y \operatorname{Im} Z_I' + A \\ \sigma_y &= \operatorname{Re} Z_I + y \operatorname{Im} Z_I' - A \\ \tau_{xy} &= -y \operatorname{Re} Z_I' \end{aligned} \qquad (2.49)$$

and

$$\begin{aligned} 2\mu u &= \frac{\kappa - 1}{2} \operatorname{Re} \tilde{Z}_I - y \operatorname{Im} Z_I + Ax \\ 2\mu v &= \frac{\kappa + 1}{2} \operatorname{Im} \tilde{Z}_I - y \operatorname{Re} Z_I - Ay \end{aligned} \qquad (2.50)$$

where it was put

$$2\phi'(z) = Z_I \qquad (2.51)$$

and $\kappa = 3 - 4\nu$ for plane strain and $\kappa = (3 - \nu)/(1 + \nu)$ for generalized plane stress.

Equations (2.49) coincide with Equations (2.45) if $A = 0$. These equations were also derived by Eftis and Liebowitz [2.17] using the complex representation of the plane elasticity problem given by MacGregor [2.18] upon whose work Westergaard based his formulations.

Following the previous developments, the Airy function U_{II} for skew-symmetric problems with respect to the x-axis is defined by

$$U_{II} = y \operatorname{Im} \tilde{Z}_{II}. \qquad (2.52)$$

Using Equations (2.40) we obtain for the stresses

$$\begin{aligned} \sigma_x &= 2 \operatorname{Re} Z_{II} - y \operatorname{Im} Z_{II}' \\ \sigma_y &= y \operatorname{Im} Z_{II}' \\ \tau_{xy} &= -\operatorname{Im} Z_{II} - y \operatorname{Re} Z_{II}'. \end{aligned} \qquad (2.53)$$

Equations (2.53) were rederived by Sih [2.16] who used the general equations of two-dimensional elasticity (2.3). For skew-symmetric problems with respect to the x-axis for which the σ_y stress is zero for $y = 0$, Equations (2.3) give

$$\operatorname{Re}[2\phi'(z) + \bar{z}\phi''(z) + \psi'(z)] = 0. \qquad (2.54)$$

Equation (2.54) renders

$$2\phi'(z) + \bar{z}\phi''(z) + \psi'(z) + iB = 0 \qquad (2.55)$$

where B is a real constant.

Introducing the value of $\psi'(z)$ from Equation (2.55) into Equations (2.3) we obtain, for the stresses and displacements,

$$\begin{aligned} \sigma_x &= 2 \operatorname{Re} Z_{II} - y \operatorname{Im} Z_{II}' \\ \sigma_y &= y \operatorname{Im} Z_{II}' \\ \tau_{xy} &= -\operatorname{Im} Z_{II} - y \operatorname{Re} Z_{II}' - B \end{aligned} \qquad (2.56)$$

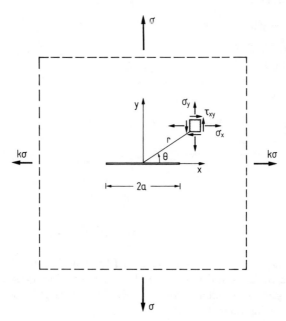

Figure 2.4. A crack of length $2a$ in an infinite plate subjected to uniform stresses σ and $k\sigma$ at infinity.

and

$$2\mu u = \frac{\kappa+1}{2} \operatorname{Re} \tilde{Z}_{II} - y \operatorname{Im} Z_{II} - By$$
$$2\mu v = \frac{\kappa-1}{2} \operatorname{Im} \tilde{Z}_{II} - y \operatorname{Re} Z_{II} - Bx \qquad (2.57)$$

where it was put

$$2\phi'(z) = Z_{II}. \qquad (2.58)$$

(b) *Opening-mode crack problems*

Consider a crack of length $2a$ which occupies the segment $-a \leq x \leq a$ along the x-axis in an infinite plate subjected to uniform stresses σ and $k\sigma$ along the y and x directions at infinity (Figure 2.4). The boundary conditions of the problem may be stated as follows:

$$\sigma_y + i\tau_{xy} = 0 \quad \text{for } y = 0, \quad -a < x < a \qquad (2.59)$$

and

$$\sigma_x = k\sigma, \quad \sigma_y = \sigma, \quad \tau_{xy} = 0 \quad \text{for } (x^2+y^2)^{1/2} \to \infty. \qquad (2.60)$$

Equations (2.49) for $y = 0$ in conjunction with Equation (2.59) give

$$\operatorname{Re} Z_I = A. \qquad (2.61)$$

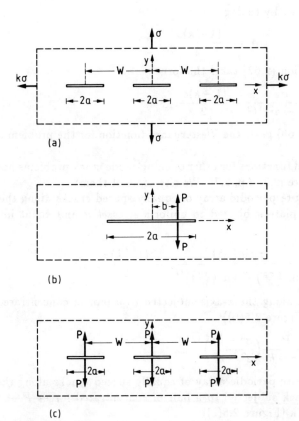

Figure 2.5. Geometrical configurations of an infinite plate with (a) a periodic array of equally spaced cracks subjected to uniform stresses σ and $k\sigma$ at infinity, (b) a single crack subjected to a pair of concentrated forces P at $x = b$ and (c) a periodic array of equally spaced cracks subjected to a pair of concentrated forces P at the center of each crack.

The function Z_I is analytic in the entire plane except for the interval $-a < x < a$. Thus, by virtue of Equation (2.61) the function Z_I may be chosen to have the form

$$Z_I = \frac{g(z)}{(z^2 - a^2)^{1/2}} + A \qquad (2.62)$$

where the function $g(z)$ is assumed to be holomorphic in the region of definition, except possibly at the point $z = \infty$. Since the denominator of the first term of Z_I is pure imaginary for $-a < x < a$, the function $g(z)$ should have zero imaginary part along the crack border in order to satisfy the boundary condition (2.61). Equations (2.49), in conjunction with Equation (2.60), render

$$(\bar{z} - z)Z_I' - 2A = (1 - k)\sigma \qquad (2.63)$$

which is satisfied by taking

$$g(z) = \sigma z, \qquad A = -\frac{(1-k)\sigma}{2}. \tag{2.64}$$

Thus Equation (2.62) takes the form

$$Z_\mathrm{I} = \frac{\sigma z}{(z^2 - a^2)^{1/2}} - \frac{(1-k)\sigma}{2}. \tag{2.65}$$

Equation (2.65) gives the Westergaard function for the problem shown in Figure 2.4.

Westergaard functions for other opening-mode crack problems have been given in the literature [2.3, 2.6, 2.7]. We cite some of them:

(a) An infinite periodic array of equally spaced cracks along the x-axis in an infinite plate subjected to uniform stresses σ and $k\sigma$ at infinity (Figure 2.5(a)):

$$Z_\mathrm{I} = \frac{\sigma \sin\left(\frac{\pi z}{W}\right)}{\left[\sin^2\left(\frac{\pi z}{W}\right) - \sin^2\left(\frac{\pi a}{W}\right)\right]^{1/2}} - \frac{(1-k)\sigma}{2}. \tag{2.66}$$

(b) A crack along the x-axis subjected to a pair of concentrated forces P at $x = b$ (Figure 2.5(b)):

$$Z_\mathrm{I} = \frac{P}{\pi(z-b)}\left(\frac{a^2 - b^2}{z^2 - a^2}\right)^{1/2}. \tag{2.67}$$

(c) An infinite periodic array of equally spaced cracks along the x-axis with each crack subjected to a pair of concentrated forces P at the center of the crack (Figure 2.5(c)):

$$Z_\mathrm{I} = \frac{P \sin(\pi a/W)}{W(\sin(\pi z/W))^2}\left[1 - \left(\frac{\sin(\pi a/W)}{\sin(\pi z/W)}\right)^2\right]^{-1/2}. \tag{2.68}$$

(c) *Sliding-mode crack problems*

Consider a crack of length $2a$ which occupies the segment $-a \leq x \leq a$ along the x-axis in an infinite plate subjected to uniform in-plane shear stresses τ at infinity (Figure 2.6). The boundary conditions of the problem may be stated as

$$\begin{aligned}\sigma_y + i\tau_{xy} &= 0 \quad \text{for } y = 0, \quad -a < x < a \\ \sigma_x = \sigma_y &= 0, \quad \tau_{xy} = \tau \quad \text{for } (x^2 + y^2)^{1/2} \to \infty.\end{aligned} \tag{2.69}$$

Following the same procedure as in the previous case of an opening mode, the Westergaard function for this case takes the form

$$Z_\mathrm{II} = \frac{-i\tau z}{(z^2 - a^2)^{1/2}} + i\tau. \tag{2.70}$$

The function Z_II satisfies the boundary conditions (2.69) with the stresses $\sigma_x, \sigma_y, \tau_{xy}$ defined from Equations (2.56) and $B = -\tau$.

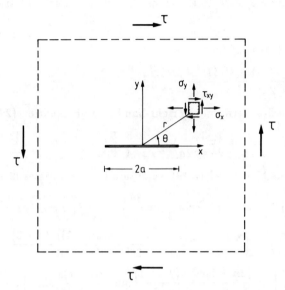

Figure 2.6. A crack of length $2a$ in an infinite plate subjected to uniform in-plane shear stresses τ at infinity.

Westergaard functions for other sliding-mode crack problems have been given in the literature. Thus, the Westergaard function for the problem of Figure 2.5(a) subjected to a uniform in-plane shear stress τ at infinity is given by

$$Z_{\text{II}} = \frac{-i\tau \sin\left(\frac{\pi z}{W}\right)}{\left[\sin^2\left(\frac{\pi z}{W}\right) - \sin^2\left(\frac{\pi a}{W}\right)\right]^{1/2}} + i\tau. \tag{2.71}$$

The Westergaard method for solving crack problems is simple but it suffers limitations, the most serious being that it is restricted to infinite bodies with cracks placed along a single straight line. For more complicated crack problems other more powerful methods will be developed.

2.5. Singular stress and displacement fields

The study of stress and displacement fields in the vicinity of the crack tip is very important because these fields govern the fracture process that takes place at the crack tip. In this section we shall make a thorough study of the stresses and displacements near the crack tip for the three deformation modes.

(a) Opening mode

The Westergaard function for an infinite plate with a crack of length $2a$ subjected to stresses σ and $k\sigma$ at infinity (Figure 2.4) is given by Equation (2.65). If we place the coordinate system to the crack tip $z = a$ through the transformation

$$\varsigma = z - a \tag{2.72}$$

Equation (2.65) takes the form

$$Z_{\mathrm{I}} = \frac{\sigma(\varsigma + a)}{[\varsigma(\varsigma + 2a)]^{1/2}} - \frac{(1-k)\sigma}{2}. \tag{2.73}$$

Expanding the first term on the right-hand side of Equation (2.73) we obtain

$$Z_{\mathrm{I}} = \frac{\sigma(\varsigma + a)}{(2a\varsigma)^{1/2}} \left[1 - \frac{1}{2}\frac{\varsigma}{2a} + \frac{1 \cdot 3}{2 \cdot 4}\left(\frac{\varsigma}{2a}\right)^2 - \frac{1 \cdot 3 \cdot 5}{2 \cdot 4 \cdot 6}\left(\frac{\varsigma}{2a}\right)^3 + \cdots \right] - \frac{(1-k)\sigma}{2}. \tag{2.74}$$

From Equations (2.74) and (2.49) we obtain for the stresses [2.19]

$$\frac{\sigma_x}{\sigma} = \left(\frac{2r}{a}\right)^{-1/2}\cos\frac{\theta}{2}\left(1 - \sin\frac{\theta}{2}\sin\frac{3\theta}{2}\right) + \frac{3}{4\sqrt{2}}\left(\frac{r}{a}\right)^{1/2}\cos\frac{\theta}{2}\left(1 + \sin^2\frac{\theta}{2}\right) + $$
$$+ \left(\frac{r}{2a}\right)^{1/2}\left\{\sum_{n=1}^{\infty}(-1)^n\frac{[1 \cdot 3 \cdot 5 \cdots (2n-1)](2n+3)}{2^{2(n+1)}(n+1)!}\right.$$
$$\left. \times \left(\frac{r}{a}\right)^n\left[\cos\frac{(2n+1)\theta}{2} - \frac{(2n+1)}{2}\sin\frac{(2n-1)\theta}{2}\sin\theta\right]\right\} - (1-k)\sigma$$

$$\frac{\sigma_y}{\sigma} = \left(\frac{2r}{a}\right)^{-1/2}\cos\frac{\theta}{2}\left(1 + \sin\frac{\theta}{2}\sin\frac{3\theta}{2}\right) + \frac{3}{4\sqrt{2}}\left(\frac{r}{a}\right)^{1/2}\cos\frac{\theta}{2}\left(1 - \sin^2\frac{\theta}{2}\right) + $$
$$+ \left(\frac{r}{2a}\right)^{1/2}\left\{\sum_{n=1}^{\infty}(-1)^n\frac{[1 \cdot 3 \cdot 5 \cdots (2n-1)](2n+3)}{2^{2(n+1)}(n+1)!}\right. \tag{2.75}$$
$$\left. \times \left(\frac{r}{a}\right)^n\left[\cos\frac{(2n+1)\theta}{2} + \frac{(2n+1)}{2}\sin\frac{(2n-1)\theta}{2}\sin\theta\right]\right\}$$

$$\frac{\tau_{xy}}{\sigma} = \left(\frac{2r}{a}\right)^{-1/2}\cos\frac{\theta}{2}\sin\frac{\theta}{2}\cos\frac{3\theta}{2} - \frac{3}{4\sqrt{2}}\left(\frac{r}{a}\right)^{1/2}\cos^2\frac{\theta}{2}\sin\frac{\theta}{2} +$$
$$+ \left(\frac{r}{2a}\right)^{1/2}\sum_{n=1}^{\infty}(-1)^{(n+1)}\frac{1 \cdot 3 \cdot 5 \cdots (2n+3)}{2^{(2n+3)}(n+1)!}$$
$$\times \left(\frac{r}{a}\right)^n\cos\frac{(2n-1)\theta}{2}\sin\theta$$

where $\varsigma = r\,e^{i\theta}$.

For very small values of r/a the first terms in the above equations dominate the others. Taking into account the constant term in σ_x we obtain for the stresses the approximations

$$\sigma_x = \frac{K_{\mathrm{I}}}{\sqrt{2\pi r}}\cos\frac{\theta}{2}\left(1 - \sin\frac{\theta}{2}\sin\frac{3\theta}{2}\right) - (1-k)\sigma$$
$$\sigma_y = \frac{K_{\mathrm{I}}}{\sqrt{2\pi r}}\cos\frac{\theta}{2}\left(1 + \sin\frac{\theta}{2}\sin\frac{3\theta}{2}\right) \tag{2.76}$$
$$\tau_{xy} = \frac{K_{\mathrm{I}}}{\sqrt{2\pi r}}\cos\frac{\theta}{2}\sin\frac{\theta}{2}\cos\frac{3\theta}{2}$$

where it was put

$$K_{\mathrm{I}} = \sigma\sqrt{\pi a}. \tag{2.77}$$

Linear elastic stress field in cracked bodies

The quantity K_I is the opening-mode stress intensity factor and expresses the strength of the singular elastic stress field. As was shown by Irwin [2.6, 2.7], Equations (2.76) have a general applicability to all crack-tip stress fields independently of crack/body geometry and the loading conditions. The constant term entering σ_x takes different values depending on the applied loads and the geometry of the cracked plate. The stress intensity factor depends linearly on the applied load and is a function of the crack length and the geometrical configuration of the cracked body. Methods of determining stress intensity factors will be discussed later in this chapter.

Introducing the value of the Westergaard function Z_I from Equation (2.74) into Equations (2.50) and retaining the singular and the constant terms we obtain for the displacements

$$u = \frac{K_I}{2\mu}\sqrt{\frac{r}{2\pi}} \cos\frac{\theta}{2}(\kappa - \cos\theta) - \frac{\sigma}{E}(1-k)r\cos\theta$$
$$v = \frac{K_I}{2\mu}\sqrt{\frac{r}{2\pi}} \sin\frac{\theta}{2}(\kappa - \cos\theta) + \frac{\nu\sigma}{E}(1-k)r\sin\theta$$
(2.78)

Equations (2.76) and (2.78) express the stress and displacement fields for plane strain and plane stress conditions in the vicinity of the crack tip. The constant term refers only to the case of the infinite plate loaded by stresses σ_x and σ_y at infinity and takes different values for other crack/body geometries and loading conditions. It is observed that the stresses and strains always have an inverse square root singularity at the crack tip. For plane strain conditions the stress σ_z $(= \nu(\sigma_x + \sigma_y))$ normal to the plane of the plate is given by

$$\sigma_z = \frac{2\nu K_I}{\sqrt{2\pi r}} \cos\frac{\theta}{2}. \tag{2.79}$$

When the Westergaard function Z_I of a crack problem is known it can always, as in the previous case, be put in the form

$$Z_I = \frac{f(\varsigma)}{\varsigma^{1/2}} \tag{2.80}$$

where the function $f(\varsigma)$ is well behaved for small $|\varsigma|$. Thus, for $|\varsigma| \to 0$ Equation (2.80) takes the form

$$Z_I = \frac{f(0)}{\varsigma^{1/2}}. \tag{2.81}$$

If now the stress σ_y along the x-axis computed from Equation (2.81) is compared with that given by Equations (2.76), we obtain

$$K_I = \lim_{|\varsigma| \to 0} \sqrt{2\pi\varsigma}\, Z_I. \tag{2.82}$$

Equation (2.82) can be used to determine the K_I stress intensity factor when the function Z_I is known. Thus, for the case of a periodic array of equally spaced cracks along the x-axis in an infinite plate subjected to uniform stresses σ and

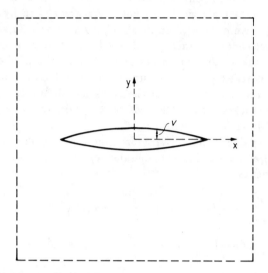

Figure 2.7. Crack deformation shape for mode-I loading.

$k\sigma$ at infinity (Figure 2.5(a)), the function Z_I is given by Equation (2.66) which, in conjunction with Equation (2.82), gives for K_I

$$K_I = \sigma\sqrt{\pi a}\sqrt{\frac{W}{\pi a} \tan \frac{\pi a}{W}}. \tag{2.83}$$

For $a/W \to 0$, K_I becomes equal to $\sigma\sqrt{\pi a}$, which corresponds to a crack of length $2a$ in an infinite plate (Equation (2.77)).

Similarly, for the case of a crack along the x-axis subjected to a pair of concentrated forces P (Figure 2.5(b)), for which the function Z_I is given by Equation (2.67), we obtain

$$K_I = \frac{P}{\sqrt{\pi a}}\sqrt{\frac{a+b}{a-b}} \tag{2.84}$$

where K_I refers to the crack tip $z = a$.

From Equations (2.76), by omitting the constant term, we obtain the principal singular stresses σ_1 and σ_2

$$\begin{aligned}\sigma_1 &= \frac{K_I}{\sqrt{2\pi r}} \cos \frac{\theta}{2}\left(1 + \sin \frac{\theta}{2}\right) \\ \sigma_2 &= \frac{K_I}{\sqrt{2\pi r}} \cos \frac{\theta}{2}\left(1 - \sin \frac{\theta}{2}\right).\end{aligned} \tag{2.85}$$

It is important to determine the displacement v (Figure 2.7) of the crack faces. From Equations (2.50), this is obtained as

$$2\mu v = \frac{\kappa + 1}{2} \operatorname{Im} \tilde{\tilde{Z}}_I \tag{2.86}$$

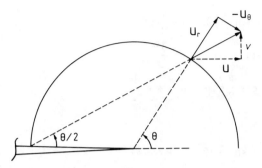

Figure 2.8. Rectangular and polar displacement components around the tip of a mode-I crack.

where for a crack of length $2a$ in an infinite plate subjected to stresses σ and $k\sigma$ at infinity the function Z_I is given by Equation (2.65). Omitting the constant term in Z_I we obtain

$$v = \frac{\kappa+1}{4\mu}\sigma\sqrt{a^2-x^2}, \quad -a \leq x \leq a. \tag{2.87}$$

Equation (2.87) can be put in the form

$$\frac{v^2}{t^2} + x^2 = a^2, \quad t = \frac{\kappa+1}{4\mu}\sigma \tag{2.88}$$

which shows the shape of the crack, after deformation, to be elliptic.

The vertical displacement v becomes maximum at the center of the crack. This is obtained from Equation (2.87):

$$v_{\max} = \frac{\kappa+1}{4\mu}\sigma a. \tag{2.89}$$

From Equations (2.21) and (2.78) which express the polar and rectangular components of the displacement in the vicinity of the crack tip, and by omitting the constant terms in Equation (2.78), we obtain

$$u_r = u, \quad u_\theta = -v. \tag{2.90}$$

This property of the displacement components is shown in Figure 2.8.

(b) Sliding mode

The Westergaard function Z_{II} for the case of a crack of length $2a$ in an infinite plate subjected to uniform in-plane shear stress τ at infinity (Figure 2.6) is given by Equation (2.70) and the stresses and displacements are obtained from Equations (2.56) and (2.57). Following the same procedure as in the previous case, and recognizing the general applicability of the singular solution for all sliding-mode crack problems, we obtain the following equations for the stresses and displacements:

$$\sigma_x = \frac{K_{II}}{\sqrt{2\pi r}} \sin\frac{\theta}{2}\left(2 + \cos\frac{\theta}{2}\cos\frac{3\theta}{2}\right)$$
$$\sigma_y = \frac{K_{II}}{\sqrt{2\pi r}} \sin\frac{\theta}{2}\cos\frac{\theta}{2}\cos\frac{3\theta}{2} \tag{2.91}$$
$$\tau_{xy} = \frac{K_{II}}{\sqrt{2\pi r}} \cos\frac{\theta}{2}\left(1 - \sin\frac{\theta}{2}\sin\frac{3\theta}{2}\right)$$

and

$$u = \frac{K_{II}}{2\mu}\sqrt{\frac{r}{2\pi}} \sin\frac{\theta}{2}(2 + \kappa + \cos\theta)$$
$$v = \frac{K_{II}}{2\mu}\sqrt{\frac{r}{2\pi}} \cos\frac{\theta}{2}(2 - \kappa - \cos\theta) \tag{2.92}$$

The quantity K_{II} is the sliding-mode stress intensity factor and, as in the previous case of the opening mode, it expresses the strength of the singular elastic stress field. When the Westergaard function Z_{II} is known, K_{II} is determined, following the same procedure as previously, by

$$K_{II} = \lim_{|\varsigma|\to 0} i\sqrt{2\pi\varsigma}\, Z_{II}. \tag{2.93}$$

For a crack of length $2a$ in an infinite plate subjected to in-plane shear stress τ at infinity, we obtain from Equations (2.93) and (2.70)

$$K_{II} = \tau\sqrt{\pi a}. \tag{2.94}$$

In a similar manner, for the case of an infinite periodic array of equally spaced cracks in an infinite plate subjected to in-plane shear stress τ at infinity, we obtain from Equations (2.93) and (2.71)

$$K_{II} = \tau\sqrt{\pi a}\sqrt{\frac{W}{\pi a}\tan\frac{\pi a}{W}}. \tag{2.95}$$

(c) *Tearing mode*

As shown in Section (2.3), the nonzero out-of-plane displacement w for the tearing (or antiplane) crack problem satisfies the Laplace equation (Equation (2.29)) and, therefore, can be put in the form

$$w = \frac{1}{\mu}\operatorname{Im} Z_{III} \tag{2.96}$$

where Z_{III} is an analytic function. From Hooke's law and the strain–displacement equations

$$\tau_{xz} = \mu\frac{\partial w}{\partial x}, \qquad \tau_{yz} = \mu\frac{\partial w}{\partial y} \tag{2.97}$$

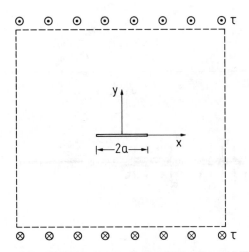

Figure 2.9. A crack of length 2a in an infinite plate subjected to uniform out-of-plane shear stress τ at infinity.

we obtain

$$\tau_{xz} = \text{Im } Z'_{\text{III}}, \qquad \tau_{yz} = \text{Re } Z'_{\text{III}}. \tag{2.98}$$

Consider now a crack of length $2a$ which occupies the segment $-a \leq x \leq a$ along the x-axis in an infinite plate subjected to uniform out-of-plane shear stress τ at infinity (Figure 2.9). The boundary conditions of the problem may be stated as

$$\begin{aligned}\tau_{yz} &= 0 \quad \text{for } y = 0, \quad -a < x < a \\ \tau_{xz} &= 0, \quad \tau_{yz} = \tau \quad \text{for } (x^2 + y^2)^{1/2} \to \infty.\end{aligned} \tag{2.99}$$

Following the same procedure as in the case of the opening mode (Section 2.4(b)), the function Z'_{III}, defined by

$$Z'_{\text{III}} = \frac{\tau z}{(z^2 - a^2)^{1/2}}, \tag{2.100}$$

satisfies the boundary conditions (2.99).

Referring the function Z'_{III} in the vicinity of the crack tip, we obtain for the stresses τ_{xz}, τ_{yz} and the displacement w

$$\tau_{xz} = -\frac{K_{\text{III}}}{\sqrt{2\pi r}} \sin \frac{\theta}{2}, \qquad \tau_{yz} = \frac{K_{\text{III}}}{\sqrt{2\pi r}} \cos \frac{\theta}{2},$$
$$w = \frac{2K_{\text{III}}}{\mu} \sqrt{\frac{r}{2\pi}} \sin \frac{\theta}{2} \tag{2.101}$$

where it was put

$$K_{\text{III}} = \tau \sqrt{\pi a}. \tag{2.102}$$

Equations (2.101) have general applicability for all tearing-mode crack problems when attention is paid in the neighborhood of the crack tip. The quantity

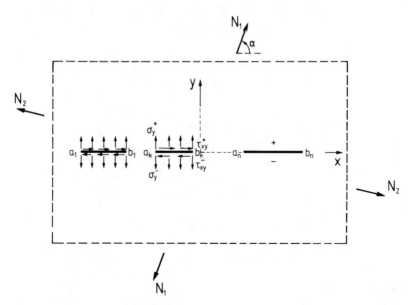

Figure 2.10. A system of cracks $L_k = a_k b_k$ $(k = 1, 2, \ldots, n)$ lying along the x-axis in an infinite plate subjected to principal stresses N_1 and N_2 at infinity.

K_{III} is the tearing-mode stress intensity factor and expresses the strength of the singular elastic stress field. When the function Z_{III} is given, K_{III} is determined by

$$K_{\text{III}} = \lim_{|\varsigma| \to 0} i\sqrt{2\pi\varsigma}\, Z_{\text{III}}. \tag{2.103}$$

In cases where two or all three deformation modes exist in a crack problem, the singular elastic stress field in the neighborhood of the crack tip is obtained by superimposing the solutions corresponding to each of the above three deformation modes and it is characterized by the respective stress intensity factors.

2.6. Method of complex potentials

The method of complex potentials, mainly developed by Muskhelishvili [2.12], provides a potential and attractive method for solving crack problems. The method is straightforward and deals with a broader class of problems than the Westergaard semi-inverse method. In this section the methodology which reduces the crack problem to that of Hilbert–Riemann is developed.

Consider a system of cracks $L_k = a_k b_k$ $(k = 1, 2, \ldots, n)$ lying along the x-axis in an infinite plate subjected to principal stresses N_1 and N_2 at infinity and let α be the angle subtended between N_1 and the x-axis (Figure 2.10). Assume also that the crack faces are subjected to the stress σ_y^+, τ_{xy}^+, σ_y^-, τ_{xy}^-, where the (+) and (−) signs refer to the boundary values on the left and right edges of the cracks when moving along the positive x-axis. Introducing the functions $\Phi(z)$ and $\Psi(z)$ from

$$\Phi(z) = \phi'(z), \qquad \Psi(z) = \psi'(z), \tag{2.104}$$

Equations (2.3) which express the stresses and displacements can be put in the form

$$\begin{aligned}
2\mu(u+iv) &= \kappa\phi(z) - z\,\overline{\phi'(z)} - \overline{\psi(z)} \\
\sigma_x + \sigma_y &= [2\Phi(z) + \overline{\Phi(z)}] \\
\sigma_y - \sigma_x + 2i\tau_{xy} &= 2[\bar{z}\Phi'(z) + \Psi(z)]
\end{aligned} \tag{2.105}$$

The functions $\Phi(z)$ and $\Psi(z)$ are holomorphic in the region S', which is the infinite plane cut along the n cracks L_k, and for large z take the form [2.12]

$$\begin{aligned}
\Phi(z) &= \Gamma - \frac{X+iY}{2\pi(1+\kappa)}\frac{1}{z} + O\!\left(\frac{1}{z^2}\right) \\
\Psi(z) &= \Gamma' + \frac{\kappa(X-iY)}{2\pi(1+\kappa)}\frac{1}{z} + O\!\left(\frac{1}{z^2}\right)
\end{aligned} \tag{2.106}$$

where (X, Y) is the resultant vector of the external forces applied to the union L of the crack faces L_k and

$$\Gamma = \tfrac{1}{4}(N_1 + N_2), \qquad \Gamma' = -\tfrac{1}{2}(N_1 - N_2)\,e^{-2i\alpha}. \tag{2.107}$$

In dealing with crack problems, a function $\Omega(z)$ is introduced from

$$\Omega(z) = \overline{\Phi}(z) + z\,\overline{\Phi'}(z) + \overline{\Psi}(z) \tag{2.108}$$

where the following notation applies for a function $F(z)$:

$$\overline{F}(z) = \overline{F(\bar{z})}. \tag{2.109}$$

From Equation (2.108) we get

$$\Psi(z) = \overline{\Omega}(z) - \Phi(z) - z\Phi'(z) \tag{2.110}$$

while from Equations (2.105), in conjunction with Equation (2.110), we have

$$\sigma_y - i\tau_{xy} = \Phi(z) + \Omega(\bar{z}) + (z - \bar{z})\,\overline{\Phi'(z)}. \tag{2.111}$$

Using Equations (2.106), the function $\Omega(z)$ has for large $|z|$ the form

$$\Omega(z) = \Gamma + \overline{\Gamma}' + \frac{\kappa(X+iY)}{2\pi(1+\kappa)}\frac{1}{z} + O\!\left(\frac{1}{z^2}\right). \tag{2.112}$$

It will be assumed that the functions $\Phi(z)$ and $\Psi(z)$ are sectionally holomorphic with a line of discontinuity L and that for all points t on L which do not coincide with an end

$$\lim_{y \to 0} y\Phi'(t + iy) = 0. \tag{2.113}$$

From Equations (2.111) and (2.113) the boundary conditions along the crack faces take the form

$$\Phi^+(t) + \Omega^-(t) = \sigma_y^+ - i\tau_{xy}^+ \\ \Phi^-(t) + \Omega^+(t) = \sigma_y^- - i\tau_{xy}^- \quad (2.114)$$

where $F^+(t)$ and $F^-(t)$ denote the boundary values of the function $F(z)$ at the point t of L from the left and right respectively when moving along the positive x-axis.

Addition and subtraction of Equations (2.114) leads to

$$[\Phi(t) + \Omega(t)]^+ + [\Phi(t) + \Omega(t)]^- = 2p(t) \\ [\Phi(t) - \Omega(t)]^+ - [\Phi(t) - \Omega(t)]^- = 2q(t) \quad (2.115)$$

where

$$2p(t) = \sigma_y^+ + \sigma_y^- - i(\tau_{xy}^+ + \tau_{xy}^-) \\ 2q(t) = \sigma_y^+ - \sigma_y^- - i(\tau_{xy}^+ - \tau_{xy}^-). \quad (2.116)$$

The first and second of Equations (2.115) constitute a Hilbert–Riemann problem [2.12]. The second of Equations (2.115) gives

$$\Phi(z) - \Omega(z) = \frac{1}{\pi i}\int_L \frac{q(t)\,dt}{t-z} - \overline{\Gamma}' \quad (2.117)$$

since, from Equations (2.106) and (2.112),

$$\Phi(\infty) - \Omega(\infty) = -\overline{\Gamma}'.$$

The first of Equations (2.115) gives

$$\Phi(z) + \Omega(z) = \frac{1}{\pi i X(z)}\int_L \frac{X(t)p(t)\,dt}{t-z} + \frac{2P_n(z)}{X(z)} \quad (2.118)$$

where

$$X(z) = \prod_{k=1}^{j}(z-a_k)^{1/2}(z-b_k)^{1/2} \quad (2.119)$$

and

$$P_n(z) = C_0 z^n + C_1 z^{n-1} + \cdots + C_n, \qquad C_0 \neq 0 \quad (2.120)$$

and $X(t)$ represents $X^+(t)$.

From Equations (2.117) and (2.118) the values of the complex potentials $\Phi(z)$ and $\Omega(z)$ are obtained. We have

$$\Phi(z) = \Phi_0(z) + \frac{P_n(z)}{X(z)} - \tfrac{1}{2}\overline{\Gamma}' \\ \Omega(z) = \Omega_0(z) + \frac{P_n(z)}{X(z)} + \tfrac{1}{2}\overline{\Gamma}' \quad (2.121)$$

where
$$\Phi_0(z) = \frac{1}{2\pi i X(z)} \int_L \frac{X(t)p(t)dt}{t-z} + \frac{1}{2\pi i} \int_L \frac{q(t)dt}{t-z} \qquad (2.122)$$
$$\Omega_0(z) = \frac{1}{2\pi i X(z)} \int_L \frac{X(t)p(t)dt}{t-z} - \frac{1}{2\pi i} \int_L \frac{q(t)dt}{t-z} \qquad (2.123)$$

The polynomial $X(z)$ has for large $|z|$ the form [2.12]
$$X(z) = z^n + a_{n-1}z^{n-1} + \cdots \qquad (2.124)$$

Adding equations and observing that $\Phi(\infty) = \Gamma$ we obtain
$$C_0 = \Gamma + \frac{\overline{\Gamma}'}{2}. \qquad (2.125)$$

The coefficients C_j ($j = 1, 2, \ldots, n$) of the polynomial $P_n(z)$ are determined from the condition of single-valuedness of the displacements. The first of Equations (2.105) takes the form
$$2\mu(u+iv) = \kappa\phi(z) - \omega(\bar{z}) - (z-\bar{z})\overline{\Phi(z)} + \text{const} \qquad (2.126)$$
where
$$\omega(z) = \int \Omega(z) = z\overline{\Phi}(z) + \overline{\psi}(z) + \text{const}. \qquad (2.127)$$

Equation (2.126) suggests that the expression $\kappa\phi(z) - \omega(\bar{z})$ should take its original value as the point z describes a contour surrounding the crack L_k. By contracting this contour to L_k the following system of n linear equations for the constants C_j ($j = 1, 2, \ldots, n$) is obtained:

$$2(\kappa+1)\int_{L_k} \frac{P_n(t)\,dt}{X(t)} + \kappa \int_{L_k} [\Phi_0^+(t) - \Phi_0^-(t)]dt +$$
$$+ \int_{L_k} [\Omega_0^+(t) - \Omega_0^-(t)]dt = 0, \quad k = 1, 2, \ldots, n \qquad (2.128)$$

When the crack faces are free of traction, $\Phi_0(z) = \Omega_0(z) = 0$ (Equations (2.122) and (2.123)), Equations (2.121) take the form
$$\Phi(z) = \frac{P_n(z)}{X(z)} - \frac{\overline{\Gamma}'}{2}, \qquad \Omega(z) = \frac{P_n(z)}{X(z)} + \frac{\overline{\Gamma}'}{2}, \qquad (2.129)$$

and the coefficients C_j are determined by Equation (2.125) and the conditions
$$\int_{L_k} \frac{P_n(t)\,dt}{X(t)} = 0, \quad k = 1, 2, \ldots, n. \qquad (2.130)$$

For a crack of length $2a$ occupying the segment $|x| \leq a$ along the x-axis, we obtain from Equations (2.129) and (2.130) that
$$\Phi(z) = \frac{(2\Gamma + \overline{\Gamma}')z}{2\sqrt{z^2-a^2}} - \frac{\overline{\Gamma}'}{2}, \qquad \Omega(z) = \frac{(2\Gamma + \overline{\Gamma}')z}{2\sqrt{z^2-a^2}} + \frac{\overline{\Gamma}'}{2}. \qquad (2.131)$$

For the case of uniform stresses $N_1 = \sigma$ and $N_2 = k\sigma$ at infinity along the y and x directions ($\alpha = 90°$), Equation (2.131) takes the form

$$\Phi(z) = \frac{\sigma z}{2\sqrt{z^2 - a^2}} - \frac{\sigma(1-k)}{4} \tag{2.132}$$

and gives the Westergaard function Z_I of Equation (2.65) by taking into account Equation (2.51).

When the complex potential $\Phi(z)$ for a given crack problem is known, the complex stress intensity factor K, defined by

$$K = K_I - iK_{II}, \tag{2.133}$$

may be obtained in the following manner suggested by Sih et al. [2.20].

From Equations (2.76) and (2.91) we obtain

$$\sigma_x + \sigma_y = \frac{2}{\sqrt{2\pi r}} \left(K_I \cos\frac{\theta}{2} - K_{II} \sin\frac{\theta}{2} \right) \tag{2.134}$$

which may be written as

$$\sigma_x + \sigma_y = \frac{\sqrt{2}}{\sqrt{\pi}} \operatorname{Re}\left[\frac{K}{\sqrt{z - z_1}} \right] \tag{2.135}$$

where

$$z = z_1 + r e^{i\theta} \tag{2.136}$$

with z_1 defining the location of the crack tip.

Then Equations (2.105) and (2.135) render

$$K = 2\sqrt{2\pi} \lim_{z \to z_1} [\sqrt{z - z_1}\, \Phi(z)]. \tag{2.137}$$

In crack problems with complicated geometry it is convenient to use conformal mapping of the physical problem on the unit circle or the half plane. The method has been used extensively by Bowie [2.21] who developed polynomial mapping approximations to complicated configurations involving cracks in finite plates and emanating from the boundary of circular holes. When the mapping is accomplished through the function

$$z = \omega(\varsigma), \tag{2.138}$$

where ς is the variable in the transformed plane, Equation (2.137) takes the form

$$K = 2\sqrt{2\pi} \lim_{\varsigma \to \varsigma_1} \left[\sqrt{\omega(\varsigma) - \omega(\varsigma_1)}\, \frac{\Phi(\varsigma)}{\omega'(\varsigma)} \right]. \tag{2.139}$$

Equation (2.139) is simplified by Sih [2.9] using L'Hopital's rule to

$$K = 2\sqrt{\pi}\, \frac{\Phi'(\varsigma_1)}{\sqrt{\omega''(\varsigma_1)}}. \tag{2.140}$$

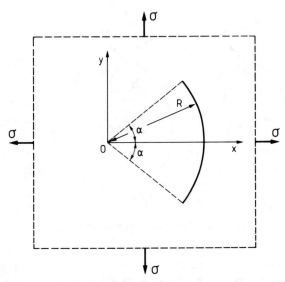

Figure 2.11. A circular crack in an infinite plate subjected to biaxial stress σ at infinity.

As an application of Equation (2.137), consider the problem of a circular crack of radius R and arc angle 2α in an infinite plate subjected to uniform biaxial stress σ (Figure 2.11). The complex potential $\Phi(z)$ is given by [2.12]

$$\Phi(z) = \frac{\sigma\sqrt{R}}{2(1+\sin^2 \alpha/2)} \left[\frac{z - R\cos\alpha}{\sqrt{R^2 - 2Rz\cos\alpha + z^2}} + \sin^2 \frac{\alpha}{2} \right]. \quad (2.141)$$

Using the transformation

$$z = i\,e^{i\alpha}\left(\varsigma - i - \sin\alpha\cos\alpha\right) \quad (2.142)$$

which relocates the crack tip on the x-axis, we obtain from Equations (2.137) and (2.141) [2.20]

$$\begin{aligned} K_{\mathrm{I}} &= \frac{\sigma\sqrt{\pi R}}{1+\sin^2(\alpha/2)}\sqrt{\frac{\sin\alpha(1+\cos\alpha)}{2}} \\ K_{\mathrm{II}} &= \frac{\sigma\sqrt{\pi R}}{1+\sin^2(\alpha/2)}\sqrt{\frac{\sin\alpha(1-\cos\alpha)}{2}}. \end{aligned} \quad (2.143)$$

2.7. Numerical methods

The methods presented in the previous sections for solving crack problems are generally restricted in elastic plates of infinite extent with simple geometrical configuration of cracks and boundary conditions. For more complicated situations one must resort to numerical methods. In what follows some of the more frequently used methods for obtaining approximate solutions of crack problems

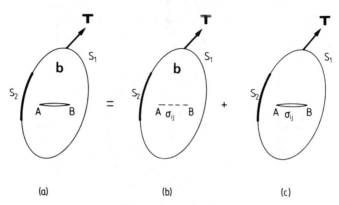

Figure 2.12. Illustration of Bueckner's principle.

are briefly described. Particular emphasis is usually paid to determining the stress intensity factors which characterize the strength of the singular elastic crack tip stress fields. For more information on these methods the reader is referred to Sih [2.14]. Before describing these methods the principle introduced by Bueckner [2.22], which greatly facilitates the solution of many crack problems, will be presented.

(a) *Bueckner's principle*

Consider an elastic body that contains a crack AB and is subjected to a body force distribution $\mathbf{b}$, prescribed tractions $\mathbf{T}$ along part of its boundary S_1 and prescribed displacements $\mathbf{u}$ along its remaining boundary S_2 (Figure 2.12(a)). Let the solution of this problem be expressed by the stress intensity factor at crack tips A, B as $K^a_{A,B}$. Consider now the uncracked body subjected to the same tractions $\mathbf{T}$, body forces $\mathbf{b}$ and displacements $\mathbf{u}$ and let σ_{ij} denote the stress distribution along the site of the crack AB (Figure 2.12(b)). If now the cracked body is subjected to the stresses σ_{ij} along the crack surfaces only ($\mathbf{T} = \mathbf{b} = \mathbf{u} = 0$) (Figure 2.12(c)) then Bueckner's principle states that $K^c_{A,B}$ is equal to $K^a_{A,B}$. Thus, any crack problem can be reduced to one in which the applied loads appear in the form of distributed tractions along the crack faces. This reduction not only greatly simplifies the solution of the problem but it allows a comparison of different load systems by comparing the corresponding systems of traction along the crack faces. Bueckner's principle forms the basis of many methods of solving crack problems, as will be seen in the sequel.

(b) *Green's function and weight functions*

The solution of a variety of crack problems can be obtained by using Bueckner's principle in conjunction with the problem of a crack with a concentrated force applied at some point on the crack boundary. The latter result can be used as a

Linear elastic stress field in cracked bodies

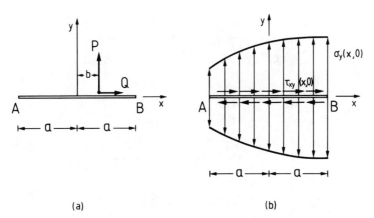

Figure 2.13. A crack of length $2a$ subjected (a) to concentrated forces P and Q and (b) to distributed forces $\sigma_y(x,0)$ and $\tau_{xy}(x,0)$ along the crack faces.

Green's function for obtaining the solution for any stress distribution along the crack surfaces. Thus, for the case of a crack AB in an infinite plate subjected to concentrated normal and tangential forces P and Q applied to the upper crack surface at $x = b$, the stress intensity factor $K = K_I - iK_{II}$ at tip B is ([2.11], Figure 2.13)

$$K = \frac{Q + iP}{2\sqrt{\pi a}} \left(\frac{\kappa - 1}{\kappa + 1} - i\sqrt{\frac{a+b}{a-b}} \right). \qquad (2.144)$$

For arbitrary distributions of tractions $\sigma_y(x, 0)$ and $\tau_{xy}(x, 0)$ across the crack boundaries, Equation (2.144) can be used as the Green's function for determining the stress intensity factor K. This is obtained in the form

$$\begin{aligned} K_I &= \frac{1}{2\sqrt{\pi a}} \int_{-a}^{a} \sigma_y(x, 0) \sqrt{\frac{a+x}{a-x}} \, dx + \frac{1}{2\sqrt{\pi a}} \frac{\kappa - 1}{\kappa + 1} \int_{-a}^{a} \tau_{xy}(x, 0) \, dx \\ K_{II} &= -\frac{1}{2\sqrt{\pi a}} \frac{\kappa - 1}{\kappa + 1} \int_{-a}^{a} \sigma_y(x, 0) \, dx + \frac{1}{2\sqrt{\pi a}} \int_{-a}^{a} \tau_{xy}(x, 0) \sqrt{\frac{a+x}{a-x}} \, dx. \end{aligned} \qquad (2.145)$$

For equal and opposite forces on the upper and lower crack faces Equations (2.145) become

$$\begin{aligned} K_I &= \frac{1}{\sqrt{\pi a}} \int_{-a}^{a} \sigma_y(x, 0) \sqrt{\frac{a+x}{a-x}} \, dx \\ K_{II} &= \frac{1}{\sqrt{\pi a}} \int_{-a}^{a} \tau_{xy}(x, 0) \sqrt{\frac{a+x}{a-x}} \, dx. \end{aligned} \qquad (2.146)$$

As an application, consider a crack in an infinite plate subjected to a linear stress distribution at infinity (Figure 2.14). Applying Bueckner's principle and using Equations (2.146) with $\sigma_y(x, 0) = \sigma_0(x/a - 1)$ we obtain K_I at the tip B

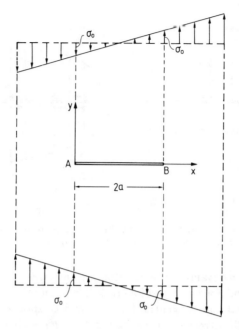

Figure 2.14. A crack of length 2a in an infinite plate subjected to a linear stress distribution at infinity.

of the crack, in the form

$$K_\text{I} = \frac{1}{\sqrt{\pi a}} \int_0^{2a} \sigma_0 \left(\frac{x}{a} - 1\right) \sqrt{\frac{x}{2a - x}} \, dx = \frac{\sigma_0 \sqrt{\pi a}}{2}. \tag{2.147}$$

The method of weight functions developed by Bueckner [2.23, 2.24] and Rice [2.25] is based on the following principle. Let $v(a, x)$ and K_I^* represent the crack face displacement and the opening-mode stress intensity factor for a crack of length a in a plate subjected to a load system. Then, the stress intensity factor K_I for another load system is given by

$$K_\text{I} = \int_0^a \sigma(x) h(a, x) \, dx \tag{2.148}$$

where $h(a, x)$ is the weight function, which is defined as

$$h(a, x) = \frac{8\mu}{1 + \kappa} \frac{1}{K^*} \frac{\partial v(a, x)}{\partial a} \tag{2.149}$$

and $\sigma(x)$ represents the stress distribution in the crack site in the uncracked body subjected to the load system which is associated with K_I.

Applying the weight function method to the previous case (Figure 2.14) and considering as an auxiliary stress system a uniform stress σ perpendicular to the

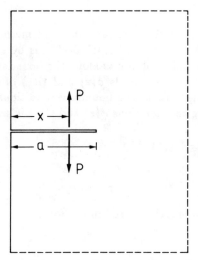

Figure 2.15. An edge crack of length a subjected to a pair of concentrated forces P in an infinite plate.

crack line, we have (Equations (2.77) and (2.87))

$$K_I^* = \sigma\sqrt{\pi a}, \qquad v(a,x) = \frac{\kappa+1}{4\mu}\sigma\sqrt{x(2a-x)}. \tag{2.150}$$

Equation (2.149) gives for the weight function

$$h(a,x) = \frac{1}{\sqrt{\pi a}}\sqrt{\frac{x}{2a-x}} \tag{2.151}$$

and substituting it into Equation (2.148) with $\sigma(x) = \sigma_0(x/a - 1)$ the value of the stress intensity factor K_I given by Equation (2.147) is obtained.

Green's functions for a variety of geometrical configurations have been obtained by various investigators. Thus, for the case of an edge crack of length a perpendicular to the boundary of a semi-infinite plate (Figure 2.15), Bueckner [2.24] gave the following expression for the stress intensity factor K_I:

$$K_I = P\sqrt{\frac{2}{\pi(a-x)}}\left[1 + 0.6147\left(1 - \frac{x}{a}\right) + 0.2502\left(1 - \frac{x}{2}\right)^2\right]. \tag{2.152}$$

The stress intensity factor for the same problem has also been determined by Hartranft and Sih [2.26]. The results of both solutions are in close agreement. Stress intensity factor solutions which can be used as Green's functions for various crack problems can be found in references [2.6], [2.11], [2.15], [2.20] and [2.27–2.29]. The method of weight functions has extensively been used in the literature for obtaining stress intensity factors. For more information the reader is referred to references [2.30–2.32].

(c) Boundary collocation

Boundary collocation provides a useful numerical method for determining the coefficients of the elastic solution of crack problems by satisfying the boundary conditions at a finite number of points along the boundary of the body. Taking as unknowns the complex potentials $\phi(z)$ and $\Omega(z)$ of Section 2.6, it can be shown that for stress-free cracks the boundary conditions across the crack faces are satisfied by putting the functions $\phi(z)$ and $\overline{\Omega}(z)$ in the form

$$\phi(z) = \frac{f(z)}{\sqrt{z^2 - a^2}} + g(z)$$
$$\overline{\Omega}(z) = \frac{\overline{f}(z)}{\sqrt{z^2 - a^2}} - \overline{g}(z) \tag{2.153}$$

for an internal crack of length $2a$, and in the form

$$\phi(z) = \frac{f(z)}{\sqrt{z}} + g(z)$$
$$\overline{\Omega}(z) = \frac{\overline{f}(z)}{\sqrt{z}} - \overline{g}(z) \tag{2.154}$$

for an edge crack, where the functions $f(z)$ and $g(z)$ are holomorphic everywhere in the body, including the crack line. Equations (2.153) are in agreement with the expressions given by Equations (2.129) for an infinite plate, where the functions $f(z)$ and $g(z)$ express the influence of the boundaries of the body on the complex potentials. The functions $f(z)$ and $g(z)$ must be determined such that the boundary conditions along the external boundaries (not including the crack faces) are satisfied. If we assume that the Laurent series expansions of the functions $f(z)$ and $g(z)$ are of the form

$$\left. \begin{array}{l} f(z) = a_0 + a_1 z + \cdots + a_n z^n \\ g(z) = b_0 + b_1 z + \cdots + b_n z^n \end{array} \right\} \quad n \geq 0 \tag{2.155}$$

the coefficients a_i, b_i $(i = 1, 2, \ldots, n)$ are determined by matching the boundary conditions at an appropriate number of discrete points on the boundary. Negative exponent integers in the series (2.155) have been excluded from the condition of finiteness of displacements at the crack tip. Usually, the number of selected points is more than that required to establish a system of algebraic equations for the determination of the unknown coefficients. The overdeterminate system of equations obtained is satisfied in a least-squares sense. For the determination of the stress intensity factor (Equation (2.137)) only the coefficient a_0 is needed.

The method of boundary collocation has been used by Gross et al. [2.33–2.35] for determining the stress intensity factors for fracture testing specimens containing single edge cracks and subjected to bending and tension. They used the eigenvalue series expansion developed by Williams [2.5], which is equivalent to the complex representation presented here. The method has also been used by Kobayashi et al. [2.36], among others, for determining the stress intensity factor in the problem of a central crack in a finite width strip.

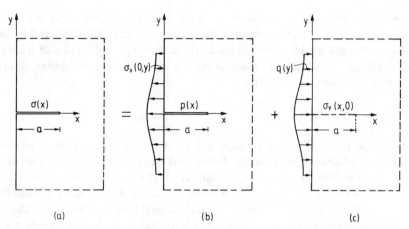

Figure 2.16. Illustration of the principle of the alternating method.

(d) Alternating method

This method, which is usually referred to as the Schwarz–Neumann alternating method, is based on the following principle: the solution of a problem is obtained by superimposing the solutions of simpler component problems, each of which satisfies the boundary conditions on some part of the boundary, so that the boundary conditions of the given problem are satisfied through an iterative process. The method is fully described by Kantorovich and Krylov [2.37] who solved potential problems by using a successive, iterative, superposition of sequences of solutions.

In order to present the method briefly, consider the case of a semi-infinite plate with an edge crack of length a perpendicular to the plate boundary and subjected to equal and opposite tractions $\sigma(x)$ along its upper and lower surfaces (Figure 2.16(a)). The method requires that the solution of the following two component problems are known: first, the problem of a crack of length $2a$ in an infinite plate subjected to tractions $p(x)$ along the crack faces (Figure 2.16(b)); and, second, the problem of a semi-infinite plate subjected to stresses $q(y)$ perpendicular to the plate boundary (Figure 2.16(c)). The solutions of these two problems are successively superimposed so that the boundary conditions of the original problem are satisfied.

The first sequence of solutions leads from the tractions $p(x)$ along the crack length to the stresses $\sigma_x(0, y)$ along the plate boundary. The second sequence of solutions leads from the stresses $q(y) = -\sigma_x(0, y)$ along the plate boundary to the tractions $\sigma_y(x, 0)$ along the perspective crack site. When the method is applied for the first time residual stresses along the perspective crack site arise. These stresses are then cancelled by solving once more the problem of Figure 2.16(b) which gives rise to normal stresses along the boundary of the semi-infinite plate. This process is repeated until the residual stresses on the crack faces and the boundary of the semi-infinite plate approach zero with any desired degree of accuracy.

The alternating method has been applied to two-dimensional crack problems by Irwin [2.38], Lachenbruch [2.39], Hartranft and Sih [2.26] and to three-dimensional crack problems by Smith et al. [2.40, 2.41], and Shah and Kobayashi [2.42, 2.43], among others. In these works the effect of the interference of a crack and a free boundary was studied.

(e) Integral transforms

This method is based on the transformation of the governing partial differential equations of elasticity to ordinary differential equations by performing an integral transform (Fourier, Laplace, Mellin, etc.) [2.44]. In order to illustrate the method consider the case of a crack of length $2a$ in an infinite plate subjected to symmetric pressure $p(x)$ along its boundaries. It is assumed that the crack occupies the segment $|x| \leq a$ along the x-axis and that the pressure is an even function of x. This problem can be formulated by specifying the following mixed boundary conditions on the boundary of the half-plane

$$\begin{aligned}\sigma_y(x,0) &= -p(x), & 0 \leq x \leq a \\ v(x,0) &= 0, & x > a\end{aligned} \tag{2.156}$$

and $\tau_{xy} = 0$ for all values of x.

The boundary conditions lead, by application of Fourier transform, to the following dual integral equations

$$\begin{aligned}\sqrt{\frac{2}{\pi}} \frac{d}{dx} \mathcal{F}_s[\Psi(\xi); x] &= p(x) & 0 \leq x \leq a \\ \mathcal{F}_c[\Psi(\xi); x] &= 0 & x > a\end{aligned} \tag{2.157}$$

for the determination of the unknown function $\Psi(\xi)$ from which the stress and displacement components are defined. The operators $\mathcal{F}_s$ and $\mathcal{F}_c$ denote the sine and cosine Fourier transforms defined by

$$\begin{aligned}\mathcal{F}_s[f(\xi,y); \xi \to x] &= \sqrt{\frac{2}{\pi}} \int_0^\infty f(\xi,y) \sin(\xi x)\, d\xi \\ \mathcal{F}_c[f(\xi,y); \xi \to x] &= \sqrt{\frac{2}{\pi}} \int_0^\infty f(\xi,y) \cos(\xi x)\, d\xi\end{aligned} \tag{2.158}$$

The unknown function $\Psi(\xi)$ is expressed in terms of another function $g(t)$ by

$$\Psi(\xi) = \int_0^c tg(t) J_0(\xi t)\, dt \tag{2.159}$$

where J_0 is the Bessel function of the first kind of zero order. The function $g(t)$ is now determined from the Abel integral equation

$$\frac{2}{\pi} \frac{d}{dx} \int_0^x \frac{tg(t)\, dt}{\sqrt{x^2 - t^2}} = p(x), \quad 0 < x < a \tag{2.160}$$

which gives

$$g(t) = \int_0^t \frac{p(x)\,dx}{\sqrt{t^2 - x^2}}, \qquad 0 < t < a. \tag{2.161}$$

The stress $\sigma_y(x,0)$ and the displacement $v(x,0)$ are defined by

$$\begin{aligned}\sigma_y(x,0) &= -\frac{2}{\pi}\frac{d}{dx}\int_0^a \frac{tg(t)\,dt}{\sqrt{x^2 - t^2}}, \quad x > a \\ v(x,0) &= \frac{4(1-\nu^2)}{\pi E}\int_x^c \frac{tg(t)\,dt}{\sqrt{t^2 - x^2}}, \quad 0 \le x \le a.\end{aligned} \tag{2.162}$$

From Equations (2.162) the stress intensity factor can be determined.

The method of integral transforms has been used by Sneddon [2.45] for the solution of a variety of crack problems.

(f) Continuous dislocations

This method is based on the representation of a crack by a continuous distribution of dislocation singularities. A brief description of the method is given for the case of a crack occupying the segment $-a \le x \le a$ of the x-axis and subjected to surface tractions $\sigma_y(x,0) = -p(x)$.

Consider an edge dislocation at $x = t$, $y = 0$ with Burger's vector b_y. The stresses along the x-axis are given by

$$\sigma_x = -\frac{2\mu b_y}{\pi(\kappa+1)}\frac{1}{x-t}, \quad \sigma_y = \frac{2\mu b_y}{\pi(\kappa+1)}\frac{1}{x-t}, \quad \tau_{xy} = 0. \tag{2.163}$$

The crack is modeled by a continuous distribution of edge dislocations with the density function $\mu(t)$, so that $\mu(t)\,dt$ represents the infinitesimal Burger's vector of a dislocation at location $x = t$. Thus, using Equations (2.163), the stresses along the x-axis for the crack are

$$\sigma_x = -\frac{2\mu}{\pi(\kappa+1)}\int\frac{\mu(t)\,dt}{x-t}, \quad \sigma_y = \frac{2\mu}{\pi(\kappa+1)}\int\frac{\mu(t)\,dt}{x-t}, \quad \tau_{xy} = 0. \tag{2.164}$$

The boundary condition $\sigma_y(x,0) = -p(x)$ along the crack leads to the equation

$$\frac{2\mu}{\pi(\kappa+1)}\int_{-a}^{a}\frac{\mu(t)\,dt}{t-x} = p(x) \tag{2.165}$$

and the condition of single-valued displacements requires that the net Burger's vector around the crack be zero, that is

$$\int_{-a}^{a}\mu(t)\,dt = 0. \tag{2.166}$$

Equation (2.165) constitutes a singular integral equation with the $1/(t-x)$ type of kernel. For the solution of such equations the reader is referred to [2.12] and [2.46]. Equations (2.165) and (2.166) lead to the well-known solution, obtained previously, of a crack in an infinite plate subjected to a uniform pressure $p(x)$. Further information on this subject can be found in [2.47].

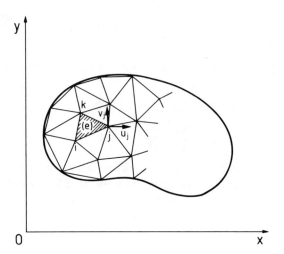

Figure 2.17. Finite element discretization of a continuum.

(g) *Finite elements*

The finite element method [2.48, 2.49] constitutes a powerful procedure for obtaining approximate solutions to continuum problems. The main advantage of the method is its ability to treat, in a unified manner, complex geometric configurations and boundary conditions in continuum bodies with general material properties. The method leads to the development of general computer programs which are applicable to the solution of a wide range of engineering problems. The following gives a brief description of the basic principles of the finite element method, with particular emphasis on two-dimensional crack problems.

(i) *Basic principles.* The two-dimensional body of interest is divided into a number of subregions, referred to as elements (Figure 2.17). An approximate form of the solution within each element expressed in terms of parameters at discrete points (nodes) of the element boundary is assumed. These parameters may be chosen to be the displacement components, the averaged boundary tractions or a combination of these two and the method is referred to as the displacement method, the force method or the hybrid method, respectively. For the most frequently used displacement method the displacement field $\{u\}$ within each element, e, is expressed in terms of the m discrete displacements $\{\delta_e\} = \{\delta_1, \ldots, \delta_m\}$ by

$$\{u\} = [N]\{\delta_e\} \tag{2.167}$$

where $[N]$ is a function of spatial coordinates and describes the assumed displacement interpolation function for the element e. From the displacement field $\{u\}$ the strain $\{\epsilon\}$ and stress $\{\sigma\}$ fields are obtained by

$$\{\epsilon\} = [C]\{u\} \tag{2.168}$$

Linear elastic stress field in cracked bodies

$$\{\sigma\} = [D]\{\epsilon\} \tag{2.169}$$

where $[C]$ is a partial differential operator and $[D]$ is the elasticity matrix.

The potential energy of the element e is

$$\Pi_e = \tfrac{1}{2} \iiint_{V_e} [\sigma]^T \{\epsilon\} \, dV - \iint_{S_e} \{u\}^T \{X\} \, dS \tag{2.170}$$

where V_e and S_e denote the volume and surface area of the element, $\{X\}$ is the surface traction vector and the superscript T denotes the transpose of the matrix to which it is applied. Using Equations (2.168) and (2.169), Equation (2.170) takes the form

$$\Pi_e = \tfrac{1}{2} \{\delta_e\}^T [k^e] \{\delta_e\} - \{\delta_e\}^T \{f^e\} \tag{2.171}$$

where the element stiffness matrix $[k^e]$ and the element force vector $\{f^e\}$ are given by

$$[k^e] = \iiint_{V_e} [N]^T [C]^T [D]^T [C][N] \, dV \tag{2.172}$$

$$\{f^e\} = \iint_{S_e} [N]^T \{X\} \, dS. \tag{2.173}$$

If the displacement interpolation functions $[N]$ ensure continuity of the displacement components across the element boundaries and the displacement boundary conditions are satisfied, then the unknown displacements δ_i ($i = 1, 2, \cdots, n$) are determined from the minimization of the potential energy Π of the body. Thus we have

$$\frac{\partial \Pi}{\partial \{\delta\}} = \begin{bmatrix} \frac{\partial \Pi}{\partial \delta_1} \\ \vdots \\ \frac{\partial \Pi}{\partial \delta_n} \end{bmatrix} = 0 \tag{2.174}$$

where

$$\Pi = \sum_{e=1}^{l} \Pi_e \tag{2.175}$$

with l representing the number of elements of the body.

From Equation (2.171) we obtain

$$\frac{\partial \Pi_e}{\partial \{\delta_e\}} = [k^e]\{\delta_e\} - \{f^e\}. \tag{2.176}$$

Thus, Equation (2.174) renders

$$[K]\{\delta\} = \{F\} \tag{2.177}$$

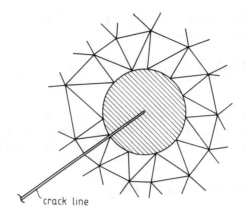

Figure 2.18. A special crack-tip element connected to regular elements.

where

$$[K_{ij}] = \sum_{e=1}^{l} [k_{ij}^e] \qquad (2.178)$$

$$\{F_i\} = \sum_{e=1}^{l} \{f_i^e\}. \qquad (2.179)$$

Equations (2.178) and (2.179) give the total stiffness matrix and the force vector of the body.

From the solution of Equation (2.177) the unknown displacements $\{\delta\}$ are obtained and the strains and stresses are calculated from Equations (2.168) and (2.169).

(*ii*) *Application to crack problems.* The finite element method was first directly applied to cracked structures in the same manner as for the analysis of non-cracked structures. The body was divided into conventional elements with a high element concentration near the crack tip to account for the stress singularity. The stress intensity factor, which measures the strength of the crack tip singularity, is calculated by substituting the value of a displacement or stress component at a node in the vicinity of the crack tip to the singular stress/displacement equations (Equations (2.76) and (2.78)). The results obtained by this straightforward application of the finite element method to crack problems are unreliable and highly inaccurate. This is attributed to the fact that the polynomial approximation for the displacement field assumed within the elements is incapable of accounting for the square root variation of the displacement components in the close neighborhood of the crack tip. By increasing the number of elements in the vicinity of the crack tip, more accurate results can be obtained.

In order to eliminate the drawbacks of the direct application of the finite element method to crack problems, and to explicitly account for the crack tip singularity, singular elements have been developed [2.50, 2.51]. With this type of element, the square root displacement field is given by the elasticity solution. The idea is to combine the singular elastic solution, which is *highly accurate* in the vicinity of the crack tip, with the finite element method, which can give excellent results everywhere *except* in the vicinity of the crack tip. For the general case of a crack in a mixed-mode stress field, the parameters which characterize the behavior of the singular element (which has a circular shape centered at the crack tip) are the stress intensity factors K_I and K_{II} and the crack tip displacements δ_I and δ_{II}. This special element is connected to standard elements along its boundary (Figure 2.18) by giving the displacement components at the nodal points of the boundary the values obtained from the singular solution. The potential energy of the body is now expressed in terms of K_I, K_{II}, δ_I, δ_{II} and the nodal displacements δ_i of the finite element grid pattern. Minimization of the potential energy Π of the cracked body with respect to these quantities results in the equations

$$\frac{\partial \Pi}{\partial K_I} = \frac{\partial \Pi}{\partial K_{II}} = \frac{\partial \Pi}{\partial \delta_I} = \frac{\partial \Pi}{\partial \delta_{II}} = \frac{\partial \Pi}{\partial \delta_i} = 0 \qquad (2.180)$$

and from their solution we obtain the values of K_I, K_{II}, δ_I, δ_{II} and δ_i.

Using this approach the displacement components are not, in general, continuous along the boundary of the singular element and the compatibility requirement which ensures convergence of the method is violated there. The displacement incompatibility vanishes as the number of nodes along the element boundary becomes infinite. The accuracy of the method depends upon the radius of the singular element and the number of elements and nodes along the element boundary. A circular singular element with a displacement function incorporating higher order terms of the asymptotic expansion of the elasticity displacement field in the vicinity of the crack tip was developed by Wilson [2.50].

A special triangular element, which has a nodal point at the crack tip and contains a displacement field that varies as the square root of the distance from the crack tip, has been developed by Tracey [2.52]. This element is made by requiring two nodes of a quadrilateral isoparametric element to coincide. The crack tip triangular elements are surrounded by ordinary isoparametric elements and interelement displacement compatibility is guaranteed. Other types of elements of major importance in crack problems are the eight-noded plane quadrilateral isoparametric element and the twenty-noded three-dimensional isoparametric element with midside nodes near the crack tip placed at the quarter point. In these elements an inverse square root stress singularity develops only along the element edges. Barsoum [2.53] showed that the triangular element that results from the quadrilateral by making two nodes coincide with the crack tip has the above singularity for all rays starting from the crack tip. The same property of the inverse square root stress singularity is found to hold for the twelve-noded quadrilateral isoparametric element by placing the intermediate nodes to the 1/9 and 4/9 positions nearest the crack tip [2.54]. A very important feature of these elements is that they satisfy the necessary convergence requirements.

In comparing the various types of special crack tip elements it is observed that the embedded singularity element results in a stress field which is asymptotically exact in the vicinity of the crack tip, but it lacks compatibility with its neighboring elements. This element has the advantage of direct calculation of stress intensity factors. The Tracey element satisfies compatibility but uses a piecewise linear function for the approximation of the circumferential variation of the displacement components. Furthermore, this element does not pass the patch test [2.48] and does not possess rigid body motion. The quadrilateral isoparametric elements with appropriately placed midside nodes pass the patch test, possess rigid body motion, interelement compatibility and displacement continuity. Thus, they satisfy the convergence requirements.

(iii) *Determination of stress intensity factors.* With the exception of the embedded singularity element where the stress intensity factor is obtained directly, in all other elements presented above it should be calculated from the obtained nodal displacements and stresses in the neighborhood of the crack tip. Thus, the stress intensity factor for mixed-mode loading conditions is determined from the values of the displacements and stresses in the vicinity of the crack tip in conjunction with the singular elastic solution. For opening-mode crack problems the stress intensity factor K_I can also be obtained by calculating the elastic strain energy of the body for two or more slightly different crack lengths. As we shall show later, the strain energy release rate, which is the energy available for unit crack extension, is given by

$$G = \pm \frac{\partial U}{\partial A} \qquad (2.181)$$

where U is the elastic strain energy of the body containing the crack and A is the crack area. The plus-and-minus sign in Equation (2.181) refers to crack extensions under constant load or constant displacement conditions. The strain energy U stored in the cracked body is obtained from the finite element solution by summing over the elements.

On the other hand, as will be shown later, the strain energy release rate is

$$G = \beta \frac{K_\mathrm{I}^2}{E} \qquad (2.182)$$

where $\beta = 1$ or $(1 - \nu^2)$ for generalized plane stress or plane strain conditions. The stress intensity factor K_I is obtained from Equations (2.181) and (2.182).

Better results for opening-mode problems are obtained by numerically calculating from the finite element solution the line integral

$$J = \int_\Gamma W \, dy - \mathbf{T} \frac{\partial \mathbf{u}}{\partial x} \, ds \qquad (2.183)$$

where Γ is an arbitrary curve surrounding the crack tip. In Equation (2.183), W is the strain energy density, $\mathbf{T}$ is the traction vector defined by $T_i = \sigma_{ij} n_j$ where n_j $(j = 1, 2, 3)$ is the outward normal to Γ, and $\mathbf{u}$ is the displacement vector. The line integral J is calculated in a counterclockwise sense.

We shall show later that

$$J = G, \tag{2.184}$$

which allows us to determine the stress intensity factor K_I.

For further details on the solution of crack problems by the finite element method the reader is referred to references [2.55–2.58].

2.8. Experimental methods

The stress field in cracked bodies can also be determined by using experimental methods. Difficulties arise in estimating the stress intensity factor, which cannot be measured directly in an experiment but is determined from its relationship to experimentally measurable quantities. Optical methods of stress analysis, such as photoelasticity, moiré, interferometry, holography, etc., have been used in determining stress intensity factors in cracked bodies. The direct application of these methods to estimate stress intensity factors needs particular attention for the evaluation of the experimental results. This results from the steep variation of the fringes in the corresponding optical patterns due to the existing singularity at the crack tip. Thus, particular methods and techniques for extracting stress intensity factors from optical patterns have been developed. In the present section the optical methods of photoelasticity and caustics which are particularly relevant in determining stress intensity factors will be briefly presented.

(a) The method of photoelasticity

The photoelastic method of stress analysis is based on the phenomenon of a temporary or artificial birefringence that is observed in certain noncrystalline and initially optically isotropic materials when they are subjected to a stress field [2.59]. In such circumstances a light ray, when entering a body, splits into two rays that are linearly polarized at right angles to each other and propagate with different velocities. The principal axes of the optical birefringence for a two-dimensional model coincide with the principal stress axes, and the optical birefringence δ is

$$\delta = C(\sigma_1 - \sigma_2)t \tag{2.185}$$

where σ_1 and σ_2 are the two principal stresses, t is the model thickness and C is the stress optical constant. Equation (2.185) can be stated in the form

$$2\tau_m = \sigma_1 - \sigma_2 = \frac{Nf}{t} \tag{2.186}$$

where $f = \lambda/C$ with λ the wavelength of the monochromatic light used in the optical polariscope, τ_m is the maximum in-plane shear stress and N is the isochromatic fringe order.

If now Equations (2.76) are used to determine $\tau_{\max}$ for opening-mode crack problems, where the term $(1-k)\sigma$ on the right-hand side of the stress σ_x in

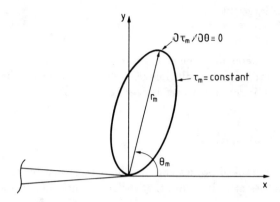

Figure 2.19. A crack-tip isochromatic fringe loop.

Equations (2.76) is substituted by the more general term σ_{0x} which accounts for distant field stresses, we obtain

$$(2\tau_m)^2 = \frac{K_I}{2\pi r} \sin^2 \theta + \frac{2\sigma_{0x} K_I}{\sqrt{2\pi r}} \sin \theta \sin \frac{3\theta}{2} + \sigma_{0x}^2. \tag{2.187}$$

A typical isochromatic fringe loop is shown in Figure 2.19. The position on this loop of the farthest point from the crack tip is dictated by

$$\frac{\partial \tau_m}{\partial \theta} = 0 \tag{2.188}$$

which gives

$$\sigma_{0x} = \frac{-K_I}{\sqrt{2\pi r_m}} \frac{\sin \theta_m \cos \theta_m}{(\cos \theta_m \sin(3\theta_m/2) + \frac{3}{2} \sin \theta_m \cos(3\theta_m/2))} \tag{2.189}$$

where r_m is the distance from the crack tip to the farthest point on a given loop and θ_m is its angle of inclination to the crack plane.

From Equations (2.186) and (2.189) the two unknown quantities K_I and σ_{0x} are determined as

$$K_I = \frac{Nf\sqrt{2\pi r_m}}{t \sin \theta_m} \left[1 + \left(\frac{2}{3 \tan \theta_m}\right)^2\right]^{1/2} \left(1 + \frac{2 \tan(3\theta_m/2)}{3 \tan \theta_m}\right) \tag{2.190}$$

$$\sigma_{0x} = -\frac{Nf}{t} \frac{\cos \theta_m}{\cos(3\theta_m/2)(\cos^2 \theta_m + \frac{9}{4} \sin^2 \theta_m)^{1/2}}. \tag{2.191}$$

The quantities N, r_m and θ_m, which are required for the determination of K_I and σ_{0x}, are measured from a single ischromatic loop. A typical isochromatic pattern in the vicinity of the crack tip is presented in Figure 2.20. The data N, r_m and θ_m should be taken from a zone dominated by the singular stresses very close to the crack tip. As we recede from the crack tip additional terms to those associated with the singular solution are needed. However, if one measures too close to the crack tip, invalid data may result. In analyzing crack problems a 'process' or 'core' region surrounding the crack tip must be introduced. This

Figure 2.20. A mode-I isochromatic fringe pattern in the vicinity of a crack tip.

arises from the inability of the analytical solution to describe the state of affairs in the immediate vicinity of the crack tip. The material within the core region is highly strained, may even be inhomogeneous, and the continuum mathematical model fails to describe its physical behavior there. The zone of valid data should, therefore, extend outside the core region. Also, additional factors incorporating the geometry of the notch-end which simulates the crack, the stress singularity and the triaxiality of the stress field put further restrictions on the determination of the data zone. An analysis of the factors which influence the isochromatic fringe pattern in the neighborhood of the crack tip can be found in the work by Gdoutos [2.60], where it is shown that measurements in the close vicinity of the crack tip are prone to the deleterious influence of various factors which alter the physical meaning of isochromatics which differ from those corresponding to the respective purely elastic problem. It is therefore attractive to gather data far enough away from the crack tip where the influence of all these factors on the isochromatic pattern becomes negligible. This requires additional terms for the description of the stress field than those associated with the singular stress components. Such studies can be found in references [2.60–2.65].

Photoelasticity has also been applied to the determination of stress intensity factors in mixed-mode two-dimensional crack problems. By incorporating the constant stress term σ_{0x} we obtain from Equations (2.76) and (2.91) the following expression for the maximum in-plane shear stress τ_m:

$$\tau_m^2 = \frac{1}{8\pi r}\left[\sin^2\theta\, K_I^2 + 2\sin 2\theta\, K_I K_{II} + (4 - 3\sin^2\theta) K_{II}^2\right] +$$
$$+ \frac{\sigma_{0x}}{2\sqrt{2\pi r}}\sin\frac{\theta}{2}\left[\sin\theta(1 + 2\cos\theta) K_I + (1 + 2\cos^2\theta + \cos\theta) K_{II}\right] +$$
$$+ \frac{\sigma_{0x}^2}{4}. \tag{2.192}$$

Equation (2.192) can be used to determine the unknown quantities K_I, K_{II}, σ_{0x} from measurements on the isochromatic fringe pattern. Further details on the use of photoelasticity in mixed-mode crack problems can be found in references [2.66–2.69].

(b) *Method of caustics*

The method of caustics, known also as the method of shadow patterns, constitutes a simple and powerful method for measuring stress intensity factors in crack problems. Generally speaking, the stress singularity in the vicinity of the crack tip is transformed by simple optical means (based on geometric optics) into an optical singularity by defining a singular boundary around the crack tip. The light rays reflected from this boundary form an optical singularity which is strongly illuminated and corresponds to a caustic. The dimensions of the caustic are directly related to the intensity of the stress field near the crack tip which is controlled by the value of the stress intensity factor. The method was first introduced by Manogg [2.70] for the simple case of a propagating crack submitted to pure tension, but it was further developed and demonstrated as an effective tool in stress analysis by Theocaris in a series of publications [2.71–2.73]. In this section the method of caustics, developed by Theocaris and Gdoutos [2.74], for the determination of the opening-mode, K_I, and sliding-mode, K_{II}, stress intensity factors in two-dimensional stress fields will be briefly presented.

The experimental arrangement for the application of the method is very simple. The cracked specimen is illuminated by a coherent monochromatic light beam emitted by a laser. The reflected and/or transmitted light rays are received on a viewing screen placed parallel and at some distance from the specimen. The caustic formed on the viewing screen is a highly illuminated curve and its dimensions can be determined accurately.

The equation of the caustic curve formed on the viewing screen is given by

$$W(X + iY) = r_0\left(e^{i\theta} + \frac{2}{3}\frac{K_I + iK_{II}}{\sqrt{K_I^2 + K_{II}^2}} e^{3i\theta/2}\right) \tag{2.193}$$

where r_0 is the radius of the circular initial curve formed on the specimen which generates the caustic and θ is the polar angle measured from the crack axis. r_0 is given by

$$r_0 = \left(\frac{3C\sqrt{K_I^2 + K_{II}^2}}{2\sqrt{2\pi}}\right)^{2/5} \tag{2.194}$$

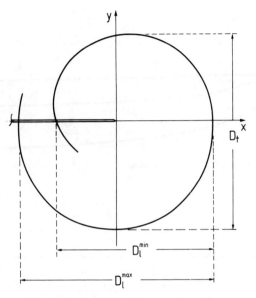

Figure 2.21. A mixed-mode caustic with $\mu = 1$.

where C is a global optical constant depending on the stress-optical properties and the thickness of the model material and also on the distance between the model and the viewing screen. C takes different values, depending on whether the caustics are formed by being transmitted or reflected from the front or the rear face of the model light rays.

Introducing an angle ω by

$$\frac{K_I + iK_{II}}{\sqrt{K_I^2 + K_{II}^2}} = e^{i\omega}, \qquad \mu = \tan \omega = \frac{K_{II}}{K_I} \qquad (2.195)$$

Equation (2.193) takes the following parametric form:

$$\begin{aligned} X &= r_0 \left[\cos \theta + \tfrac{2}{3} \cos \left(\tfrac{3\theta}{2} + \omega \right) \right] \\ Y &= r_0 \left[\sin \theta + \tfrac{2}{3} \sin \left(\tfrac{3\theta}{2} + \omega \right) \right] \end{aligned} \qquad (2.196)$$

If these equations are referred to a new system of Cartesian axes passing through the crack tip and subtending an angle -2ω with the original axes, the following equations are obtained

$$\begin{aligned} X' &= r_0 \left(\cos \phi + \tfrac{2}{3} \cos \tfrac{3\phi}{2} \right) \\ Y' &= r_0 \left(\sin \phi + \tfrac{2}{3} \sin \tfrac{3\phi}{2} \right) \\ \phi &= \theta + 2\omega. \end{aligned} \qquad (2.197)$$

From the latter equations it is concluded that the caustic is symmetrical about an axis subtending an angle -2ω with respect to the crack axis. A graphical rep-

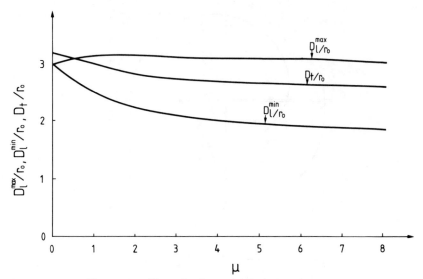

Figure 2.22. Normalized caustic diameters versus μ.

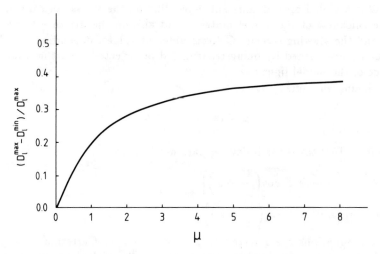

Figure 2.23. Variation of $(D_l^{\max} - D_l^{\min})/D_l^{\max}$ versus μ.

resentation of the caustic curve, which has the form of a generalized epicycloid, is shown in Figure 2.21 for $\mu = K_{II}/K_I = 1$. From the geometrical properties of the caustic, the variation of the quantities $D_l^{\max}/r_0$, $D_l^{\min}/r_0$ and D_t/r_0 – where $D_l^{\max}$, $D_l^{\min}$ denote the maximum and minimum diameters of the caustic along the crack axis and D_t is the transverse diameter of the caustic (Figure 2.21) – is shown versus μ in Figure 2.22. Furthermore, Figure 2.23 presents the variation of $(D_l^{\max} - D_l^{\min})/D_l^{\max}$ versus μ.

Linear elastic stress field in cracked bodies

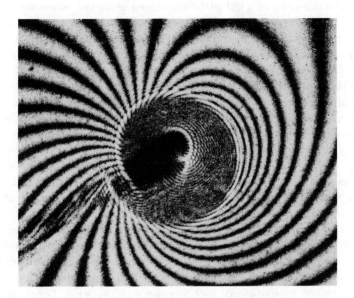

Figure 2.24. Experimental mixed-mode caustic.

From Equation (2.194) we get

$$K_I = \frac{2\sqrt{2\pi}}{3C} \frac{1}{\sqrt{1+\mu^2}} \left(\frac{D}{\delta}\right)^{5/2} \quad (2.198)$$

$$K_{II} = \mu K_I$$

where δ denotes the ratio D/r_0 and D represents any of the three diameters $D_l^{\max}$, $D_l^{\min}$ or D_t of the caustic.

Equation (2.198), in conjunction with Figures 2.22 and 2.23, is used for the experimental determination of the stress intensity factors K_I and K_{II}. The diameters $D_l^{\max}$, $D_l^{\min}$ and D_t of the caustic obtained from the experiment (Figure 2.24) are measured and the diagram of Figure 2.23 is first used to determine the ratio $\mu = K_{II}/K_I$. The quantity δ is then obtained from Figure 2.22 and K_I, K_{II} are determined from Equations (2.198).

Further information on the method of caustics and its application to crack problems can be found in the two review papers by Theocaris [2.72, 2.73].

Besides the methods of photoelasticity and caustics, other experimental stress analysis methods have also been used for analyzing crack problems. For a survey of these methods the reader is referred to reference [2.75].

2.9. Three-dimensional crack problems

(a) Introductory remarks

In the preceding sections of this chapter the problem of a plane crack extending through the thickness of a flat plate was solved using the two-dimensional theory of elasticity. However, many embedded cracks or flaws in engineering structures have irregular shapes. These flaws are three-dimensional and for purposes of analysis are usually idealized as planes of discontinuities bounded by smooth curves. The basic shapes that are most suitable for analysis are the circular or penny-shaped and the elliptical embedded cracks. In the elliptical crack, various degrees of crack-border curvedness may be obtained by varying the ellipticity.

Three-dimensional surface and embedded cracks are frequently encountered in engineering structures. Thus, surface cracks are usually initiated from the interior of pressure vessels and pipelines which are commonly used in the nuclear industry. Because of their importance in the design of a variety of structures, three-dimensional cracks have attracted the interest of engineers and researchers. A great amount of effort has been spent on the determination of the stress distribution in bodies with three-dimensional cracks. Sneddon [2.4] and Sack [2.76] were the first to find the solution for the case of a penny-shaped crack in an infinite solid subjected to a uniform tension normal to the plane of the crack. The case when the solid is loaded by a uniform shear parallel to the crack plane was studied by Segedin [2.77]. The problem of an ellipsoidal cavity in an infinite solid was solved by Sadowsky and Sternberg [2.78], while Green and Sneddon [2.79] found the stress distribution near an elliptical crack. Irwin [2.80], using the results of [2.79], calculated the stress intensity factor at any location along the crack border.

Since these pioneering works a large number of publications have appeared in the literature. Sih and coworkers [2.15, 2.81–2.84] performed a thorough study of three-dimensional crack problems. They expressed the local stress field near the crack front in a form analogous to the two-dimensional case in terms of three stress intensity factors which are independent of the local coordinates, and are depending only on the crack geometry, the form of loading and the location of the point along the crack border. This result is fundamental in analyzing the fracture behavior of cracks and provides uniform expressions for the local stresses under various geometrical and loading conditions where only the values of stress intensity factors differ. A great variety of stress solutions for internal and external cracks in three dimensions, including also the effects of material anisotropy and nonhomogeneity, is provided by Kassir and Sih [2.15].

The above analytical solutions are mainly concerned with bodies of infinite extent. However, when the dimensions of the body become finite, mathematical difficulties are experienced and numerical or experimental methods are used. Kobayashi and coworkers [2.40, 2.42, 2.43, 2.85, 2.86] determined stress intensity factor variations along the periphery of a semi-elliptical crack in a plate of finite thickness under various loading conditions. The finite element method was also extensively used for the same problem. Special three-dimensional elements were introduced, in which the inverse square root singularity of the stress field in the vicinity of the crack front was embedded [2.87, 2.88]. Results using the finite element method for semi-elliptical surface cracks were obtained by Raju and Newman [2.89–2.90], and solutions of three-dimensional crack problems using numerical techniques can also be found in references [2.41, 2.91–2.93]. From

Linear elastic stress field in cracked bodies

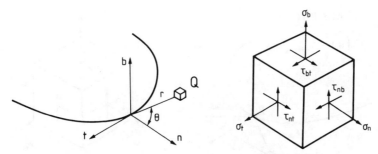

Figure 2.25. Crack front stresses in the normal plane.

the experimental point of view, the stress-frozen photoelastic method has been extensively used by Smith and coworkers [2.61, 2.62, 2.94] to determine the stress intensity factors in surface crack problems.

In the following the local stress field and the stress intensity factors for an elliptical crack will be discussed briefly.

(b) Local stress field

Let P be a fixed point of the ellipse defining the front of an elliptical crack and let **n**, **b** and **t** be the normal, binormal and tangent unit vectors at P (Figure 2.25). The plane defined by the vectors **n** and **b** is called the normal plane of the elliptical crack at P. The crack exists in a body subjected to a general type of loading. In such circumstances the singular stress field at a point Q lying in the normal plane and in the neighborhood of the point P can be expressed as a combination of the stress fields for the three plane modes of crack deformation – namely, the opening mode, the sliding mode and the tearing mode. Using Equations (2.76), (2.91) and (2.101) the result is

$$\sigma_n = \frac{K_I}{\sqrt{2\pi r}} \cos\frac{\theta}{2}\left(1 - \sin\frac{\theta}{2}\sin\frac{3\theta}{2}\right) - \frac{K_{II}}{\sqrt{2\pi r}} \sin\frac{\theta}{2}\left(2 + \cos\frac{\theta}{2}\cos\frac{3\theta}{2}\right)$$

$$\sigma_b = \frac{K_I}{\sqrt{2\pi r}} \cos\frac{\theta}{2}\left(1 + \sin\frac{\theta}{2}\sin\frac{3\theta}{2}\right) + \frac{K_{II}}{\sqrt{2\pi r}} \sin\frac{\theta}{2}\cos\frac{\theta}{2}\cos\frac{3\theta}{2}$$

$$\sigma_t = 2\nu\left(\frac{K_I}{\sqrt{2\pi r}} \cos\frac{\theta}{2} - \frac{K_{II}}{\sqrt{2\pi r}} \sin\frac{\theta}{2}\right) \quad (2.199)$$

$$\tau_{nb} = \frac{K_I}{\sqrt{2\pi r}} \sin\frac{\theta}{2}\cos\frac{\theta}{2}\cos\frac{3\theta}{2} + \frac{K_{II}}{\sqrt{2\pi r}} \cos\frac{\theta}{2}\left(1 - \sin\frac{\theta}{2}\sin\frac{3\theta}{2}\right)$$

$$\tau_{nt} = -\frac{K_{III}}{\sqrt{2\pi r}} \sin\frac{\theta}{2}$$

$$\tau_{bt} = \frac{K_{III}}{\sqrt{2\pi r}} \cos\frac{\theta}{2}$$

where r, θ are the polar coordinates of point Q in the normal plane (Figure 2.25), K_I, K_{II} and K_{III} are the opening-mode, sliding-mode and tearing-mode stress intensity factors and ν is the Poisson's ratio of the material. The stress intensity

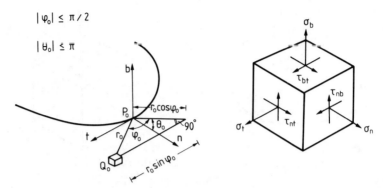

Figure 2.26. Crack front stresses in an arbitrary position.

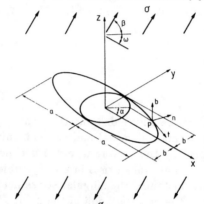

Figure 2.27. Elliptical crack at an angle to a uniform applied stress in an infinite medium.

factors K_I, K_{II}, K_{III} are defined, as in the two-dimensional case, such that they are independent of the coordinate variables r, θ and depend only on the position of the point P at the crack front, the nature of loading and the crack geometry. Note that σ_t is equal to Poisson's ratio ν times the sum of σ_n and σ_b, indicating a state of plane strain as in the two-dimensional case.

It should be emphasized that Equations (2.199) give only the stress field in the normal plane nb. The stresses in other planes are different. It was found, however, by Hartranft and Sih [2.83] that Equations (2.199) can still be used to determine the stresses at a point Q_0 defined by the spherical coordinates r_0, θ_0, ϕ_0 as in Figure 2.26 and not lying in the normal plane when r in Equations (2.199) is substituted by $r_0 \cos \phi_0$. In other words, the stresses on the element at a point Q_0 are the same as those on an element in the normal plane nb located at a distance $r = r_0 \cos \phi_0$ from P_0 and at an angle $\theta = \theta_0$ from n.

The stress field given by (2.199) is strictly valid for the elliptical crack. However, although not proven, it seems quite likely that similar results would be found for arbitrary cracks. The crack front may be a general curve in space described implicitly by its curvature and torsion, and it will be assumed that the osculating plane of the crack front curve coincides with the tangent to the free

Linear elastic stress field in cracked bodies

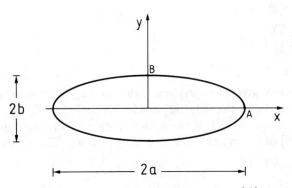

Figure 2.28. Elliptical crack with axes $2a$ and $2b$.

surfaces of the crack at each point.

(c) Stress intensity factors

Consider an elliptical crack with semi-axes a and b embedded in an infinite body which is subjected to a uniform uniaxial stress σ. The crack lies in the xy-plane and the stress σ is applied at an angle β with reference to the crack plane and at an angle ω with the xz-plane, as shown in Figure 2.27. For this case the stress intensity factors K_I, K_II and K_III entering into Equations (2.199) for point P defined by angle α are given by [2.83]

$$K_\text{I} = \frac{\sigma \sin^2 \beta}{E(k)} \left(\frac{\pi b}{a}\right)^{1/2} (a^2 \sin^2 \alpha + b^2 \cos^2 \alpha)^{1/4}$$

$$K_\text{II} = \frac{\sigma \sin \beta \cos \beta}{(a^2 \sin^2 \alpha + b^2 \cos^2 \alpha)^{1/4}} \left(1 - \frac{b^2}{a^2}\right) \left(\frac{\pi b}{a}\right)^{1/2}$$

$$\times \left[\frac{a \sin \alpha \sin \omega}{(k^2 + \nu k'^2) E(k) - \nu k'^2 K(k)} + \frac{b \cos \alpha \sin \omega}{(k^2 - \nu) E(k) + \nu k'^2 K(k)}\right] \quad (2.200)$$

$$K_\text{III} = \frac{(1 - \nu)\sigma \sin \beta \cos \beta}{(a^2 \sin^2 \alpha + b^2 \cos^2 \alpha)^{1/4}} \left(1 - \frac{b^2}{a^2}\right) \left(\frac{\pi b}{a}\right)^{1/4}$$

$$\times \left[\frac{a \sin \alpha \cos \omega}{(k^2 - \nu) E(k) + \nu k'^2 K(k)} - \frac{b \cos \alpha \sin \omega}{(k^2 + \nu k'^2) E(k) - \nu k'^2 K(k)}\right]$$

where $k = 1 - b^2/a^2$, $k' = b/a$, ν is Poisson's ratio and $K(k)$ and $E(k)$ are complete elliptic integrals of the first and second kind, respectively, defined as

$$K(k) = \int_0^{\pi/2} \frac{d\phi}{(1 - k^2 \sin^2 \phi)^{1/2}} \quad (2.201)$$

$$E(k) = \int_0^{\pi/2} (1 - k^2 \sin^2 \phi)^{1/2} \, d\phi. \quad (2.202)$$

For the case when the applied stress σ is normal to the crack plane, $\beta = 90°$,

Equations (2.200) render

$$K_I = \frac{\sigma\sqrt{\pi b}}{E(k)} \left(\sin^2 \alpha + \frac{b^2}{a^2} \cos^2 \alpha\right)^{1/4} \qquad (2.203)$$
$$K_{II} = K_{III} = 0.$$

Equation (2.203) gives the variation of the opening-mode stress intensity factor K_I along the border of the elliptical crack. We find that K_I takes its maximum value at point B ($\alpha = 90°$) of the minor crack axis and its minimum value at point A ($\alpha = 0$) of the major crack axis (Figure 2.28). These values are given by

$$K_{IA} = \frac{\sigma\sqrt{\pi b}}{E(k)} \sqrt{\frac{b}{a}}, \qquad K_{IB} = \frac{\sigma\sqrt{\pi b}}{E(k)}. \qquad (2.204)$$

For a circular crack of radius a in an infinite solid, Equation (2.203) gives

$$K_I = \frac{2}{\pi}\sigma\sqrt{\pi a}. \qquad (2.205)$$

In the limit as $b/a \to 0$, $k \to 1$ and $E(k) \to 1$, Equation (2.203) for $\alpha = 90°$ gives

$$K_I = \sigma\sqrt{\pi b}$$

which is the stress intensity factor at the tip of a through the thickness crack of length $2b$.

The above formulas for the stress intensity factor refer to an embedded elliptical crack in an infinite solid. Results for semi-elliptical and quarter-elliptical cracks in plates of finite width have been obtained using numerical methods and can be found in references [2.40–2.43 and 2.85–2.92].

2.10. Cracks in bending plates and shells

(a) *Introductory remarks*

This section presents a brief study of the stress field developed in the neighborhood of the edge of through cracks in plates subjected to bending. The related problem of the same plate configuration under extensional loads has been considered previously in this chapter. Shells in which the through cracks extend mainly along two dimensions and generaly have curved (cylindrical, spheroidal, etc.) surfaces are also examined. The predominant deformation of a shell consists of bending stresses in combination with in-plane membrane stresses. Problems of cracked plates and shells are, strictly speaking, three-dimensional in nature and their solution should satisfy the equations of the three-dimensional theory of elasticity in conjunction with the appropriate boundary conditions. Usually however, in order to make the problem manageable, simplifying assumptions are made to relax some of the differential equations and/or boundary conditions. Along these lines several theories of bending plates and shells have developed and have been used to study problems of cracked plates and shells. A concise discussion of the results of the solution of cracked plates and shells by the application of the classical Kirchhoff theory and the more accurate Reissner theory follows. For further information on this section the reader is referred to reference [2.95].

(b) *Kirchhoff theory of plate bending*

Consider a cracked plate of thickness h referred to a system of Cartesian coordinates $Oxyz$ with the Oxy-plane coinciding with the plane that bisects the plate thickness. The plate contains a through crack placed along the Oxz-plane and is subjected to bending (Figure 2.29). The in-plane stresses σ_x, σ_y and τ_{xy} on the element located at (x, y, z) are assumed to vary linearly with z, i.e.

$$\sigma_x = \frac{12z}{h^3} M_x, \qquad \sigma_y = \frac{12z}{h^3} M_y, \qquad \tau_{xy} = \frac{12z}{h^3} M_{xy} \tag{2.206}$$

while the shear stresses τ_{xz} and τ_{yz} vary parabolically with z, i.e.

$$\tau_{xz} = \frac{3}{2h}\left[1 - \left(\frac{2z}{h}\right)^2\right] Q_x, \qquad \tau_{yz} = \frac{3}{2h}\left[1 - \left(\frac{2z}{h}\right)^2\right] Q_y \tag{2.207}$$

where M_x and M_y are the bending moments, M_{xy} is the twisting moment and Q_x and Q_y are the shearing forces per unit length. They are given by

$$M_x(x, y) = \int_{-h/2}^{h/2} \sigma_x(x, y, z) z \, dz \tag{2.208}$$

$$M_y(x, y) = \int_{-h/2}^{h/2} \sigma_y(x, y, z) z \, dz \tag{2.209}$$

$$M_{xy}(x, y) = \int_{-h/2}^{h/2} \tau_{xy}(x, y, z) z \, dz \tag{2.210}$$

$$Q_x(x, y) = \int_{-h/2}^{h/2} \tau_{xz}(x, y, z) \, dz \tag{2.211}$$

$$Q_y(x, y) = \int_{-h/2}^{h/2} \tau_{yz}(x, y, z) \, dz. \tag{2.212}$$

The boundary conditions along the crack plane are

$$M_y(x, 0) = M_{xy}(x, 0) = Q_y(x, 0) = 0. \tag{2.213}$$

The Kirchhoff theory of plate bending neglects the effect of transverse shear deformation and thus replaces the last two boundary conditions on M_{xy} and Q_y by a single one on $Q_y + \partial M_{xy}/\partial x$. Thus the three boundary conditions of Equation (2.213) are replaced by the following two:

$$M_y(x, 0) = Q_y(x, 0) + \frac{\partial M_{xy}(x, 0)}{\partial x} = 0. \tag{2.214}$$

Based on Kirchhoff theory, and using the method of eigenfunction expansion, Williams [2.96] obtained the asymptotic behavior of the bending stresses in the

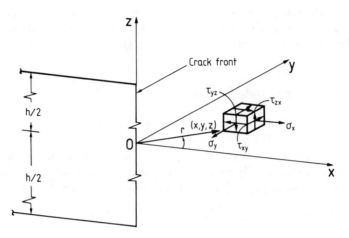

Figure 2.29. Stresses on an element ahead of a through crack in a thin plate.

vicinity of the crack. The same problem was treated later by Sih [2.97] who formulated it as a Hilbert boundary problem. The stress field is given by

$$\sigma_x = -\frac{1-\nu}{2(3+\nu)} \frac{K_I z}{h(2\pi r)^{1/2}} \left[3\cos\frac{\theta}{2} + \cos\frac{5\theta}{2}\right] +$$
$$+ \frac{1-\nu}{2(3+\nu)} \frac{K_{II} z}{h(2\pi r)^{1/2}} \left[\sin\frac{5\theta}{2} - \left(\frac{9+7\nu}{1-\nu}\right)\sin\frac{\theta}{2}\right] + O(r^0)$$

$$\sigma_y = \frac{1-\nu}{2(3+\nu)} \frac{K_I z}{h(2\pi r)^{1/2}} \left[\cos\frac{5\theta}{2} + \left(\frac{11+5\nu}{1-\nu}\right)\cos\frac{\theta}{2}\right] +$$
$$+ \frac{1-\nu}{2(3+\nu)} \frac{K_{II} z}{h(2\pi r)^{1/2}} \left[\sin\frac{\theta}{2} - \sin\frac{5\theta}{2}\right] + O(r^0)$$

$$\tau_{xy} = -\frac{1-\nu}{2(3+\nu)} \frac{K_I z}{h(2\pi r)^{1/2}} \left[\sin\frac{5\theta}{2} + \left(\frac{7+\nu}{1-\nu}\right)\sin\frac{\theta}{2}\right] + \qquad (2.215)$$
$$+ \frac{1-\nu}{2(3+\nu)} \frac{K_{II} z}{h(2\pi r)^{1/2}} \left[\frac{5+3\nu}{1-\nu}\cos\frac{\theta}{2} - \cos\frac{5\theta}{2}\right] + O(r^0)$$

$$\sigma_z = 0$$

$$\tau_{yz} = -\frac{1}{2(3+\nu)} \frac{(h^2-4z^2)\pi}{h(2\pi r)^{3/2}} \left[K_I \sin\frac{3\theta}{2} + K_{II}\cos\frac{3\theta}{2}\right] + O(r^{-1})$$

$$\tau_{xz} = \frac{1}{2(3+\nu)} \frac{(h^2-4z^2)\pi}{h(2\pi r)^{3/2}} \left[-K_I \cos\frac{3\theta}{2} + K_{II}\sin\frac{3\theta}{2}\right] + O(r^{-1})$$

where K_I and K_{II} are bending stress intensity factors that depend on the load and geometry. They have been defined such that for the problem of a plate with a crack of length $2a$ subjected to all-round moment M_0 at infinity the value of K_I coincides with its corresponding two-dimensional counterpart. For such case the all-round stress σ_0 in the surface layer at infinity is $6M_0/h^2$ and the value of K_I is $(\sigma M_0/h^2)\sqrt{\pi a}$.

It is observed from Equations (2.215) that the in-plane stresses σ_x, σ_y and τ_{xy}

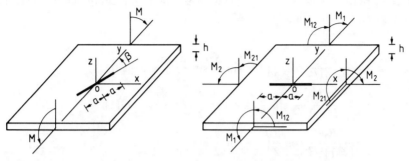

(a) Angle crack under bending   (b) Combined bending and twisting

Figure 2.30. An angle crack under bending and analysis to bending and twisting moments with respect to the crack plane.

have an inverse square root singularity at the crack edge, as in the extensional case. Equations (2.215), however, present the following drawbacks:

(i) The angular variations of the in-plane stresses σ_x, σ_y and τ_{xy} depend on Poisson's ratio ν, which is in contrast to the general three-dimensional solution (Equations (2.199)).

(ii) The transverse shear stresses τ_{yz} and τ_{xz} have a singularity of the order $r^{-3/2}$, which is inconsistent with the singularity $r^{-1/2}$ of the three-dimensional solution (Equations (2.199)).

(iii) The normal stress σ_z is everywhere zero, while along the crack front in most of the interior region, except near the free surface, it should satisfy the plane strain condition $\sigma_z = \nu(\sigma_x + \sigma_y)$.

(iv) For symmetrical loading ($K_I = 0$) the ratio of the normal stresses σ_x and σ_y along the crack plane ($\theta = 0$)

$$\frac{\sigma_x(r,0)}{\sigma_y(r,0)} = -\left(\frac{1-\nu}{3+\nu}\right) + \cdots \qquad (2.216)$$

is far beyond equal to unity as in the extensional case. This means that for bending loads relatively more yielding would take place along the crack plane.

(c) *Reissner theory of plate bending*

The above drawbacks indicate the inadequacy of the Kirchhoff theory of plate bending which replaces the three boundary conditions along the crack plane (Equation (2.213)) by two (Equation (2.214)). However, since the most important region in a cracked plate is that around the crack, it is of utmost importance to preserve the exact nature of the crack plane boundary conditions in the approximate solution of the problem. Thus, Knowles and Wang [2.98] used the sixth-order theory of Reissner [2.99, 2.100] and obtained the asymptotic solution for a plate of vanishingly small thickness. The complete solution of the plate bending problem was provided by Hartranft and Sih [2.101] who studied the effect of plate thickness on the local stress field employing the Reissner theory. They gave, for the singular stresses,

$$\sigma_x = \frac{K_I}{(2\pi r)^{1/2}} \cos \frac{\theta}{2} \left(1 - \sin \frac{\theta}{2} \sin \frac{3\theta}{2}\right) -$$
$$- \frac{K_{II}}{(2\pi r)^{1/2}} \sin \frac{\theta}{2} \left(2 + \cos \frac{\theta}{2} \cos \frac{3\theta}{2}\right) + O(1)$$
$$\sigma_y = \frac{K_I}{(2\pi r)^{1/2}} \cos \frac{\theta}{2} \left(1 + \sin \frac{\theta}{2} \sin \frac{3\theta}{2}\right) +$$
$$+ \frac{K_{II}}{(2\pi r)^{1/2}} \sin \frac{\theta}{2} \cos \frac{\theta}{2} \cos \frac{3\theta}{2} + O(1)$$
$$\sigma_z = 2\nu \left[\frac{K_I}{(2\pi r)^{1/2}} \cos \frac{\theta}{2} - \frac{K_{II}}{(2\pi r)^{1/2}} \sin \frac{\theta}{2}\right] + O(1) \qquad (2.217)$$
$$\tau_{xy} = \frac{K_I}{(2\pi r)^{1/2}} \sin \frac{\theta}{2} \cos \frac{\theta}{2} \cos \frac{3\theta}{2} +$$
$$+ \frac{K_{II}}{(2\pi r)^{1/2}} \cos \frac{\theta}{2} \left(1 - \sin \frac{\theta}{2} \sin \frac{3\theta}{2}\right) + O(1)$$
$$\tau_{xz} = -\frac{K_{III}}{(2\pi r)^{1/2}} \sin \frac{\theta}{2} + O(1)$$
$$\tau_{yz} = \frac{K_{III}}{(2\pi r)^{1/2}} \cos \frac{\theta}{2} + O(1)$$

where K_I and K_{II} are the moment intensity factors and K_{III} is the shear force intensity factor. Note that the $r^{-1/2}$ singularity and the θ-dependence in Equations (2.217) are the same as for a crack subjected to the three deformation modes considered in Section 2.5 and in accordance with the three-dimensional solution (Equations (2.199)). This is consistent with the view that each layer of the plate experiences extension or compression, with the aggregate of layers producing bending.

Consider now the case of a through crack in a bent plate subjected to a bending moment of magnitude M whose plane makes an angle β with the crack plane (Figure 2.30(a)). This loading is equivalent to the application of bending moments M_1, M_2 and twisting moment M_{12} (Figure 2.30(b)) given according to the transformation properties of stresses by:

$$M_1 = M \sin^2 \beta$$
$$M_2 = M \cos^2 \beta \qquad (2.218)$$
$$M_{12} = M \sin \beta \cos \beta.$$

For this problem, Hartranft and Sih [2.102] and Wang [2.103] gave the following expressions for the moment K_I, K_II and shear force K_III intensity factors:

$$K_\mathrm{I} = \frac{12z}{h^3}\Phi(1)M\sqrt{\pi a}\sin^2\beta$$

$$K_\mathrm{II} = \frac{12z}{h^3}\Psi(1)M\sqrt{\pi a}\sin\beta\cos\beta \qquad (2.219)$$

$$K_\mathrm{III} = -\frac{3\sqrt{10}}{2(1+\nu)h^2}\left[1 - \left(\frac{2z}{h}\right)^2\right]\Omega(1)M\sqrt{\pi a}\sin\beta\cos\beta$$

where the functions $\Phi(1)$, $\Psi(1)$ and $\Omega(1)$ have been computed in references [2.102, 2.103] as a function of h/a for various values of Poisson's ratio.

(d) Shells

The presence of curvature in shell structures results in the development of both extensional and bending stresses under either stretching or bending loads. Studies of cracked shells [2.104–2.109], based on the classical shell theory of Reissner, resulted in a singular stress field in the vicinity of the crack tip which is a combination of the respective stress fields of the Kirchhoff plate bending analysis (Equations (2.215)) and the familiar extensional problem (Equations (2.76) and (2.91)). Thus, the local stress field is described by the extensional and the bending stress intensity factors with different angular distributions for each case. This behavior is inconsistent with the general three-dimensional solution described by Equations (2.199). The difficulty was resolved by Sih and Hagendorf [2.110] who developed a tenth-order system of shell equations which took into account the effect of transverse shear. The individual physical boundary conditions along the crack plane were satisfied in contrast to previous results where they were only approximately satisfied in the Kirchhoff sense. Using such analysis the same angular stress distribution for both the membrane and bending stress fields was obtained, which is in accordance with the three-dimensional solution. For more information on these and related results the reader is referred to the works of Folias, Erdogan and to Sih and Hagendorf in reference [2.95].

References

2.1. Inglis, C. E., Stresses in a plate due to the presence of cracks and sharp corners, *Transactions of the Institute of Naval Architects* **55**, 219–241 (1913).
2.2. Westergaard, H. M., Stresses at a crack, size of the crack and the bending of reinforced concrete, *Proceedings of the American Concrete Institute* **30**, 93–102 (1934).
2.3. Westergaard, H. M., Bearing pressures and cracks, *Journal of Applied Mechanics, Trans. ASME* **6**, A.49–A.53 (1937).
2.4. Sneddon, I. N., The distribution of stress in the neighborhood of a crack in an elastic solid, *Proceedings of the Royal Society of London* **187A**, 229–260 (1946).
2.5. Williams, M. L., On the stress distribution at the base of a stationary crack, *Journal of Applied Mechanics, Trans. ASME* **24** 109–114 (1975).
2.6. Irwin, G. R., Analysis of stresses and strains near the end of a crack traversing a plate, *Journal of Applied Mechanics, Trans. ASME* **24**, 361-364 (1957).
2.7. Irwin, G. R., Fracture, in *Encyclopedia of Physics*, Vol. VI, *Elasticity and Plasticity* (ed. S. Flügge), Springer–Verlag, pp. 551–590 (1958).

2.8. Sih, G. C., Strength of stress singularities at crack tips for flexural and torsional problems, *Journal of Applied Mechanics, Trans. ASME* 30, 419–425 (1963).
2.9. Sih, G. C., Stress distribution near internal crack tips for longitudinal shear problems, *Journal of Applied Mechanics, Trans. ASME* 32, 51–58 (1965).
2.10. Sih, G. C., Heat conduction in the infinite medium with lines of discontinuity, *Journal of Applied Mechanics, Trans. ASME* 32, 293–298 (1965).
2.11. Paris, P. C. and Sih, G. C., Stress analysis of cracks, in *Fracture Toughness Testing and its Applications*, American Society for Testing and Materials, Philadelphia, pp. 30–81 (1965).
2.12. Muskhelishvili, N. I., *Some Basic Problems of the Mathematical Theory of Elasticity* (transl. from the Russian by J. R. M. Radok), (2nd ed.), Noordhoff Int. Publ., The Netherlands (1975).
2.13. Sih, G. C. and Liebowitz, H., Mathematical theories of brittle fracture, in *Fracture-An Advanced Treatise*, Vol. II, *Mathematical Fundamentals* (ed. H. Liebowitz), Pergamon Press, pp. 67–190 (1968).
2.14. Sih, G. C. (ed.), *Mechanics of Fracture* Vol. 1, *Methods of Analysis and Solutions of Crack Problems*, Noordhoff Int. Publ., The Netherlands (1973).
2.15. Kassir, M. K. and Sih, G. C., *Mechanics of Fracture*, Vol. 2, *Three Dimensional Crack Problems*, Noordhoff Int. Publ., The Netherlands (1975).
2.16. Sih, G. C., On the Westergaard method of crack analysis, *Journal of Fracture Mechanics* 2, 628–630 (1966).
2.17. Eftis, J. and Liebowitz, H., On the modified Westergaard equations for certain plane crack problems, *International Journal of Fracture Mechanics* 8, 383–392 (1972).
2.18. MacGregor, C. W., The potential function method for the solution of two-dimensional stress problems, *Transactions of the American Mathematical Society* 38, 177–186 (1935).
2.19. Theocaris, P. S. and Gdoutos, E. E., Limitations of the Westergaard equation for experimental evaluations of stress intensity factors, Discussion, *Journal of Strain Analysis* 12, 349–350 (1977).
2.20. Sih, G. C., Paris P. C. and Erdogan, F., Crack-tip stress-intensity factors for plane extension and plate bending problems, *Journal of Applied Mechanics, Trans. ASME* 29, 306–312 (1962).
2.21. Bowie, O. L., Solution of plane crack problems by mapping technique, in *Mechanics of Fracture*, Vol. 1, *Methods of Analysis and Solutions of Crack Problems* (ed. G. C. Sih), Noordhoff Int. Publ., The Netherlands, pp. 1–55 (1973).
2.22. Bueckner, H. F., The propagation of cracks and the energy of elastic deformation, *Transactions of the American Society of Mechanical Engineers* 80, 1225–1230 (1958).
2.23. Bueckner, H. F., A novel principle for the computations of stress intensity factors, *Zeitschrift für Angewandte Mathematik und Mechanik* 50, 529–546 (1970).
2.24. Bueckner, H. F., Weight functions for the notched bar, *Zeitschrift für Angewandte Mathematik und Mechanik* 51, 97–109 (1971).
2.25. Rice, J. R., Some remarks on the elastic crack-tip stress fields, *International Journal of Solids and Structures* 8, 751–758 (1972).
2.26. Hartranft, R. J. and Sih, G. C., Alternating method applied to edge and surface crack problems, in *Mechanics of Fracture*, Vol. 1, *Methods of Analysis and Solutions of Crack Problems* (ed. G. C. Sih), Noordhoff Int. Publ., The Netherlands, pp. 179–238 (1973).
2.27. Sih, G. C., *Handbook of Stress Intensity Factors*, Institute of Fracture and Solid Mechanics, Lehigh University, Bethlehem, Pennsylvania (1973).
2.28. Hsu, T. M. and Rudd, J. L., Green's function for thru-crack emanating from fastener holes, *Advances in Research on the Strength and Fracture of Materials, Fourth Int. Conf. Fract. Ontario*, (ed. D. M. R. Taplin), Pergamon Press, Vol. 3A, pp. 139–148 (1977).
2.29. Cartwright, D. J. and Rooke, D. P., Green's functions in fracture mechanics, in *Fracture Mechanics* (ed. R. A. Smith), Pergamon Press, pp. 91–123 (1979).
2.30. Paris, P. C., McMeeking, R. M. and Tada, H., The weight function method for determining stress intensity factors, in *Crack and Fracture*, STP 601, American Society for Testing and Materials, Philadelphia, pp. 471–489 (1976).
2.31. Grandt, A. F., Stress intensity factor for some through-cracked fastener holes, *International Journal of Fracture* 11, 283–294 (1975).
2.32. Petroski, H. J., Computation of the weight function from a stress intensity factor, *Engineering Fracture Mechanics* 10, 257–266 (1978).
2.33. Gross, B., Srawley, J. E. and Brown, W. F. Jr, Stress intensity factors for a single-edge-notch tension specimen by boundary collocation of a stress function, NASA NT D-2395

(1964).
2.34. Gross, B. and Srawley, J. E., Stress-intensity factors for single-edge-notch specimens in bending or combined bending and tension by boundary collocation of a stress function, NASA NT D-2603 (1965).
2.35. Gross, B. and Srawley, J. E., Stress-intensity factors for three-point bend specimens by boundary collocation, NASA NT D-3092 (1965).
2.36. Kobayashi, A. S., Cherepy, R. B. and Kinsel, W. C., A numerical procedure for estimating the stress intensity factor of a crack in a finite plate, *Journal of Basic Engineering* **86**, 681–684 (1964).
2.37. Kantorovich, L. V. and Krylov, V. I., *Approximate Methods of Higher Analysis*, Interscience, New York (1964).
2.38. Irwin, G. R., The crack extension force for a crack at a free surface boundary, U.S. Naval Research Laboratory Report 5120 (1958).
2.39. Lachenbruch, A. H., Depth and spacing of tension cracks, *Journal of Geophysical Research* **66**, 4273–4292 (1961).
2.40. Smith, F. W., Emery, A. F. and Kobayashi, A. S., Stress intensity factors for semi-circular cracks, part II–Semi-infinite solid, *Journal of Applied Mechanics, Trans. ASME* **34**, 953–959 (1967).
2.41. Smith, F. W. and Alavi, M. J., Stress intensity factors for a penny shaped crack in a half-space, *Engineering Fracture Mechanics* **3**, 241–254 (1971).
2.42. Shah, R. C. and Kobayashi, A. S., Stress intensity factor for an elliptical crack approaching the surface of a plate in bending, *Stress Analysis and Growth of Cracks*, STP 513, American Society for Testing and Materials, Philadelphia, pp. 3–21 (1972).
2.43. Shah, R. C. and Kobayashi, A. S., Stress intensity factor for an elliptical crack approaching the free surface of a semi-infinite solid, *International Journal of Fracture* **9**, 133–146 (1973).
2.44. Sneddon, I. N., *The Use of Integral Transforms*, McGraw-Hill (1972).
2.45. Sneddon, I. N., Integral transform methods, in *Mechanics of Fracture*, Vol. 1, *Methods of Analysis and Solutions of Crack Problems* (ed. G. C. Sih), Noordhoff Int. Publ., The Netherlands, pp. 315–357 (1973).
2.46. Erdogan, F. Gupta, G. D. and Cook, T. S., Numerical solution of singular integral equations, in *Mechanics of Fracture*, Vol. 1, *Methods of Analysis and Solutions of Crack Problems* (ed. G. C. Sih), Noordhoff Int. Publ., The Netherlands, pp. 368–425 (1973).
2.47. Bilby, B. A. and Eshelby, J. D., Dislocations and the theory of fracture, in *Fracture–An Advanced Treatise*, Vol. 1 (ed. H. Liebowitz), Academic Press, pp. 99–182 (1968).
2.48. Zienkiewicz, O. C., *The Finite Element Method*, McGraw-Hill (1967).
2.49. Zienkiewicz, O. C. and Cheung, Y. K., *The Finite Element Method in Structural and Continuum Mechanics*, McGraw-Hill (1967).
2.50. Wilson, W. K., Finite element methods for elastic bodies containing cracks, in *Mechanics of Fracture*, Vol. 1, *Methods of Analysis and Solutions of Crack Problems* (ed. G. C. Sih), Noordhoff Int. Publ., The Netherlands, pp. 484–515 (1973).
2.51. Hilton, P. D. and Hutchinson, J. W., Plastic intensity factors for cracked plates, *Engineering Fracture Mechanics* **3**, 435–451 (1971).
2.52. Tracey, D. M., Finite elements for determination of crack tip elastic stress intensity factors, *Engineering Fracture Mechanics* **3**, 255–265 (1971).
2.53. Barsoum, R. S., On the use of isoparametric finite elements in linear fracture mechanics, *International Journal for Numerical Methods in Engineering* **10**, 25–37 (1976).
2.54. Pu, S. L. and Hussain, M. A., The collapsed cubic isoparametric element as a singular element for crack problems, *International Journal for Numerical Methods in Engineering* **12**, 1727–1742 (1978).
2.55. Hilton P. D. and Sih, G. C., Applications of the finite element method to the calculations of stress intensity factors, in *Mechanics of Fracture*, Vol. 1, *Methods of Analysis and Solutions of Crack Problems* (ed. G. C. Sih), Noordhoff Int. Publ., The Netherlands, pp. 426–483 (1973).
2.56. Rice, J. R. and Tracey, D. M., Computational fracture mechanics, in *Numerical and Computer Methods in Structural Mechanics* (eds. S. J. Fenves et al.), Academic Press, New York, pp. 585–623 (1973).
2.57. Gallagher, P. H., A review of finite element techniques in fracture mechanics, in *Numerical Methods in Fracture Mechanics* (eds. A. R. Luxmoore and D. R. J. Owen), Pineridge Press, Swansea, U.K., pp. 1–25 (1978).
2.58. Hellen, T. K., Numerical methods in fracture mechanics, in *Developments in Fracture*

Mechanics, Vol. 1 (ed. G. G. Chell), Applied Science Publishers, London, pp. 145–181 (1979).
2.59. Theocaris, P. S. and Gdoutos, E. E., *Matrix Theory of Photoelasticity*, Springer–Verlag (1979).
2.60. Gdoutos, E. E., Photoelastic study of crack problems, in *Photoelasticity in Engineering Practice* (eds. S. A. Paipetis and G. S. Hollister), Elsevier Applied Science Publishers, pp. 181–204 (1985).
2.61. Smith, C. W., Use of three-dimensional photoelasticity in fracture mechanics, *Experimental Mechanics* **13**, 539–544 (1973).
2.62. Schroedl, M. A., McGowan, J. J. and Smith, C. W., Determination of stress intensity factors from photoelastic data with applications to surface-flaw problems, *Experimental Mechanics* **14**, 392–399 (1974).
2.63. Theocaris, P. S. and Gdoutos, E. E., A photoelastic determination of K_I stress intensity factors, *Engineering Fracture Mechanics* **7**, 331–339 (1975).
2.64. Etheridge, J. M. and Dally, J. W., A three-parameter method for determining stress intensity factors from isochromatic fringe patterns, *Journal of Strain Analysis* **13**, 91–94 (1978).
2.65. Etheridge, J. M., Dally, J. W. and Kobayashi, T., A new method of determining the stress intensity factor K from isochromatic fringe loops, *Engineering Fracture Mechanics* **10**, 81–93 (1978).
2.66. Smith, D. G. and Smith, C. W., Photoelastic determination of mixed mode stress intensity factors, *Engineering Fracture Mechanics* **4**, 357–366 (1972).
2.67. Gdoutos, E. E. and Theocaris, P. S., A photoelastic determination of mixed-mode stress-intensity factors, *Experimental Mechanics* **18**, 87–96 (1978).
2.68. Dally, J. W. and Sanford, R. J., Classification of stress-intensity factors from isochromatic fringe patterns, *Experimental Mechanics* **18**, 441–448 (1978).
2.69. Sanford, R. J. and Dally, J. W., A general method for determining mixed-mode stress intensity factors from isochromatic fringe patterns, *Engineering Fracture Mechanics* **11**, 621–633 (1979).
2.70. Manogg, P., Die Lichtablenkung durch eine elastisch beanspruchte Platte und die Schattenfiguren von Kreiss- und Risskerbe, *Glastechnische Berichte* **39**, 229–329 (1966).
2.71. Theocaris, P. S., Local yielding around a crack tip in plexiglas, *Journal of Applied Mechanics, Trans. ASME* **37**, 409–415 (1970).
2.72. Theocaris, P. S., The method of caustics applied to elasticity problems, in *Developments in Stress Analysis* (ed. G. S. Hollister), Applied Science Publishers, pp. 27–63 (1979).
2.73. Theocaris, P. S., Elastic stress intensity factors evaluated by caustics, in *Mechanics of Fracture*, Vol. 7, *Experimental Evaluation of Stress Concentration and Intensity Factors* (ed. G. C. Sih), Martinus Nijhoff Publ., pp. 189–252 (1981).
2.74. Theocaris, P. S. and Gdoutos, E. E., An optical method for determining opening-mode and edge sliding-mode stress-intensity factors, *Journal of Applied Mechanics, Trans. ASME* **39**, 91–97 (1972).
2.75. Kobayashi, A. S. (ed.), *Experimental Techniques in Fracture Mechanics*, Vols 1 and 2, Society for Experimental Stress Analysis of U.S.A. (1973, 1975).
2.76. Sack, R. A., Extension of Griffith theory of rupture to three dimensions, *Proceedings of the Physical Society of London* **58**, 729–736 (1946).
2.77. Segedin, C. M., Note on a penny-shaped crack under shear, *Proceedings of the Cambridge Philosophical Society* **47**, 396–400 (1950).
2.78. Sadowsky, M. A. and Sternberg, E., Stress concentration around a triaxial ellipsoidal cavity, *Journal of Applied Mechanics, Trans. ASME* **16**, 149–157 (1949).
2.79. Green, A. E. and Sneddon, I. N., The distribution of stress in the neighbourhood of a flat elliptical crack in an elastic solid, *Proceedings of the Cambridge Philosophical Society* **46**, 159–163 (1950).
2.80. Irwin, G. R., Crack-extension force for a part-through crack in a plate, *Journal of Applied Mechanics, Trans. ASME* **29**, 651–654 (1962).
2.81. Kassir, M. K. and Sih, G. C., Three-dimensional stress distribution around an elliptical crack under arbitrary loading, *Journal of Applied Mechanics, Trans. ASME* **33**, 601–611 (1966).
2.82. Hartranft, R. J. and Sih, G. C., The use of eigenfunction expansions in the general solution of three-dimensional crack problems, *Journal of Mathematics and Mechanics* **19**, 123–138 (1969).
2.83. Hartranft, R. J. and Sih, G. C., Stress singularity for a crack with an arbitrarily curved

front, *Engineering Fracture Mechanics* 9, 705–718 (1977).
2.84. Sih, G. C. and Chen, E. P., *Mechanics of Fracture*, Vol. 6, *Cracks in Composite Materials*, Martinus Nijhoff Publ., The Netherlands (1981).
2.85. Shah, R. C. and Kobayashi, A. S., Stress intensity factor for an elliptical crack under arbitrary normal loading, *Engineering Fracture Mechanics* 3, 71–96 (1971).
2.86. Kobayashi, A. S., Surface flaws in plates in bending, *Proceedings of the 12th Annual Meeting of the Society of Engineering Science, Austin, Texas*, pp. 343–352 (1975).
2.87. Tracey, D. M., Finite element for three-dimensional elastic crack analysis, *Nuclear Engineering and Design* 26, 282–290 (1974).
2.88. Barsoum, R. S., On the use of isoparametric finite elements in linear fracture mechanics, *International Journal of Numerical Methods in Engineering* 10, 25–37 (1976).
2.89. Raju, I. S. and Neuman, J. C. Jr, Stress-intensity factors for a wide range of semi-elliptical surface cracks in finite-thickness plates, *Engineering Fracture Mechanics* 11, 817–829 (1979).
2.90. Neuman, J. C. Jr and Raju, I. S., An empirical stress-intensity factor equation for the surface crack, *Engineering Fracture Mechanics* 15, 185–192 (1981).
2.91. Tan, C. L. and Fenner, R. T., Stress intensity factors for semi-elliptical surface cracks in pressurized cylinders using the boundary integral equation method, *International Journal of Fracture* 16, 233–245 (1980).
2.92. Isida, M., Noguchi, H. and Yoshida, T., Tension and bending of finite thickness plates with a semi-elliptical surface crack, *International Journal of Fracture* 26, 157–188 (1984).
2.93. Wu, X. R., Stress intensity factors for half-elliptical surface cracks subjected to complex crack face loadings, *Engineering Fracture Mechanics* 19, 387–405 (1984).
2.94. Smith, C. W., Use of photoelasticity in fracture mechanics, in *Mechanics of Fracture*, Vol. 7, *Experimental Evaluation of Stress Concentration and Intensity Factors* (ed. G. C. Sih), Martinus Nijhoff Publ., pp. 163–187 (1981).
2.95. Sih, G. C. (ed.), *Mechanics of Fracture*, Vol. 3, *Plates and Shells with Cracks*, Noordhoff Int. Publ., The Netherlands (1977).
2.96. Williams, M. L., The bending stress distribution at the base of a stationary crack, *Journal of Applied Mechanics, Trans. ASME* 28, 78–82 (1961).
2.97. Sih, G. C., Flexural problems of cracks in mixed media, *Proceedings of the First International Conference on Fracture* (eds T. Yokobori, T. Kawasaki and J. L. Swedlow), Sendai, Japan, Vol. 1, pp. 391–409 (1965).
2.98. Knowles, J. K. and Wang, N. M., On the bending of an elastic plate containing a crack, *Journal of Mathematics and Physics* 39, 223–236 (1960).
2.99. Reissner, E., The effect of transverse shear deformation on the bending of elastic plates, *Journal of Applied Mechanics, Trans. ASME* 67, A69–77 (1945).
2.100. Reissner, E., On bending of elastic plates, *Quarterly Journal of Applied Mathematics* 5, 55–68 (1974).
2.101. Hartranft, R. J. and Sih, G. C., Effect of plate thickness on the bending stress distribution around through cracks, *Journal of Mathematics and Physics* 47, 276–291 (1968).
2.102. Hartranft, R. J. and Sih, G. C., An approximate three-dimensional theory of plates with application to crack problems, *International Journal of Engineering Science* 8, 711–729 (1970).
2.103. Wang, N. M., Twisting of an elastic plate containing a crack, *International Journal of Fracture Mechanics* 6, 367–378 (1970).
2.104. Folias, E. S., The stresses in a cracked spherical shell, *Journal of Mathematics and Physics* 44, 164–176 (1965).
2.105. Folias, E. S., An axial crack in a pressurized cylindrical shell, *International Journal of Fracture Mechanics* 1, 104–113 (1965).
2.106. Folias, E. S., A circumferential crack in a pressurized cylindrical shell, *International Journal of Fracture Mechanics* 3, 1–12 (1967).
2.107. Copley, L. G. and Sanders, J. L. Jr, A longitudinal crack in a cylindricall shell under internal pressure, *International Journal of Fracture Mechanics* 5, 113–131 (1969).
2.108. Erdogan, F. and Ratwani, M., A circumferential crack in a cylindrical shell under torsion, *International Journal of Fracture Mechanics* 8, 87–95 (1972).
2.109. Sih, G. C. and Dobreff, P. S., Crack-like imperfections in a spherical shell, *Glasgow Mathematical Journal* 12, 65–88 (1971).
2.110. Sih, G. C. and Hagendorf, H. C., A new theory of spherical shells with cracks, *Thin-Shell Structures: Theory, Experiment, and Design* (Eds Y. C. Fung and E. E. Sechler), Prentice-Hall, New Jersey, pp. 519–545 (1974).

3

Elastic–plastic stress field in cracked bodies

3.1. Introduction

The analysis of the stress field in cracked bodies by the linear elastic theory, dealt with in the preceding chapter, strictly speaking only concerns the ideal situation of brittle materials where the amount of inelastic deformation in the vicinity of the crack tip is negligible. In most cases, however – depending on material and conditions – some inelasticity in the neighborhood of the crack tip is always present in the form of rate-independent plasticity, creep or phase change. A study of the local stress fields for the three modes of loading showed that they have general applicability and are governed by the values of three stress intensity factors. In other words, the applied loading, the crack length and the geometrical configuration of the cracked bodies influence the strength of these fields only through the stress intensity factors. In two cracked bodies of different geometry, crack length and applied loads producing the same mode, the stress and deformation fields near the crack tip will be the same if the stress intensity factors are equal.

The singular stress fields represent the asymptotic fields as the distance from the crack tip tends to zero, and their realm of applicability is confined to a very small region around the crack tip. Evans and Luxmoore [3.1], and Theocaris and Gdoutos [3.2] established the percentage error of the singular stresses and displacements compared to the exact solution as we recede from the crack tip. The latter authors gave also the number of terms of the series expansion of the stress field which are needed for the convergence of stresses with a given accuracy. Let the singular solution dominate inside a circle of radius D surrounding the crack tip (Figure 3.1). Consider also that the region of inelastic deformation attending the crack tip is represented by R. When R is sufficiently small compared to D and any other characteristic geometric dimension such as notch radius, plate thickness, crack ligament, etc., the singular stress field governed by the stress intensity factors forms a useful approximation of the elastic field in the ring enclosed by radii R and D. This situation has been termed 'small-scale yielding'.

A mathematical formulation of the small-scale yielding condition that is useful for the determination of the near tip elastic–plastic field has been given by Rice

Elastic–plastic stress field in cracked bodies

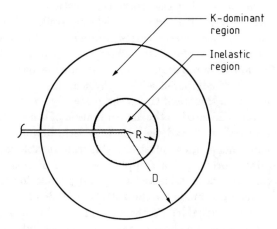

Figure 3.1. Inelastic and K-dominant regions around a crack tip.

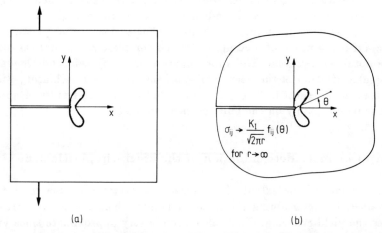

Figure 3.2. (a) Small-scale yielding around a crack tip and (b) replacement of the actual crack with a semi-infinite crack and appropriate boundary conditions.

[3.3]. The situation is that elastic singularity governs the stress field at distances from the crack tip that are large compared to the plastic zone but simultaneously small with regard to other characteristic dimensions such as the crack length. Thus, the actual configuration (Figure 3.2(a)) may be replaced by a semi-infinite crack in an infinite plate (Figure 3.2(b)) which is subjected at infinity to a stress distribution given by the singular elastic solution, that is,

$$\sigma_{ij} = \frac{K}{\sqrt{2\pi r}} f_{ij}(\theta) \quad \text{for } r \to \infty \tag{3.1}$$

where $f_{ij}(\theta)$ expresses the angular distribution of the singular stress field and K is the stress intensity factor of the associated elastic crack problem.

Small-scale yielding, however, should be interpreted with care when applied to the elastic–plastic solution of a crack in an infinite medium. Hilton and

Hutchinson [3.4] have shown that the size of the plastic zone can be many times larger than the crack length. The extent of yielding would still be considered as small scale because the plastic energy corresponding to a finite region will always be small compared to that of the elastic energy stored in an infinite region. A more precise definition of small-scale yielding should be based on the relative size of the plastic zone as compared with a physical dimension of the specimen such as the distance from the crack tip to the plate edge or the thickness.

The present chapter is devoted to the determination of the elastic–plastic stress and displacement distribution and the elastic–plastic boundary accompanying the crack tip for time-independent plasticity. Insurmountable mathematical difficulties have, to date, prevented a complete solution of this problem for the most important case of cracks subjected to opening-mode loading. Substantial progress, however, has been made for the antiplane mode of loading and a number of solutions are now available. Furthermore, for real bodies the plastic zone surrounding the crack tip is of a three-dimensional nature, which further complicates the situation. This chapter starts with an approximate calculation of the plastic zone for 'small' applied loads and describes the actual plastic enclaves revealed by experiments in plates of finite thickness. The elastic–plastic analysis for the antiplane mode of loading, for which complete mathematical solutions can be obtained, is given. The approximate models of Irwin and Dugdale and the singular solution of the near crack-tip field given by Hutchinson, Rice and Rosengren are presented. The chapter concludes with the results of numerical solutions based mainly on the finite element method.

3.2. Approximate determination of the crack-tip plastic zone

A first estimate of the extent of the plastic zone attending the crack tip can be obtained by determining the locus of points where the elastic stress field satisfies the yield criterion. This calculation is very approximate since yielding results in stress redistribution and modifies the size and shape of the plastic zone. Strictly speaking, the plastic zone should be determined from an elastic–plastic analysis of the stress field around the crack tip. However, from the approximate calculation some useful results regarding the shape of the plastic zone can be obtained.

For opening-mode loading introducing the expressions for the singular principal stresses given by Equations (2.85) into the von Mises yield criterion expressed by Equation (1.2), we obtain the following expression for the radius of the plastic zone

$$r_p(\theta) = \frac{1}{4\pi} \left(\frac{K_I}{\sigma_Y}\right)^2 \left(\tfrac{3}{2} \sin^2 \theta + 1 + \cos \theta\right) \tag{3.2}$$

for plane stress, and

$$r_p(\theta) = \frac{1}{4\pi} \left(\frac{K_I}{\sigma_Y}\right)^2 \left[\tfrac{3}{2} \sin^2 \theta + (1 - 2\nu)^2 (1 + \cos \theta)\right] \tag{3.3}$$

Elastic-plastic stress field in cracked bodies

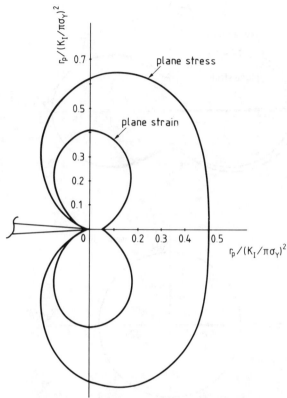

Figure 3.3. Approximate estimation of the crack-tip plastic zones for mode-I loading under plane stress and plane strain. $\nu = 1/3$.

for plane strain, where σ_Y is the yield stress.

The extent of the plastic zone along the crack axis ($\theta = 0$) is given by

$$r_p(0) = \frac{1}{2\pi}\left(\frac{K_I}{\sigma_Y}\right)^2 \tag{3.4}$$

for plane stress, and

$$r_p(0) = \frac{1}{18\pi}\left(\frac{K_I}{\sigma_Y}\right)^2 \tag{3.5}$$

for plane strain, with $\nu = 1/3$.

Figure 3.3 shows the shapes of the plastic zones for plane stress and plane strain with $\nu = 1/3$. Observe that the plane stress zone is much larger than the plane strain zone because of the higher constraint for plane strain. From Equations (3.4) and (3.5) it is shown that the extent of the plastic zone along the crack axis for plane strain is equal to 1/9 of that of plane stress.

For sliding-mode loading we obtain, in an analogous way,

$$r_p(\theta) = \frac{1}{4\pi}\left(\frac{K_{II}}{\sigma_Y}\right)^2 (25 - \cos\theta - 18\sin^2\theta) \tag{3.6}$$

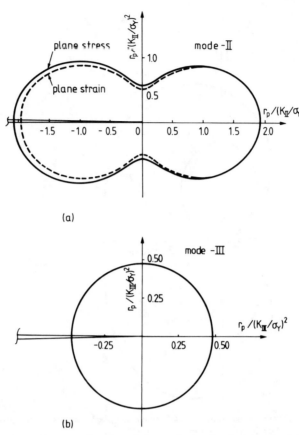

Figure 3.4. Approximate estimation of the crack-tip plastic zones for (a) mode-II loading under plane stress and plane strain and (b) mode-III loading. $\nu = 1/3$.

for plane stress, and

$$r_p(\theta) = \frac{1}{4\pi}\left(\frac{K_{II}}{\sigma_Y}\right)^2 [24 + (1-2\nu)^2(1-\cos\theta) - 18\sin^2\theta] \qquad (3.7)$$

for plane strain.

For antiplane-mode loading the plastic zone is a circle centered at the crack tip with radius

$$r_p = \frac{3}{2\pi}\left(\frac{K_{III}}{\sigma_Y}\right)^2. \qquad (3.8)$$

The plastic zones for modes II and III are shown in Figure 3.4.

Results for the plastic zones for pressure modified yield criteria applicable in glassy polymers for mixed-mode loading were obtained by Gdoutos [3.5, 3.6].

The above results obtained for the idealized cases of plane stress and plane strain give only a rough estimate of the shape and size of crack-tip plastic zones. Before we proceed to more accurate calculations, a few remarks concerning the qualitative nature of the plastic zones in plates of finite thickness are in order.

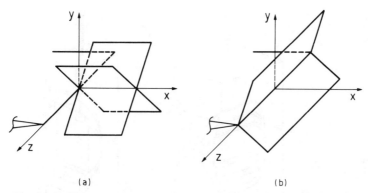

Figure 3.5. Planes of maximum shear stress in front of a mode-I crack for (a) plane stress and (b) plane strain.

Consider first a material element in front of the crack ($\theta = 0$) for mode-I loading and assume conditions of plane stress or plane strain. The in-plane principal stresses σ_1 and σ_2 are equal ($\sigma_1 = \sigma_2 = \sigma$) (Equations (2.85)), while the transverse stress σ_z is a principal stress and is equal to zero or $2\nu\sigma$ for plane stress or plane strain, respectively. Assuming a Tresca yield criterion, the maximum shear stress for plane stress is equal to $\sigma/2$ and occurs in planes making 45° with the plane of the plate (Figure 3.5(a)). On the other hand, the maximum shear stress for plane strain is equal to $\sigma/3$ (assuming $\nu = 1/3$) and occurs in planes normal to the plane of the plate and making 45° with the directions of σ_1 and σ_2 (Figure 3.5(b)). Thus, for plane strain, not only is much more stress required to yield a material element than in plane stress, but also the planes of yielding are different. Analogous conditions hold for different θ angles.

Conditions of plane stress dominate in very thin plates where it can be assumed that the transverse stress σ_z is zero through the plate thickness. On the other hand, for thick plates the state of stress is primarily one of plane strain. The type of plastic deformation associated with these two cases is shown in Figure 3.6. Under plane stress, slip takes place on planes at 45° to the plate surface, producing a rather large strain through the thickness (Figure 3.6(a)), while, in plane strain, slip occurs on planes perpendicular to the plate surface, giving a hinge-type deformation pattern (Figure 3.6(b)).

In cracked plates conditions of plane stress dominate at the traction-free surfaces, while plane strain prevails in the interior. This results in a variation of the plastic zone through the plate thickness, which decreases from the surface to the interior of the plate (Figure 3.7). Although the state of stress is always a combination of plane stress and plane strain, some guidelines for determining the predominant type can be established. This is achieved by comparing the size of the plastic zone to the thickness, B, of the plate. When the length of the plastic zone c in front of the crack is of the order of B then plane stress dominates. On the other hand, if c is much less than B the greatest part of the thickness is under plane strain. According to the American Society for Testing

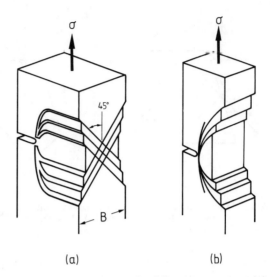

Figure 3.6. Slip-planes around a mode-I crack for (a) plane stress and (b) plane strain.

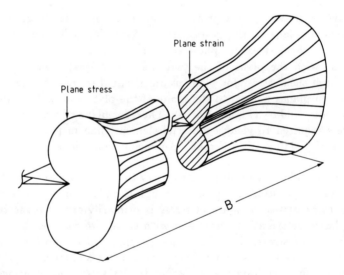

Figure 3.7. Schematic representation of the three-dimensional nature of the crack-tip plastic zone around a crack tip in a finite thickness plate.

and Materials, Standard E399-81 [3.7], plane strain dominates when $c < B/25$ where c is the length of the plane strain plastic zone along the crack axis.

The above qualitative predictions on the characteristics of the crack tip plastic zones were experimentally verified by Hahn, Rosenfield and Dai [3.8, 3.9]. They performed experiments on steel cracked specimens, and by etching their polished surfaces revealed the three-dimensional character of the plastic zone in front of the crack. The plastic zones in the interior were obtained by sectioning, repolishing and etching. Figure 3.8 shows the plastic zones (appearing as dark

Elastic-plastic stress field in cracked bodies

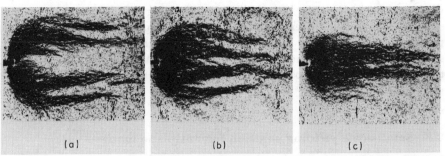

Figure 3.8. Plastic zones, appearing as dark regions, in a cracked plate at (a) the surface of the specimen, (b) the section halfway between the surface and the midsection and (c) the midsection. (Photograph by P. N. Mincer, Battelle Memorial Institute.)

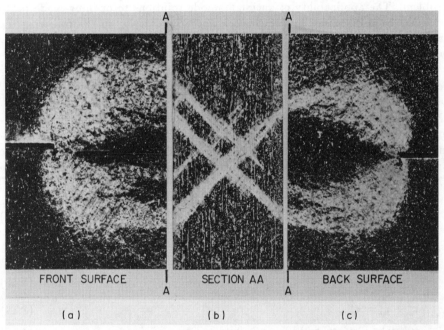

Figure 3.9. Plastic zones, appearing as light regions, in a cracked plate at (a) the front and (c) the back surfaces of the plate and (b) a section normal to the crack plane. (Photograph by P. N. Mincer, Battelle Memorial Institute.)

regions) on (a) the surface of specimen, (b) a section halfway between the surface and the midsection and (c) the midsection, at an applied load producing a net section stress equal to 0.9 of the yield stress. The specimen has a thickness of 0.232 in. and plane stress dominates. Observe that the two yielded regions on the surface of the specimen merge into a single region on the midsection. Figure 3.9 shows the plastic zones (appearing as light regions) on (a) the front surface, (b) a section normal to the crack plane and (c) the back surface, of a 0.197 in.

thickness specimen at the same stress level as in Figure 3.8. The appearance of slip bands on planes subtending 45° with the crack plane, which is indicative of plane stress, is shown. For lower stress levels for which the plastic zones are much smaller than the plate thickness it was found that the yielded regions through the thickness remain almost the same, which is consistent with the character of plane strain deformation.

3.3. Small-scale yielding solution for antiplane mode

The analysis of the elastic–plastic stress and displacement distribution for the antiplane-mode of loading is important both because it represents one of the few cases available for which closed-form plastic solutions can be obtained and, historically, it provided the basis for the understanding of the opening-mode problem. The applicability of antiplane-mode solutions to opening-mode problems has been discussed by McClintock and Irwin [3.10]. In the following the cases of perfectly plastic and strain-hardening behavior for small-scale yielding will be studied separately.

(a) Perfect plasticity

The analytical solution of a crack in a perfectly plastic material under antiplane-mode loading has first been given by Hult and McClintock [3.11]. Before proceeding to the solution of this problem some general characteristics of the antiplane problem within the framework of the deformation theory of plasticity will be given. Using the following notation to save indices

$$\begin{aligned}\tau_{zx} &= \tau_x, & \tau_{zy} &= \tau_y \\ \gamma_{zx} &= \gamma_x, & \gamma_{zy} &= \gamma_y\end{aligned} \tag{3.9}$$

we have the following equations

$$\frac{\partial \tau_x}{\partial x} + \frac{\partial \tau_y}{\partial y} = 0 \tag{3.10}$$

$$\tau_x^2 + \tau_y^2 = \tau_Y^2 \tag{3.11}$$

$$\dot{\gamma}_x = \lambda \tau_x, \qquad \dot{\gamma}_y = \lambda \tau_y, \quad \lambda \geq 0 \tag{3.12}$$

which express the equilibrium condition, the yield condition and the flow rule. The dot symbol denotes differentiation with respect to time and τ_Y is the yield stress in shear.

Furthermore, we have the strain-displacement equations

$$\gamma_x = \frac{\partial w}{\partial x}, \qquad \gamma_y = \frac{\partial w}{\partial y} \tag{3.13}$$

$$\dot{\gamma}_x = \frac{\partial \dot{w}}{\partial x}, \qquad \dot{\gamma}_y = \frac{\partial \dot{w}}{\partial y} \tag{3.14}$$

where w is the displacement along the z-direction.

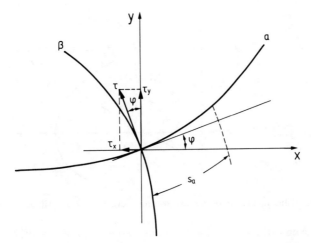

Figure 3.10. Slip lines for antiplane deformation of a perfectly plastic material.

Consider now in Figure 3.10 the slip lines at a point in the plastically deformed region and assume that the β-line is taken along the direction of the principal shear stress $\tau(\tau_x, \tau_y)$. From Figure 3.10 and Equation (3.11) it follows that

$$\begin{aligned} \tau_x &= -\tau \sin \phi = -\tau_Y \sin \phi \\ \tau_y &= \tau \cos \phi = \tau_Y \cos \phi \end{aligned} \tag{3.15}$$

where ϕ is the angle that the a-line makes with the x-axis.

By differentiating Equations (3.15) with respect to x and y and substituting into Equation (3.10) we get

$$\cos \phi \frac{\partial \phi}{\partial x} + \sin \phi \frac{\partial \phi}{\partial y} = 0. \tag{3.16}$$

Introducing the length s_α along the a-line, Equation (3.16) takes the form

$$\frac{\partial x}{\partial s_\alpha} \frac{\partial \phi}{\partial x} + \frac{\partial y}{\partial s_\alpha} \frac{\partial \phi}{\partial y} = \frac{\partial \phi}{\partial s_\alpha} = 0. \tag{3.17}$$

Equation (3.17) indicates that the a-lines are straight lines for antiplane loading.

Furthermore, from Equations (3.14), (3.12) and (3.15) we obtain

$$\frac{\partial \dot{w}}{\partial s_\alpha} = \cos \phi \frac{\partial \dot{w}}{\partial x} + \sin \phi \frac{\partial \dot{w}}{\partial y} = \lambda(\tau_x \cos \phi + \tau_y \sin \phi) = 0 \tag{3.18}$$

which shows that the antiplane displacement increments dw along the a-lines are constant, so that $w = w(\theta)$ in the plastic region for monotonic loading.

The a-lines should be perpendicular to the unloaded crack faces and pass through the crack tip. On a-lines we have

$$\tau_r = 0, \qquad \tau_\theta = \tau_Y. \tag{3.19}$$

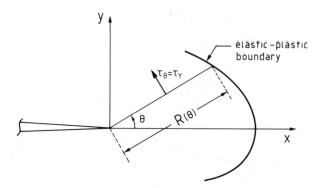

Figure 3.11. Elastic–plastic boundary around a crack tip for antiplane deformation.

Using polar coordinates we have for the value of γ_θ at $r = R(\theta)$

$$\gamma_\theta = \frac{1}{R(\theta)} \frac{dw}{d\theta} \qquad (3.20)$$

where $R(\theta)$ is the distance from the crack tip to the elastic–plastic boundary (Figure 3.11).

On the other hand, by Hooke's law we have, for γ_θ at $r = R(\theta)$,

$$\gamma_\theta = \frac{\tau_\theta}{\mu} = \frac{\tau_Y}{\mu} \qquad (3.21)$$

where μ is the shear modulus.

From Equations (3.20) and (3.21) it is found that

$$\frac{dw}{d\theta} = R(\theta) \frac{\tau_Y}{\mu} \qquad (3.22)$$

so that we have, for γ_θ in the plastic region,

$$\gamma_\theta = \frac{R(\theta)}{r} \frac{\tau_Y}{\mu}. \qquad (3.23)$$

Equation (3.23) shows that the shear strain γ_θ has a r^{-1} singularity in the plastic zone which is stronger than the $r^{-1/2}$ singularity of the elastic case (see Equations (2.101)).

The displacement w in the plastic zone is given in terms of $R(\theta)$ from Equation (3.22) by

$$w = \frac{\tau_Y}{\mu} \int_0^\theta R(\beta)\, d\beta \qquad (3.24)$$

where the displacement on the line ahead of the crack was taken equal to zero ($w(0) = 0$).

The solution of the problem is completed by determining the distance $R(\theta)$ to the elastic–plastic boundary which separates the inside plastic zone from the outside elastic material. In the elastic region it follows from Equation (2.96) that

$$\tau_y + i\tau_x = Z'_{III}(z) \qquad (3.25)$$

Elastic–plastic stress field in cracked bodies

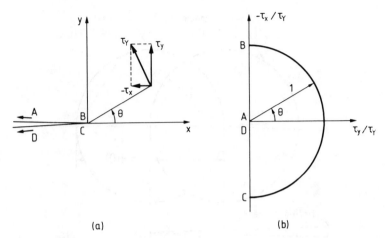

Figure 3.12. (a) Physical plane and (b) its mapping into the ξ-plane for the antiplane deformation solution of a cracked plate under small-scale yielding.

where $Z'_{III}(z)$ is an analytic function defined from Equation (2.96).

By defining the normalized quantity

$$\xi = \frac{\tau_y + i\tau_x}{\tau_Y} \tag{3.26}$$

Equation (3.25) suggests that z must be an analytic function of ξ, that is

$$z = F(\xi) \tag{3.27}$$

where, from Equation (3.11),

$$|\xi| = 1. \tag{3.28}$$

Consider now the mapping between the planes z and ξ defined by Equation (3.27) and shown in Figure 3.12. Equation (3.27) renders

$$R(\theta) = e^{-i\theta} F(e^{-i\theta}). \tag{3.29}$$

The unknown function $z = F(\xi)$ will be determined from the following three conditions:

(i) Along the crack faces ($y = 0$, $x < 0$) $\tau_y = 0$ and therefore ξ is purely imaginary. Thus, the crack faces are mapped on the imaginary axis of the ξ-plane and the function $z = F(\xi)$ must be real and negative for $\tau_y/\tau_Y = \operatorname{Re} \xi = 0$.

(ii) Equation (3.29) indicates that

$$\operatorname{Im}[e^{-i\theta} F(e^{-i\theta})] = 0. \tag{3.30}$$

(iii) Following the small-scale yielding formulation given by Equation (3.1), and using Equations (2.101), it follows that

$$z = F(\xi) \rightarrow \frac{K_{III}^2}{2\pi \tau_Y^2 \, \xi^2} \quad \text{for } |\xi| \rightarrow 0. \tag{3.31}$$

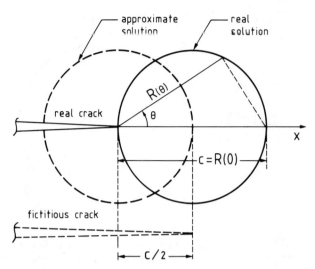

Figure 3.13. Exact and approximate solution of the plastic zone around a crack tip for small-scale yielding antiplane deformation.

A solution satisfying these three conditions is

$$z = F(\xi) = \frac{K_{III}}{2\pi \tau_Y^2}\left(1 + \frac{1}{\xi^2}\right). \tag{3.32}$$

Thus, for the distance $R(\theta)$ of the crack tip from the elastic–plastic boundary, Equation (3.29) gives

$$R(\theta) = \frac{K_{III}^2}{\pi \tau_Y^2}\cos\theta. \tag{3.33}$$

Furthermore, Equations (3.26) and (3.32) render

$$\tau_y + i\tau_x = \frac{K_{III}}{\sqrt{2\pi}}\left(z - \frac{K_{III}^2}{2\pi \tau_Y^2}\right)^{-1/2}. \tag{3.34}$$

Equation (3.33) indicates that the plastic zone is a circle passing from the crack tip with diameter

$$c = R(0) = \frac{K_{III}^2}{\pi \tau_Y^2} \tag{3.35}$$

as shown in Figure 3.13. In the same figure, the plastic zone – which, according to the approximate determination of Section 3.2, is a circle centered at the crack tip with radius $c/2$ (Equation (3.8) with $\sigma_Y = \sqrt{3}\,\tau_Y$ according to the von Mises yield criterion) – is indicated. Observe that the extent of the plastic zone along the crack direction obtained by the approximate solution is half of that predicted by the small-scale yielding elastic–plastic solution.

From Equations (2.101) of the elastic stress field we obtain that

$$\tau_y + i\tau_x = \frac{K_{III}}{\sqrt{2\pi z}}. \tag{3.36}$$

By comparing Equations (3.34) and (3.36) it is deduced that the elastic stress field outside the plastic zone is the same as that of a crack with length equal to the original crack length plus half the plastic zone length. This result motivated Irwin to propose a model for the determination of the plastic zone for opening-mode loading, as we will see in Section 3.5.

Introducing the value of $R(\theta)$ from Equation (3.33) into Equations (3.23) and (3.24) we obtain for the strain γ_θ and displacement w

$$\gamma_\theta = \frac{K_{III}^2}{\pi \mu \tau_Y} \frac{\cos \theta}{r}; \qquad w = \frac{K_{III}^2}{\pi \mu \tau_Y} \sin \theta. \tag{3.37}$$

The crack-tip opening displacement is

$$\delta = w\left(\frac{\pi}{2}\right) - w\left(-\frac{\pi}{2}\right) = \frac{2K_{III}^2}{\pi \mu \tau_Y}. \tag{3.38}$$

Equations (3.37) give the strain and displacement fields within the circular plastic zone of diameter c defined by Equation (3.35) which passes from the crack tip. Outside the plastic zone the elastic stress field is the same as that of a crack that is longer by $c/2$ than the original crack. These propositions completely define the exact elastic–plastic solution of the antiplane-mode crack problem under small-scale yielding.

(b) Strain-hardening

The solution of the antiplane-mode crack problem for strain-hardening materials under small-scale yielding has been given by Rice [3.12] within the framework of the deformation plasticity theory. In the following some general characteristic features of the solution will be given, while for further details the reader is referred to reference [3.12].

For an elastic strain-hardening material the principal antiplane shear stress τ and strain γ, defined by

$$\tau = \sqrt{\tau_x^2 + \tau_y^2}, \qquad \gamma = \sqrt{\gamma_x^2 + \gamma_y^2} \tag{3.39}$$

are related by

$$\tau = \frac{\tau_Y}{\gamma_Y}\gamma, \quad \gamma < \gamma_Y$$
$$\tau = \tau(\gamma), \quad \gamma > \gamma_Y \tag{3.40}$$

where τ_Y, γ_Y correspond to an initial yield point and the function $\tau(\gamma)$ describes the stress–strain relation in the hardening region.

The Hencky isotropic stress–strain relations are

$$\tau_x = \frac{\tau(\gamma)}{\gamma}\gamma_x, \qquad \tau_y = \frac{\tau(\gamma)}{\gamma}\gamma_y. \tag{3.41}$$

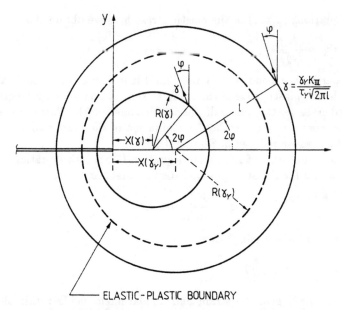

Figure 3.14. Lines of constant strain in the elastic and plastic region and the elastic–plastic boundary for antiplane deformation of a strain-hardening elastic–plastic material under small-scale yielding [3.12].

Under these conditions the basic results of the solution are shown in Figure 3.14 and may be summarized as follows:

In the plastic region lines of constant γ and τ are circles with radius $R(\gamma)$ and centers located on the x-axis at a distance $X(\gamma)$ from the crack tip. They are described by the equation

$$[x - X(\gamma)]^2 + y^2 = [R(\gamma)]^2 \qquad (3.42)$$

where

$$X(\gamma) = \frac{K_{III}^2}{2\pi\tau_Y^2}\left[2\gamma_Y\tau_Y\int_\gamma^\infty \frac{du}{u^2\tau(u)} - \frac{\gamma_Y\tau_Y}{\gamma\tau(\gamma)}\right] \qquad (3.43)$$

$$R(\gamma) = \frac{K_{III}^2}{2\pi\tau_Y^2}\frac{\gamma_Y\tau_Y}{\gamma\tau(\gamma)}. \qquad (3.44)$$

The elastic–plastic boundary is a circle with center at a distance $X(\gamma_Y)$ from the crack tip and radius $R(\gamma_Y)$ given by

$$R(\gamma_Y) = \frac{K_{III}^2}{2\pi\tau_Y^2}. \qquad (3.45)$$

Equation (3.45) shows that $R(\gamma_Y)$ is independent of the stress–strain relation in the hardening range, while from Equations (3.43) and (3.45) it is deduced that $X(\gamma_Y) < R(\gamma_Y)$, which indicates that the elastic–plastic boundary encompasses the crack tip. From Equations (3.35) and (3.45) it follows that the radius of the elastic–plastic boundary is the same for perfect plasticity and strain-hardening behavior, while for $\tau(\gamma) = \gamma_Y$ Equation (3.43) gives $X(\gamma_Y) = 0$, which was obtained previously for perfect plasticity. The angle that the principal shear strain at any point on a constant strain circle inside the plastic zone makes with the y-axis is half the angle that the radial line from the center of the circle subtends with the x-axis (Figure 3.14). The magnitude of the strain γ along a constant strain circle is determined from Equations (3.43) and (3.44) by solving

$$x = X(\gamma) + R(\gamma) = \frac{K_{III}^2 \gamma_Y}{\pi \tau_Y} \int_\gamma^\infty \frac{du}{u^2 \tau(u)} \qquad (3.46)$$

where x is the abscissa of the point where the circle intersects the x-axis.

In the elastic region lines of constant γ are also circles which are concentric with the elastic–plastic boundary. The stresses are obtained by

$$\tau_y + i\tau_x = \frac{K_{III}}{\sqrt{2\pi[z - X(\gamma_Y)]}} \qquad (3.47)$$

which, by comparing Equation (3.36), indicates that the effect of yielding is to shift the crack tip by $X(\gamma_Y)$ to the center of the plastic zone, as in the case of perfect plasticity.

The shear strain along a constant strain circle is given by (Figure 3.14)

$$\gamma = \frac{\gamma_Y}{\tau_Y} \frac{K_{III}}{\sqrt{2\pi l}}, \qquad l = \sqrt{[x - X(\gamma_0)]^2 + y^2} \qquad (3.48)$$

where l represents the distance from the center of the elastic–plastic boundary to the point under consideration.

As an example, we consider the case of a power law strain-hardening material with a stress–strain relation given by

$$\tau = \frac{\tau_Y}{\gamma_Y}\gamma, \quad \gamma < \gamma_Y; \qquad \tau = \tau_Y \left(\frac{\gamma}{\gamma_Y}\right)^N, \quad \gamma > \gamma_Y. \qquad (3.49)$$

The values $N = 0$ and $N = 1$ correspond to perfect plasticity and perfect elasticity, respectively. From Equations (3.43) and (3.44) we obtain

$$R(\gamma) = \frac{K_{III}^2}{2\pi \tau_Y^2}\left(\frac{\gamma_Y}{\gamma}\right)^{N+1} \qquad (3.50)$$

$$X(\gamma) = \frac{1-N}{1+N} R(\gamma). \qquad (3.51)$$

The plastic zone extends a distance

$$c = X(\gamma_Y) + R(\gamma_Y) = \frac{K_{III}^2}{(1+N)\pi \tau_Y^2} \qquad (3.52)$$

ahead of the crack tip, and a distance

$$R(\gamma_Y) - X(\gamma_Y) = \frac{N K_{III}^2}{(1+N)\pi \tau_Y^2} \qquad (3.53)$$

behind the crack tip.

The strains and stresses ahead of the crack in the plastic region are given by

$$\gamma_y(x,0) = \gamma_Y \left[\frac{K_{III}^2}{(N+1)\pi\tau_Y^2 x}\right]^{1/(N+1)} \tag{3.54}$$

$$\tau_y(x,0) = \tau_Y \left[\frac{K_{III}^2}{(N+1)\pi\tau_Y^2 x}\right]^{N/(N+1)} \tag{3.55}$$

Observe in Equations (3.54) and (3.55) the $r^{-1/(N+1)}$ and $r^{-N/(N+1)}$ singularity of the strain and stress fields. Equations (3.50)–(3.55) degenerate to the corresponding equations for perfect plastic and elastic behavior for $N=0$ and $N=1$, respectively.

3.4. Complete solution for antiplane mode

For applied loads resulting to large plastic zones the small-scale yielding approximation becomes invalid and a complete solution must be sought. Hult and McClintock [3.11] solved the problem of an edge crack in a semi-infinite plate under antiplane-mode loading, while the problem of an edge notch in finite-width plates has been studied by Koskinen [3.13] and Rice [3.12, 3.14]. In the following some basic results of these solutions for perfectly plastic and strain-hardening behavior will be given.

(a) Perfect plasticity

For an edge crack of length a in a semi-infinite plate subjected to a uniform remote stress τ_∞ the extent c of the plastic zone in front of the crack and the crack-opening displacement δ are given by

$$\frac{c}{a} = \frac{2}{\pi}\frac{1+s^2}{1-s^2}E_2\left(\frac{2s}{1+s^2}\right) - 1 \tag{3.56}$$

$$\frac{\delta}{a} = 2\gamma_Y\left[\frac{2}{\pi}(1+s^2)E_1(s^2) - 1\right] \tag{3.57}$$

where $s = \tau_\infty/\tau_Y$ and E_1 and E_2 are the complete elliptic integrals of the first and second type, respectively. For small applied stresses, when terms of order s^2 are negligible compared to unity, equations (3.56) and (3.57) reduce to the small-scale yielding results. For higher stresses the plastic zone elongates from the circular shape which occurs at small stresses and, at the limit load, extends to infinity in the x-direction with a height in the y-direction equal to $4a/\pi$. Results for the normalized plastic zone $(c/a)[(\pi a/2b)\operatorname{ctn}(\pi a/2b)/[1-(a/b)^2]]$ in terms of the normalized net section stress $\tau_n/\tau_Y = \tau_\infty/[(1-a/b)\tau_Y]$ for finite-width plates are shown in Figure 3.15. The small-scale yielding solution is shown by the dotted line. Observe that the results of the complete solution start to deviate significantly from those of the small-scale yielding solution for $\tau_\infty > (0.4\text{--}0.5)\tau_Y$.

Elastic–plastic stress field in cracked bodies

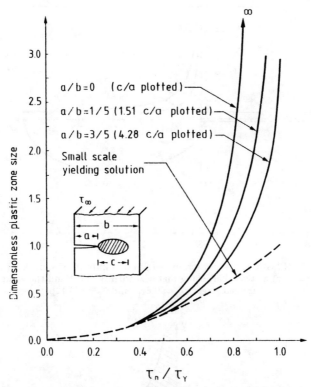

Figure 3.15. Normalized plastic zone length versus the normalized net section stress for antiplane deformation of a perfectly plastic material. a/b takes the values 0, 1/5 and 3/5 [3.3].

(b) Strain-hardening

Results for the plastic zone and the strain distribution in front of the crack in a strain-hardening material with $N = 0.1$ are shown in Figure 3.16 for two values of applied stress $\tau_\infty = 0.6\tau_Y$ and $\tau_\infty = 0.8\tau_Y$. Observe the transition from the circular plastic zone at small applied stresses to elongated plastic zones for higher load levels. As expected, the plastic zones are more elongated for smaller values of the strain-hardening exponent N for which the post-yield material behavior approaches that of a perfectly plastic material.

3.5. Irwin's model

Unlike the antiplane mode, treatment of the elastic–plastic problem for the opening mode encounters mathematical difficulties that prevent an analytical solution in this case. In this respect the simplified model proposed by Irwin [3.15] in the early days of development of fracture mechanics for the determination of the plastic zone attending the crack tip under small-scale yielding is most appropri-

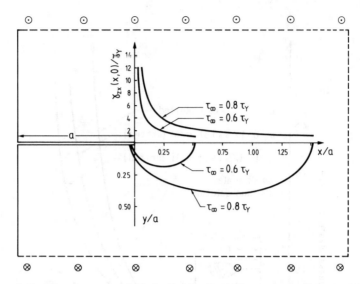

Figure 3.16. Plastic zones and strain distribution in front of a crack for antiplane deformation of a strain-hardening material with $N = 0.1$ and the values of the applied stress $\tau_\infty = 0.6\tau_Y$ and $0.8\tau_Y$ [3.12].

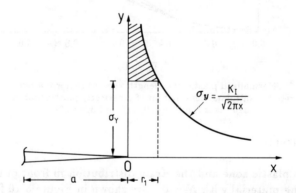

Figure 3.17. Elastic σ_y stress distribution ahead of a crack.

ate. Attention is focused only on the extent along the crack axis and not on the shape of the plastic zone for an elastic–perfectly plastic material.

To begin with, let us consider the elastic distribution of the σ_y ($=\sigma_x$) stress along the crack axis in Figure 3.17 and assume that the plate is under plane stress. An estimate of the extent of the plastic zone in front of the crack following the approximate solution of Section 3.2 is obtained by determining the distance r_1 from the crack tip to the point at which the yield stress σ_Y is exceeded. The value of r_1 determined from the condition $\sigma_y = \sigma_Y$ is given by Equation (3.4). The σ_y stress distribution along the x-axis is represented by the horizontal line $\sigma_y = \sigma_Y$ up to the point $x = r_1$ followed by the elastic singular σ_y-curve.

It is apparent in this determination that the equilibrium condition along the y-direction is violated since the actual elastic stress distribution inside the plastic

Elastic–plastic stress field in cracked bodies 95

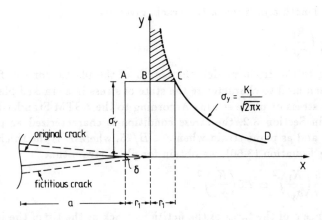

Figure 3.18. Elastoplastic σ_y stress distribution ahead of a crack according to the Irwin model

zone is replaced by a constant stress equal to σ_Y. The stresses in the shaded area in Figure 3.17 should produce a stress redistribution along the x-axis and the actual plastic zone length must be larger than r_1. Thus, as a result of the crack-tip plasticity, the displacements are larger and the stiffness of the plate is lower than in the elastic case.

These observations, in conjunction with the small-scale yielding solution for the antiplane mode (Section 3.3), led Irwin to propose that the effect of plasticity makes the plate behave as if it had a crack longer than the actual crack size. The fictitious crack length is determined as follows: The area underneath the σ_y-curve up to the point $x = r_1$ is given by

$$\int_0^{r_1} \frac{K_I}{\sqrt{2\pi x}} = 2\sigma_Y r_1 \tag{3.58}$$

where the value of r_1 was introduced from Equation (3.4). Therefore, the shaded area in Figure 3.17 is $\sigma_Y r_1$. This result suggests that in order to satisfy equilibrium along the y-direction the original crack should be extended by a length r_1, as in Figure 3.18. In this case the σ_y stress distribution is represented by the curve $ABCD$, so that the area underneath this curve is equal to the area underneath the σ_y-curve in Figure 3.17 and equilibrium is maintained. Thus, the length of the plastic zone c in front of the crack is equal to $2r_1$ and is given by

$$c = \frac{1}{\pi}\left(\frac{K_I}{\sigma_Y}\right)^2 \tag{3.59}$$

for plane stress. Equation (3.59) shows that the length of the plastic zone c, according to the Irwin model, is twice that determined from the approximate solution of Section 3.2. For plane strain Irwin [3.16] suggested a constraint factor that increases the stress required to produce yielding of $\sqrt{3}$. This results in a plastic zone length c in front of the crack, given by

$$c = \frac{1}{3\pi}\left(\frac{K_I}{\sigma_Y}\right)^2 \quad (3.60)$$

According to the Irwin model, the length of the plastic zone in front of the crack has been used to characterize the state of stress in a cracked plate as being either plane stress or plane strain. According to the ASTM Standard E399 [3.7] referred to in Section 3.2, the stress condition is characterized as plane stress when $c = B$ and as plane strain when $c < B/25$, where B is the thickness of the plate. Using Equation (3.60), we obtain for plane strain that

$$B > \frac{25}{3\pi}\left(\frac{K_I}{\sigma_Y}\right)^2 \simeq 2.5\left(\frac{K_I}{\sigma_Y}\right)^2. \quad (3.61)$$

The distance δ of the faces of the fictitious crack at the tip of the initial crack of length a is given by the use of Equation (2.87)

$$\delta = 2v = \frac{\kappa+1}{2\mu}\sigma\sqrt{(a+c)^2 - a^2} \quad (3.62)$$

which gives

$$\delta = \frac{4}{\pi E}\frac{K_I^2}{\sigma_Y} \quad (3.63)$$

for plane stress, and

$$\delta = \frac{4(1-\nu^2)}{3\pi E}\frac{K_I^2}{\sigma_Y} \quad (3.64)$$

for plane strain.

The quantity δ given by Equation (3.63) has played an important role in characterizing the propensity of a crack to extend, and will be described in detail in Chapter 5.

3.6. Dugdale's model

A simplified model for plane stress yielding which avoids the complexities of a true elastic–plastic solution has been introduced by Dugdale [3.17]. The model applies to very thin plates in which plane stress conditions dominate for materials with elastic–perfectly plastic behavior which obey the Tresca yield criterion. To analyze the model, the case of a crack of length $2a$ in an infinite plate subjected to uniaxial uniform stress σ at infinity perpendicular to the crack plane is considered (Figure 3.19), and the following hypotheses are made:

(i) All plastic deformation concentrates in a line in front of the crack.
(ii) An effective crack that is longer than the physical crack by the length of the plastic zone is considered.

The first hypothesis is justified from the fact that for plane stress, following the considerations of Section 3.2, yielding takes place on planes that subtend 45°

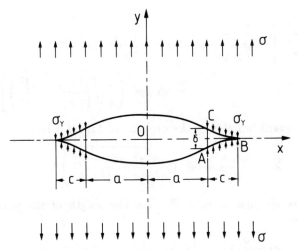

Figure 3.19. Dugdale model for a mode-I crack of length 2a.

with the plate surface and the height of the plastic zone is equal to the plate thickness. Thus, for very thin plates the plastic zones approach line segments. Following the Tresca yield criterion, stresses equal to the yield stress σ_Y should apply along the plastic zone. The length of the plastic zone c is determined from the condition that the σ_y stress at the tip of the effective crack should remain bounded and equal to the yield stress σ_Y.

Based on the above arguments the solution of the elastic–plastic problem of Figure 3.19 is, according to the Dugdale model, reduced to an elastic problem. The Westergaard function Z of the problem is obtained by adding these functions for the following two problems:

(i) A crack of length $2(a+c)$ in an infinite plate subjected to a uniform stress σ at infinity. The Westergaard function is given by (Equation (2.65) by omitting the constant term)

$$Z_1 = \frac{\sigma z}{\sqrt{z^2 - (a+c)^2}} \tag{3.65}$$

(ii) A crack of length $2(a+c)$ in an infinite plate subjected to a uniform stress distribution equal to σ_Y along the plastic zone $(a < |x| < a+c)$. The Westergaard function for a pair of concentrated forces at the points $\pm x$ is given by (Equation (2.67))

$$Z = \frac{2\sigma_Y z \sqrt{(a+c)^2 - x^2}}{\pi(z^2 - x^2)\sqrt{z^2 - (a+c)^2}} \tag{3.66}$$

and therefore this function for the problem under consideration is

$$Z_2 = \int_a^{a+c} \frac{2\sigma_Y z}{\pi(z^2 - x^2)} \frac{\sqrt{(a+c)^2 - x^2}}{\sqrt{z^2 - (a+c)^2}} \, dx \tag{3.67}$$

or

$$Z_2 = \frac{2\sigma_Y}{\pi}\left[\frac{z}{\sqrt{z^2-(a+c)^2}}\text{arc cos}\left(\frac{a}{a+c}\right)-\right.$$
$$\left.-\text{arc cot}\left(\frac{a}{z}\sqrt{\frac{z^2-(a+c)^2}{(a+c)^2-a^2}}\right)\right]. \quad (3.68)$$

The Westergaard function of the problem of Figure 3.19 is

$$Z = Z_1 - Z_2 = \frac{2\sigma_Y}{\pi}\text{arc cot}\left(\frac{a}{z}\sqrt{\frac{z^2-(a+c)^2}{(a+c)^2-a^2}}\right) \quad (3.69)$$

by zeroing the singular term of Z. For the length of the plastic zone, this condition gives

$$\frac{a}{a+c} = \cos\left(\frac{\pi}{2}\frac{\sigma}{\sigma_Y}\right). \quad (3.70)$$

Equation (3.70) for small values of σ/σ_Y gives

$$c = \frac{\pi}{8}\left(\frac{K_I}{\sigma_Y}\right)^2. \quad (3.71)$$

By comparing Equations (3.71) and (3.59) we can deduce that the Irwin model underestimates the length of the plastic zone as compared to the Dugdale model by about 20 per cent.

The displacement of the crack faces obtained by introducing the value of Z from Equation (3.69) into Equation (3.68) is given by

$$v = \frac{(a+c)\sigma_Y}{\pi E}\left[\frac{x}{a+c}\ln\frac{\sin^2(\theta_2-\theta)}{\sin^2(\theta_2+\theta)} + \cos\theta_2\ln\frac{(\sin\theta_2+\sin\theta)^2}{(\sin\theta_2-\sin\theta)^2}\right] \quad (3.72)$$

where we put

$$\theta = \text{arc cos}\frac{x}{a+c}, \quad \theta_2 = \frac{\pi}{2}\frac{\sigma}{\sigma_Y}. \quad (3.73)$$

The opening of the effective crack at the tip of the physical crack is given by

$$\delta = 2\lim_{x\to\pm a} v = \frac{8\sigma_Y a}{\pi E}\ln(\sec\theta_2). \quad (3.74)$$

By expanding Equation (3.74) and retaining the first term for small values of σ/σ_Y we obtain

$$\delta = \frac{K_I^2}{E\sigma_Y}. \quad (3.75)$$

By comparing Equations (3.74) and (3.63) we can deduce that the Irwin model overestimates δ as compared to the Dugdale model by 27 per cent. Furthermore, from Equation (3.72) it can be shown that the curve $v = v(x)$ has a vertical

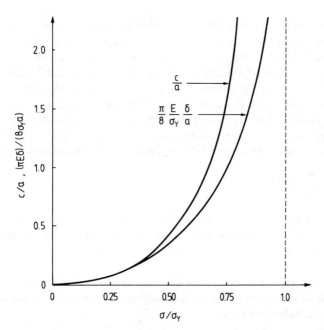

Figure 3.20. Normalized plastic zone length and crack-tip opening displacement versus normalized applied stress for the Dugdale model.

slope at the point $x = \pm a$. The variation of the dimensionless quantities c/a and $(\pi E \delta)/(8\sigma_Y a)$ versus σ/σ_Y is shown in Figure 3.20.

Experimental evidence for the Dugdale model was provided by Dugdale [3.17] for low-carbon steels, by Rosenfield et al. [3.9] for silicon iron and plain carbon steels, by Mills [3.18] and Gales and Mills [3.19] for polycarbonate, polysulfane and polyvinylchloride, and by Brinson [3.20] for polycarbonate, among others. Theocaris and Gdoutos [3.21], using the optical method of caustics, found the realm of validity of the Dugdale model for different types of steel.

The spread of plasticity in front of the crack according to the Dugdale model for modes II and III was studied by Bilby et al. [3.22] who modeled the crack and the yielded zone as arrays of dislocations. The expressions for the length of the plastic zone and the opening of the crack faces are analogous to Equations (3.70) and (3.74). Related results for other crack configurations under mode III were provided by Smith [3.23, 3.24]. Other problems for the Dugdale model under mode I are discussed in references [3.25-3.33]. Modifications of the Dugdale model to include the influence of strain hardening on the size of the plastic zone are provided in references [3.8], [3.9] and [3.34-3.37].

The Dugdale model presents many similarities although it has a completely different physical meaning from a model proposed by Barenblatt [3.38] for the study of the structure of the stress field and the shape of an elastic crack around its ends.

3.7. Singular solution for a work-hardening material

The problem of stress and strain distribution near the tip of a crack in an opening-mode field for a material with a strain-hardening behavior was solved simultaneously by Hutchinson [3.39] and by Rice and Rosengren [3.40]. This solution is sometimes referred to in the literature as the HRR solution from the names of the above investigators. The solution concerns the immediate vicinity of the crack tip in which the elastic strains are small compared to plastic strains and takes place within the framework of the deformation theory of plasticity for small strains, which is actually a nonlinear elasticity theory. The uniaxial stress–strain curve of the material is modeled by the Ramberg–Osgood relation

$$\frac{\epsilon}{\epsilon_Y} = \frac{\sigma}{\sigma_Y} + \alpha \left(\frac{\sigma}{\sigma_Y}\right)^m \tag{3.76}$$

where σ_Y and ϵ_Y are referred to the yield point, $m > 1$ is the strain-hardening exponent and α is a dimensionless material parameter. The limiting values of $m = 1$ and ∞ correspond to an elastic and an elastic–perfectly plastic material, respectively.

In the deformation theory of plasticity it is assumed that plastic deformation is independent of the hydrostatic stress component. Thus, the generalized stress–strain relation takes the form

$$\epsilon_{ij} = \frac{1+\nu}{E} s_{ij} + \frac{1-2\nu}{3E} \sigma_{pp} \delta_{ij} + \frac{3}{2} \frac{\epsilon_e^p}{\sigma_e} s_{ij} \tag{3.77}$$

where the deviatoric stress components are given by

$$s_{ij} = \sigma_{ij} - \tfrac{1}{3}\sigma_{pp}\delta_{ij} \tag{3.78}$$

and the effective stress σ_e and effective plastic strain ϵ_e^p are given by

$$\sigma_e = \left(\tfrac{3}{2} s_{ij} s_{ij}\right)^{1/2} \tag{3.79}$$

$$\epsilon_e^p = \left(\tfrac{3}{2} \epsilon_{ij}^p \epsilon_{ij}^p\right)^{1/2}. \tag{3.80}$$

From Equation (3.76) it is deduced that

$$\epsilon_e^p = \alpha \epsilon_Y \left(\frac{\sigma_e}{\sigma_Y}\right)^m \tag{3.81}$$

so that Equation (3.77) takes the form

$$\epsilon_{ij} = \tfrac{3}{2} \alpha \epsilon_Y \left(\frac{\sigma_e}{\sigma_Y}\right)^m \frac{s_{ij}}{\sigma_e} \tag{3.82}$$

by ignoring the elastic strains relatively to the plastic strains in the neighborhood of the crack tip. For $m = 1$, Equation (3.82) takes the form of Hooke's law for an incompressible material, while for $m = \infty$ perfect plasticity is obtained.

We now have to solve an elastic problem governed by the equations of equilibrium, compatibility and the nonlinear stress–strain relation expressed by Equation (3.82). Following the Airy stress function representations, the stress components in polar coordinates are given in terms of a function U by

$$\sigma_\theta = \frac{\partial^2 U}{\partial r^2}, \qquad \sigma_r = \nabla^2 U - \sigma_\theta, \qquad \tau_{r\theta} = -\frac{\partial}{\partial r}\left(\frac{1}{r}\frac{\partial U}{\partial \theta}\right). \tag{3.83}$$

For plane stress, considered first, the strains from Equation (3.82) are given by

$$\begin{aligned}\epsilon_r &= \alpha\sigma_e^{n-1}(\sigma_r - \tfrac{1}{2}\sigma_\theta) \\ \epsilon_\theta &= \alpha\sigma_e^{n-1}(\sigma_\theta - \tfrac{1}{2}\sigma_r) \\ \epsilon_{r\theta} &= \tfrac{3}{2}\alpha\sigma_e^{n-1}\tau_{r\theta}\end{aligned} \tag{3.84}$$

where the effective stress is

$$\sigma_e^2 = \sigma_r^2 + \sigma_\theta^2 - \sigma_r\sigma_\theta + 3\tau_{r\theta}^2. \tag{3.85}$$

Introducing the values of strains from Equations (3.84) into the compatibility equation

$$\frac{\partial^2 \epsilon_\theta}{\partial r^2} + \frac{2}{r}\frac{\partial \epsilon_\theta}{\partial r} - \frac{1}{r}\frac{\partial^2 \gamma_{r\theta}}{\partial r \partial \theta} - \frac{1}{r^2}\frac{\partial \gamma_{r\theta}}{\partial \theta} + \frac{1}{r^2}\frac{\partial^2 \epsilon_r}{\partial \theta^2} - \frac{1}{r}\frac{\partial \epsilon_r}{\partial r} = 0 \tag{3.86}$$

the following differential equation for the function U is obtained

$$\frac{\partial^2}{\partial r^2}\left[\sigma_e^{m-1}\left(2r\frac{\partial^2 U}{\partial r^2} - \frac{\partial U}{\partial r} - \frac{1}{r}\frac{\partial^2 U}{\partial \theta^2}\right)\right] + \frac{6}{r}\frac{\partial^2}{\partial r \partial \theta}\left[\sigma_e^{m-1} r\frac{\partial}{\partial r}\left(\frac{1}{r}\frac{\partial U}{\partial \theta}\right)\right] + $$
$$+ \frac{\partial}{\partial r}\left[\sigma_e^{m-1}\left(-\frac{2}{r}\frac{\partial U}{\partial r} - \frac{2}{r^2}\frac{\partial^2 U}{\partial \theta^2} + \frac{\partial^2 U}{\partial r^2}\right)\right] + $$
$$+ \frac{1}{r}\frac{\partial^2}{\partial \theta^2}\left[\sigma_e^{m-1}\left(-\frac{\partial^2 U}{\partial r^2} + \frac{2}{r}\frac{\partial U}{\partial r} + \frac{2}{r^2}\frac{\partial^2 U}{\partial \theta^2}\right)\right] = 0. \tag{3.87}$$

Following Williams's eigenvalue method for solving singularity problems (see reference [2.5]), the function U in the neighborhood of the crack tip is expressed by

$$U = Kr^{\lambda+1}f(\theta) \tag{3.88}$$

where K is a proportional factor.

From Equation (3.83) we obtain for the stresses

$$\begin{aligned}\sigma_r &= r^{\lambda-1}\left[(\lambda+1)f + \frac{d^2 f}{d\theta^2}\right] = Kr^{\lambda-1}\bar{\sigma}_r \\ \sigma_\theta &= r^{\lambda-1}(\lambda+1)\lambda f = Kr^{\lambda-1}\bar{\sigma}_\theta \\ \tau_{r\theta} &= -r^{\lambda-1}\lambda\frac{df}{d\theta} = Kr^{\lambda-1}\bar{\tau}_{r\theta}\end{aligned} \tag{3.89}$$

and for the effective stress

$$\sigma_e = r^{\lambda-1}\sqrt{\bar{\sigma}_r^2 + \bar{\sigma}_\theta^2 - \bar{\sigma}_r\bar{\sigma}_\theta + 3\bar{\tau}_{r\theta}^2} = r^{\lambda-1}\bar{\sigma}_e \tag{3.90}$$

where the quantities $\bar{\sigma}_r$, $\bar{\sigma}_\theta$, $\bar{\tau}_{r\theta}$ and $\bar{\sigma}_e$ are functions of the polar angle θ.

Introducing the value of U from Equation (3.88) into Equation (3.87), the following nonlinear differential equation for the function $f = f(\theta)$ is obtained:

$$\left[m(\lambda-1) - \frac{d^2}{d\theta^2}\right]\left[\bar{\sigma}_e^{m-1}\left[(\lambda+1)(\lambda-2)f - 2\frac{d^2 f}{d\theta^2}\right]\right] +$$

$$+ [m(\lambda-1)+1]m(\lambda-1)\bar{\sigma}_e^{m-1}\left[(\lambda+1)(2\lambda-1)f - \frac{d^2 f}{d\theta^2}\right] +$$

$$+ 6[m(\lambda-1)+1]\lambda\frac{d}{d\theta}\left(\bar{\sigma}_e^{m-1}\frac{df}{d\theta}\right) = 0. \qquad (3.91)$$

From the stress-free boundary conditions along the crack faces we obtain

$$f(\pm\pi) = \frac{df(\pm\pi)}{d\theta} = 0. \qquad (3.92)$$

Due to symmetry, $\tau_{r\theta}$, $\partial\sigma_r/\partial\theta$ and $\partial\sigma_\theta/\partial\theta$ should vanish for $\theta = 0$, which leads to

$$\frac{df(0)}{d\theta} = \frac{d^3 f(0)}{d\theta^3} = 0. \qquad (3.93)$$

Analogous equations can be obtained for plane strain. By solving numerically the nonlinear eigenvalue equation (Equation (3.91)) in conjunction with the boundary conditions given by (3.92) and (3.93) and its counterpart for plane strain [3.39], the values of λ fit the equation

$$\lambda = \lambda_1 = \frac{m}{m+1}. \qquad (3.94)$$

Using energy considerations, as we did in Section 2.3, we find that for $\lambda < \lambda_1$ the total work of deformation within a finite area, including the crack tip, becomes infinite, while for $\lambda > \lambda_1$ the resulting stress field is nonsingular. With $\lambda = \lambda_1$ the stresses σ_{ij} and strains ϵ_{ij} from Equations (3.89) and (3.84) take the form

$$\sigma_{ij} = Kr^{-1/(m+1)}\tilde{\sigma}_{ij} \qquad (3.95)$$

and

$$\epsilon_{ij} = \alpha\epsilon_Y K^m r^{-m/(m+1)}\tilde{\epsilon}_{ij} \qquad (3.96)$$

where K has the dimension of $(\text{length})^{1/(m+1)}$.

The value of λ given by Equation (3.94) can also be obtained from the path independent line integral J defined by Equation (2.183). For a circular path of radius r encompassing the crack tip, J takes the form

$$J = r\int_{-\pi}^{\pi}\left[w[\epsilon(r,\theta)]\cos\theta - T_k(r,\theta)\frac{\partial u_k(r,\theta)}{\partial x}\right]d\theta. \qquad (3.97)$$

In order to make the integral J path independent (i.e. does not depend on the value of r), the integrand in Equation (3.97) must exhibit an inverse r singularity. Since all terms in the integrand are of the order of stress × strain, it follows that

$$\sigma_{ij}\epsilon_{ij} = \frac{F(\theta)}{r} \quad \text{for } r \to 0. \qquad (3.98)$$

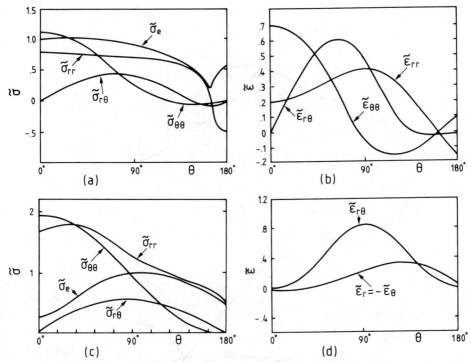

Figure 3.21. Normalized stress $\tilde{\sigma}_{ij}$ and strain $\tilde{\epsilon}_{ij}$ components and equivalent stress $\tilde{\sigma}_e$ versus polar angle θ for a strain-hardening material with $n = 3$ under (a) and (b) plane stress and (c) and (d) plane strain [3.41].

Thus, for power law hardening materials described by Equation (3.76) the stresses and strains should have the singularities of Equations (3.95) and (3.96). For linear elastic materials ($m = 1$) the inverse square root singularity is obtained, while for perfectly plastic materials ($m = \infty$) the stress field is nonsingular, while the strain field has a r^{-1} singularity. Equation (3.98) applies to the homogeneous crack stress and strain field. For crack tips in a nonhomogeneous stress and strain environment, $\sigma_{ij}\epsilon_{ij}$, can possess singularities other than the form $1/r$.

Equations (3.95) and (3.96) express the singular stress and strain fields in terms of the amplitude K and the functions $\tilde{\sigma}_{ij}(\theta)$ and $\tilde{\epsilon}_{ij}(\theta)$. The θ-variation of these functions has been calculated numerically and is shown in Figure 3.21 for $n = 3$ under conditions of plane stress and plane strain. Furthermore, these functions for perfect plasticity, as calculated by Hutchinson [3.41], are shown in Figure 3.22.

It is important to connect the amplitude factor K with the J-integral. After determining the displacements from strains and substituting into Equation (3.96) we get

$$J = \alpha \epsilon_Y \sigma_Y K^{m+1} I \tag{3.99}$$

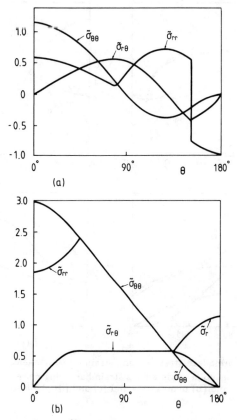

Figure 3.22. Normalized stress $\tilde{\sigma}_{ij}$ components versus polar angle θ for a perfectly plastic material under (a) plane stress and (b) plane strain [3.41].

where I depends only on the strain-hardening exponent m. Values of I for selected values of m under plane stress and plane strain are shown in Table 3.1.

In terms of the J-integral, Equations (3.95) and (3.96) take the form

$$\sigma_{ij} = \left(\frac{J}{\alpha\epsilon_Y\sigma_Y Ir}\right)^{1/(m+1)} \tilde{\sigma}_{ij} \tag{3.100}$$

$$\epsilon_{ij} = \alpha\epsilon_Y \left(\frac{J}{\alpha\epsilon_Y\sigma_Y Ir}\right)^{m/(m+1)} \tilde{\epsilon}_{ij} \tag{3.101}$$

while for the displacements u_i we obtain

$$u_i = \alpha\epsilon_Y \left(\frac{J}{\alpha\epsilon_Y\sigma_Y I}\right)^{m/(m+1)} r^{1/(m+1)} \tilde{u}_i(\theta). \tag{3.102}$$

It is interesting to compare the values of stresses in front of the crack for plane stress and plane strain. For perfect plasticity it can be shown from Equation (3.100) that the ratio of $\sigma_y(x)$ for plane strain and plane stress is equal to $1+\pi/2$, while the same ratio for linear elasticity is equal to unity. Furthermore, the ratio

Table 3.1. Values of $I(m)$

m	1	3	5	9	13	∞
Plane stress	2π	3.86	3.41	3.03	2.87	2.8
Plane strain	$2\pi(1-\nu^2)$	5.51	5.01	4.60	4.40	4.3

of stress triaxiality for perfect plasticity measured by $\sigma_{ij}/3$ for plane strain and plane stress is equal to $1 + \pi$. The high values of those stresses for plane strain that develop ahead of the crack and exceed the yield stress play a major role in the mechanism of crack growth by formation, growth and coalescence of voids.

The solution presented in this section for a strain-hardening material took place by the deformation theory of plasticity which is actually a nonlinear elasticity theory. Applicability of the results obtained is restricted to monotonically increasing stresses and no unloading is permitted. However, when the stresses at every point in the plastic zone remain in fixed proportion, the deformation theory is a good representation of the actual plastic deformation and the results of this section can offer an adequately accurate approximation of the state of affairs near the crack tip. Generally, for stationary cracks that are subjected to monotonically increasing loading, the deformation theory gives an accurate description of the state of affairs in the cracked body. On the other hand, for situations of substantial but slow crack growth before failure, unloading takes place and the stress field near the crack tip deviates from proportionality. In such cases the deformation theory is inadequate and incremental plasticity theories should be used.

Analogous results for mode-II loading were presented by Hutchinson [3.41], while the case of a power-law hardening orthotropic material was solved by Pan and Shih [3.42]. Further solutions related to the results of this section can be found in references [3.43, 3.44].

3.8. Numerical solutions

The most commonly applied numerical method for solving crack problems in the plastic range is the finite element method. The basic principles of the method for the solution of nonlinear problems can be found in many articles and books [3.45, 3.46] and they will not be referred to here. In this section a brief review and some general characteristic features of the application of the method to elastoplastic crack problems will be made.

Among the earliest solutions of the elastoplastic stress and strain distribution in plates made of work-hardening materials are those by Swedlow et al. [3.47] and Swedlow [3.48] who used conventional triangular elements. A very fine grid in the neighborhood of the crack tip was used to account for the high stress and/or strain elevation developed in this area. This method, which was discussed in Section 2.7(g) for the elastic case, gives poor results near the crack tip. The embedded singularity finite element method was developed by Hilton and Hutchinson [3.4] for antiplane and opening-mode loading. The dominant

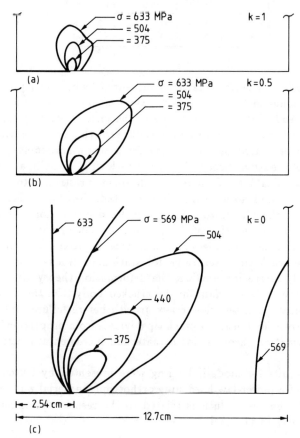

Figure 3.23. Elastic–plastic boundaries for various values of the applied biaxial stress for (a) $k = 1$, (b) $k = 0.5$ and (c) $k = 0$. (k is the biaxiality coefficient, Figure 3.26).

singularity for in-plane loading was that of the HRR solution developed in the previous section. The method was used by Hilton and Sih [3.49] for solving antiplane shear and inplane loading crack problems. The elements with the embedded singularity are assumed to be fully yielded, which is not always the case, especially behind the crack tip. This method presents the disadvantage of the incompatibility of enriched and conventional elements and requires substantial computational time to obtain convergence. As in the elastic case, isoparametric elements with a proper choice of placement of the midside nodes to obtain the $\sqrt{r}$ term of the displacement shape functions have been used [3.50–3.52]. A very important feature of using such elements is that they satisfy the compatibility condition with the adjoining conventional elements resulting in less computer time than the enriched elements.

A number of computer codes based on the use of conventional or special elements for elastic–plastic analysis of crack problems have appeared in the lit-

Elastic–plastic stress field in cracked bodies 107

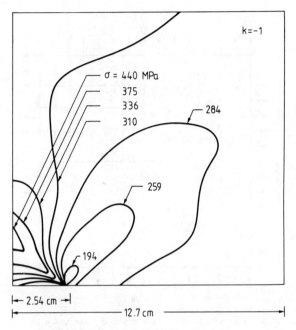

Figure 3.24. Elastic–plastic boundaries for various values of the applied biaxial stress for $k = -1$.

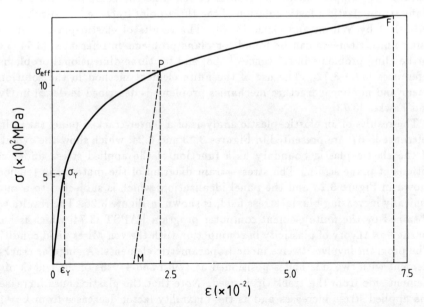

Figure 3.25. True stress–true strain diagram of the material in uniaxial tension.

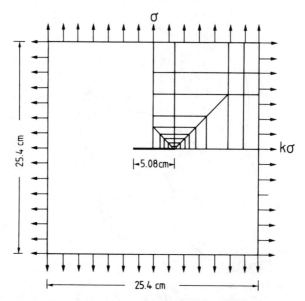

Figure 3.26. Biaxially loaded plane cracked plate and a finite element idealization of one quarter of the plate.

erature. A comparison of the results of ten different computer codes for the plane strain elastic–plastic solution of the three-point fracture test specimen was published by Wilson and Osias [3.53]. The results of elastic–plastic solutions using finite elements can be found for plane problems in references [3.54–3.67], for bending problems in references [3.68] and for three-dimensional problems in references [3.69–3.72]. The use of the finite element method in the solution of linear and nonlinear fracture mechanics problems is described in detail in Owen and Fawkes [3.73].

The results of an elastic–plastic analysis of a center-cracked panel taken from reference [3.67] are presented in Figures 3.23 and 3.24, which show the evolution of the elastic–plastic boundary as a function of the applied stress under conditions of plane strain. The stress–strain diagram of the material in tension is shown in Figure 3.25 and the panel idealization, which is subjected to a monotonically increasing biaxial stress field, is shown in Figure 3.26. The results were obtained by the finite element computer program PAPST [3.74] which is based on the flow theory of plasticity in conjunction with the von Mises yield condition. The program involves twelve mode isoparametric elements. A singular crack-tip element with two side nodes positioned at 1/9th and 4/9ths of the length of the element side from the crack tip is used. Note that the plastic zones increase as the applied stress increases and as the biaxiality factor decreases from $k = 1$ to $k = -1$.

References

3.1. Evans, W. T. and Luxmoore, A. R., Limitations of the Westergaard equations for the experimental evaluations of stress intensity factors, *Journal of Strain Analysis* 11, 177–185 (1976).
3.2. Theocaris, P. S. and Gdoutos, E. E., Discussion of the above paper, *Journal of Strain Analysis* 12, 349–350 (1977).
3.3. Rice, J. R., Mathematical analysis in the mechanics of fracture, in *Fracture-An Advanced Treatise*, Vol. II, *Mathematical Fundamentals* (ed. H. Liebowitz), Pergamon Press, pp. 191–311 (1968).
3.4. Hilton, P. D. and Hutchinson, J. W., Plastic Intensity Factors for Cracked Plates, *Engineering Fracture Mechanics*, 3, pp. 435–451 (1971).
3.5. Gdoutos, E. E., Crack tip plastic zones in glassy polymers under small scale yielding, *Journal of Applied Polymer Science* 26, 1919–1930 (1981).
3.6. Gdoutos, E. E., Plastic zones at the tips of inclined cracks in glassy polymers under small scale yielding, *Journal of Applied Polymer Science* 27, 879–892 (1982).
3.7. Standard test method for plane–strain fracture toughness of metallic materials, *Annual Book of ASTM Standards*, Part 10, E399–81, American Society for Testing and Materials, Philadelphia, pp. 592–621 (1981).
3.8. Hahn, G. T. and Rosenfield, A. R., Local yielding and extension of a crack under plane stress, *Acta Metallurgica* 13, 293–306 (1965).
3.9. Rosenfield, A. R., Dai, P. K. and Hahn, G. T., Crack extension and propagation under plane stress, *Proceedings of the First International Conference on Fracture* (eds. T. Yokobori, T. Kawasaki and J. L. Swedlow), Sendai, Japan, Vol. 1, pp. 223–258 (1966).
3.10. McClintock, F. A. and Irwin, G. R., Plasticity aspects of fracture mechanics, in *Fracture Toughness Testing and its Applications*, ASTM STP No. 381, American Society for Testing and Materials, Philadelphia, pp. 84–113 (1965).
3.11. Hult, J. A. H. and McClintock, F. A., Elastic–plastic stress and strain distributions around sharp notches under repeated shear, *Proceedings of the 9th International Congress on Applied Mechanics*, 8, 51–58 (1956).
3.12. Rice, J. R., Stresses due to a sharp notch in a work-hardening elastic–plastic material loaded by longitudinal shear, *Journal of Applied Mechanics, Trans. ASME* 34, 287–298 (1967).
3.13. Koskinen, M. F., Elastic–plastic deformation of a single grooved flat plate under longitudinal shear, *Journal of Applied Mechanics, Trans. ASME* 30, 585–594 (1963).
3.14. Rice, J. R., Contained plastic deformation near cracks and notches under longitudinal shear, *International Journal of Fracture* 2, 426–447 (1966).
3.15. Irwin, G. R., Plastic zone near a crack tip and fracture toughness, *Proceedings of the Seventh Sagamore Ordnance Material Conference*, pp. IV63–IV78 (1960).
3.16. Irwin, G. R., Linear fracture mechanics, fracture transition, and fracture control, *Engineering Fracture Mechanics* 1, 241–257 (1968).
3.17. Dugdale, D. S., Yielding of steel sheets containing slits, *Journal of the Mechanics and Physics of Solids* 8, 100–104 (1960).
3.18. Mills, N. J., Dugdale yielding zones in cracked sheets of glassy polymers, *Engineering Fracture Mechanics* 6, 537–549 (1974).
3.19. Gales, R. D. R. and Mills, N. J., The plane strain fracture of polysulfone, *Engineering Fracture Mechanics* 6, 93–104 (1974).
3.20. Brinson, H. F., The ductile fracture of polycarbonate, *Experimental Mechanics* 10, 72–77 (1970).
3.21. Theocaris, P. S. and Gdoutos, E. E., Verification of the validity of the Dugdale–Barenblatt model by the method of caustics, *Engineering Fracture Mechanics* 6, 523–535 (1974).
3.22. Bilby, B. A., Cottrell, A. H. and Swinden, K. H., The spread of plastic yield from a notch, *Proceedings of the Royal Society*, Series A, 272, 304–314 (1963).
3.23. Smith, E., The spread of plasticity between two cracks, *International Journal of Engineering Sciences* 2, 379–387 (1964).
3.24. Smith, E., Fracture at stress concentrations, *Proceedings of the First International Conference on Fracture* (eds. T. Yokobori, T. Kawasaki and J. L. Swedlow), Sendai, Japan, Vol. 1, pp. 133–151 (1966).
3.25. Goodier, J. N. and Field, F. A., Plastic energy dissipation in crack propagation, in *Fracture*

of Solids (eds. D. C. Drucker and J. J. Gilman), Wiley, New York, pp. 103–118 (1963).
3.26. Rice, J. R., Plastic yielding at a crack tip, *Proceedings of the First International Conference on Fracture* (eds. T. Yokobori, T. Kawasaki and J. L. Swedlow) Sendai, Japan, Vol. 1, pp. 283–308 (1966).
3.27. Keer, L. M. and Mura, T., Stationary crack and continuous distributions of dislocations, *Proceedings of the First International Conference on Fracture* (eds. T. Yokobori, T. Kawasaki and J. L. Swedlow), Sendai, Japan, Vol. 1, pp. 99–115 (1966).
3.28. Burdekin, F. M. and Stone, D. E. W., The fracture opening displacement approach to fracture mechanics in yielding materials, *Journal of Strain Analysis* 1, 145–153 (1966).
3.29. Kanninen, M. F., A solution for a Dugdale crack subjected to a linearly varying tensile loading, *International Journal of Engineering Sciences* 8, 85–95 (1970).
3.30. Embley, G. T. and Sih, G. C., Plastic flow around an expanding crack, *Engineering Fracture Mechanics* 4, 431–442 (1972).
3.31. Yi-Zhou Chen, A Dugdale problem for a finite internally cracked plate, *Engineering Fracture Mechanics* 17, 579–583 (1983).
3.32. Theocaris, P. S., Dugdale models for two collinear unequal cracks, *Engineering Fracture Mechanics* 18, 545–559 (1983).
3.33. Petroski, H. J., Dugdale plastic zone sizes for edge cracks, *International Journal of Fracture* 15, pp. 217–230 (1979).
3.34. Newman, J. C. Jr, Fracture of cracked plates under plane stress, *Engineering Fracture Mechanics* 1, 137–154 (1968).
3.35. Theocaris, P. S. and Gdoutos, E. E., The modified Dugdale–Barenblatt model adapted to various fracture configurations in metals, *International Journal of Fracture* 10, 549–564 (1974).
3.36. Theocaris, P. S. and Gdoutos, E. E., The size of plastic zones in cracked plates made of polycarbonate, *Experimental Mechanics* 15, 169–176 (1975).
3.37. Harrop, L. P., Application of the modified Dugdale model to the K vs COD relation, *Engineering Fracture Mechanics* 10, 807–816 (1978).
3.38. Barenblatt, G. I., The mathematical theory of equilibrium cracks in brittle fracture, in *Advances in Applied Mechanics*, Academic Press, Vol. 7, pp. 55–129 (1962).
3.39. Hutchinson, J. W., Singular behavior at the end of a tensile crack in a hardening material, *Journal of the Mechanics and Physics of Solids* 16, 13–31 (1968).
3.40. Rice, J. R. and Rosengren, G. F., Plane strain deformation near a crack tip in a power-law hardening material, *Journal of the Mechanics and Physics of Solids* 16, 1–12 (1968).
3.41. Hutchinson, J. W., Plastic stress and strain fields at a crack tip, *Journal of the Mechanics and Physics of Solids* 16, 337–347 (1968).
3.42. Pan, J. and Shih, C. F., Plane-strain crack-tip fields for power-law hardening orthotropic materials, *Mechanics of Materials* 5, 299–316 (1986).
3.43. Goldman, N. L. and Hutchinson, J. W., Fully plastic crack problems: The center-cracked strip under plane strain, *International Journal of Solids and Structures* 11, 575–591 (1975).
3.44. He, M. Y. and Hutchinson, J. W., The penny-shaped crack and the plane strain crack in an infinite body of a power-law material, *Journal of Applied Mechanics, Trans. ASME* 48, pp. 830–840 (1981).
3.45. Zienkiewicz, O. C., *The Finite Element Method* (3rd edn), McGraw–Hill (1982).
3.46. Owen, D. R. J. and Hinton, E., *Finite Elements in Plasticity–Theory and Practice*, Pineridge Press, Swansea, U.K. (1980).
3.47. Swedlow, J. L., Williams, M. L. and Yang, W. H., Elasto-plastic stresses and strains in cracked plates, *Proceedings of the First International Conference on Fracture* (eds. T. Yokobori, T. Kawasaki and J. L. Swedlow), Sendai, Japan, Vol. 1, pp. 259–282 (1966).
3.48. Swedlow, J. L., Elasto-plastic cracked plates in plane strain, *International Journal of Fracture Mechanics* 5, 33–44 (1969).
3.49. Hilton, P. D. and Sih, G. C., Applications of the finite element method for the calculations of stress intensity factors, in *Mechanics of Fracture*, Vol. 1, *Methods of Analysis and Solutions of Crack Problems* (ed. G. C. Sih), Noordhoff Int. Publ., The Netherlands, pp. 426–483 (1973).
3.50. Barsoum, R. S., Triangular quarter-point elements as elastic and perfectly-plastic crack tip elements, *International Journal for Numerical Methods in Engineering* 11, 85–98 (1977).
3.51. Barsoum, R. S., Application of triangular quarter-point elements as crack tip elements of power law hardening material, *International Journal of Fracture* 12, 463–466 (1976).
3.52. Benzley, S. E., Nonlinear calculations with quadratic quarter-point crack tip element, *International Journal of Fracture* 12, 477–480 (1976).

3.53. Wilson, W. K. and Osias, J. R., A comparison of finite element solutions for an elastic-plastic crack problem, *International Journal of Fracture* **14**, R95–R108 (1978).
3.54. Levy, N., Marcal, P. V., Ostergren, W. J. and Rice, J. R., Small scale yielding near a crack in plane strain: A finite element analysis, *International Journal of Fracture Mechanics* **7**, 143–156 (1971).
3.55. Lee, J. D. and Liebowitz, H., Considerations of crack growth and plasticity in finite element analysis, *Computers and Structures* **8**, 403–410 (1978).
3.56. Miller, K. J. and Kfouri, A. P., A comparison of elastic–plastic fracture parameters in biaxial stress states, in *Elastic-Plastic Fracture*, ASTM STP 668 (ed J. D. Landes, J. A. Begley and G. A. Clarke), American Society for Testing and Materials, Philadelphia, pp. 214–228 (1979).
3.57. Kfouri, A. P. and Miller, K. J., Crack separation energy rates for inclined cracks in an elastic–plastic material, in *Three-Dimensional Constitutive Relations and Ductile Fracture* (ed. S. Nemat-Nasser), North-Holland Publ. Co., pp. 83–109 (1981).
3.58. Sih, G. C., Mechanics of ductile fracture, *Proceedings of Conference on Fracture Mechanics and Technology* (eds. G. C. Sih and C. L. Chow), Sijthoff and Noordhoff Int. Publ., Vol. 2, pp. 767–784 (1977).
3.59. Sih, G. C. and Madenci, E., Crack growth resistance characterized by the strain energy density function, *Engineering Fracture Mechanics* **18**, 1159–1171 (1983).
3.60. Sih, G. C. and Madenci, E., Fracture initiation under gross yielding: strain energy density criterion, *Engineering Fracture Mechanics* **18**, 667–677 (1983).
3.61. Sorensen, E. P., A finite element investigation of stable crack growth in anti-plane shear, *International Journal of Fracture* **14**, 485–500 (1978).
3.62. Rice, J. R., McMeeking, R. M., Parks, D. M. and Sorensen, E. P., Recent finite element studies in plasticity and fracture mechanics, *Computer Methods in Applied Mechanics and Engineering* **17/18**, 411–442 (1979).
3.63. McMeeking, R. M., Finite deformation analysis of crack-tip opening in elastic–plastic materials and implications for fracture, *Journal of the Mechanics and Physics of Solids* **25**, 357–381 (1977).
3.64. Kim, Y. J. and Hsu, T. R., A numerical analysis on stable crack growth under increasing load, *International Journal of Fracture* **20**, 17–32 (1982).
3.65. Gdoutos E. E. and Papakaliatakis G., Crack growth initiation in elastic–plastic materials, *International Journal of Fracture* **32**, 143–156 (1986).
3.66. Gdoutos E. E. and Papakaliatakis G., The influence of plate geometry and material properties on crack growth, *Engineering Fracture Mechanics* **25**, 141–156 (1986).
3.67. Gdoutos E. E. and Papakaliatakis G., The effect of load biaxiality on crack growth in non-linear materials, *Theoretical and Applied Fracture Mechanics* **5**, 133–140 (1986).
3.68. Jones, D. P. and Swedlow, J. L., The influence of crack closure and elasto-plastic flow on the bending of a cracked plate, *International Journal of Fracture* **11**, 897–914 (1975).
3.69. Sih, G. C. and Kiefer, B. V., Nonlinear response of solids due to crack growth and plastic deformation, in *Nonlinear and Dynamic Fracture Mechanics* (eds. N. Perrone and S. N. Atluri), The American Society of Mechanical Engineering, AMD, Vol. 35, pp. 136–156 (1979).
3.70. Sih, G. C. and Chen, C., Non-self-similar crack growth in elastic–plastic finite thickness plate, *Theoretical and Applied Fracture Mechanics* **3**, 125–139, (May 1985).
3.71. Moyer, E. T. Jr and Liebowitz, H., Effect of specimen thickness on crack front plasticity characteristics in three-dimensions, *Proceedings of the Sixth International Conference on Fracture* (eds. S. R. Valluri, D. M. R. Taplin, P. R. Rao, J. F. Knott and R. Dubey), Pergamon Press, Vol. 2, pp. 889–896 (1984).
3.72. Moyer, E. T. Jr, Poulose, P. K. and Liebowitz, H., Prediction of plasticity characteristics for three-dimensional fracture specimens comparison with experiment, *Engineering Fracture Mechanics* **24**, 677–689 (1986).
3.73. Owen, D. R. J. and Fawkes, A. J., *Engineering Fracture Mechanics – Numerical Methods and Applications*, Pineridge Press, Swansea, U.K. (1983).
3.74. PAPST, Finite element computer code, Lehigh University (1981).

Crack growth based on energy balance

4.1. Introduction

During the phenomenon of fracture of solids new surfaces are created in the medium in a thermodynamically irreversible manner. Material separation is caused by the rupture of atomic bonds due to a sufficiently high elevation of local stresses. The study of the mechanics of the fracture process requires the simultaneous consideration of various factors including material behavior, rate of loading, environmental conditions and microscopic phenomena which tend to perplex and complicate the problem. The phenomenon of fracture may be approached from different points of view depending on the scale of observation. At one extreme of the scale is the atomistic approach where the interested phenomena take place in the material within distances of the order of 10^{-7} cm, and at the other extreme of the scale is the continuum approach where material behavior at distances greater than 10^{-2} cm is involved. In the atomistic approach, study of the problem takes place using the concepts of quantum mechanics, while the continuum approach uses the theories of continuum mechanics and classical thermodynamics. A different approach should be used to explain the phenomena that take place in the material between these two extreme scales involving movement of dislocations, formation of subgrain boundary precipitates, slip bands, and grain inclusions and voids. The complex nature of the phenomenon of fracture prohibits a unified treatment of the problem and the existing theories deal with the subject either from the microscopic or the macroscopic point of view. Attempts have been made to bridge the gap between these two approaches.

The continuum mechanics approach of fracture assumes the existence of defects with sufficiently large size compared to the characteristic dimensions of the microstructure and considers the material as a homogeneous continuum. Study of the problem of growth of an existing crack, void or other defect necessitates a stress analysis coupled with a postulate predicting the phenomenon of fracture itself. A number of hypotheses known as failure criteria have been advanced over the years. Each criterion assumes a quantity that has to be related with the loss of continuity and has a critical value that serves as a measure of the resistance of the material to separation. Among the critical quantities proposed in the literature one may mention the stress, the strain, the stress intensity factor, the strain energy density, the separation of the crack faces close to the crack tip and the J-integral.

In the present chapter the theory of crack growth based on the global energy balance of the entire system will be developed. Historically, this approach, which was proposed by Griffith [1.8, 1.9] more than six decades ago, constitutes the earliest attempt to formulate a linear elastic theory of crack propagation. Griffith, using the first law of thermodynamics, postulated that a necessary condition for crack growth is that the energy necessary in creating new fracture surface is supplied by the released strain energy in the elastic body. When the surface energy of the material and the crack size are known, the energy criterion can predict the minimum load for fracture. Thus, Griffith resolved the paradox arising in the Inglis solution of a sharp crack in an elastic body according to which an infinite stress occurs at the crack tip and, therefore, a body with a crack could sustain no applied load.

This chapter starts with the global energy balance in a continuum during crack growth from which the Griffith criterion is deduced as a special case. In an attempt to extend the principles of linear elastic analysis to situations of highly localized yielding at the crack front, the various irreversibilities associated with fracture are lumped together to define the fracture toughness of the material. This approach allows the applicability of Griffith's theory to metals and other engineering materials. The equivalence of the energy approach and the intensity of the local stress field are established, leading to the critical stress intensity factor fracture criterion. The crack growth resistance curve method is described and a general definition of fracture toughness for semi-brittle fracture is given. The chapter concludes with the prediction of mixed-mode crack growth based on the global energy balance approach.

4.2. Energy balance during crack growth

A general energy balance of a deformable continuum subjected to arbitrary loading and containing a crack is presented. No particular assumption regarding the constitutive equations relating stresses and strains is made. The crack is not necessarily stationary but may be propagating and it is assumed that the crack growth is described by the crack area A as a single parameter. We represent by Σ the bounding surface of the solid and by Σ_c the cracked surface which changes with time t as the crack propagates (Figure 4.1). We state this as $\Sigma(t) = \Sigma_T + \Sigma_c(t)$. The cracked continuum is subjected to surface tractions T_k ($k = 1, 2, 3$) on the bounding surface Σ_T and to body forces F_k throughout the region R occupied by the body. The crack surfaces are assumed to be stress free and the body volume V is unaffected by crack growth. All quantities are referred to a rectangular Cartesian coordinate system (x_1, x_2, x_3) and the summation subscript notation with repeated indices is used. The usual infinitesimal deformation assumption is invoked in the analysis.

According to the first law of thermodynamics the work $\dot{W}$ performed per unit time by the surface tractions T_k on Σ_T and body forces F_k in R plus the thermal

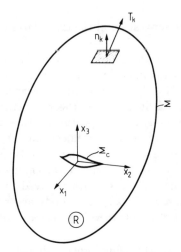

Figure 4.1. A continuum body with a crack.

energy $\dot{Q}$ applied to the body per unit time is equal to the rate of change of the internal energy E and kinetic energy K of the body and the energy $\dot{\Gamma}$ per unit time spent in increasing the crack surface Σ_c. We have,

$$\dot{W} + \dot{Q} = \dot{E} + \dot{K} + \dot{\Gamma} \tag{4.1}$$

where

$$\dot{W} = \int_{\Sigma_T} T_k \dot{u}_k \, d\Sigma + \int_R \rho F_k \dot{u}_k \, dV \tag{4.2}$$

$$\dot{Q} = \int_{\Sigma} \dot{q}_k n_k \, d\Sigma + \int_R \rho h \, dV \tag{4.3}$$

$$\dot{E} = \frac{d}{dt} \int_R \rho e \, dV = \int_R \rho \dot{e} \, dV \tag{4.4}$$

$$\dot{K} = \frac{d}{dt} \int_R \tfrac{1}{2} \rho \dot{u}_k \dot{u}_k \, dV = \int_R \rho \dot{u}_k \ddot{u}_k \, dV \tag{4.5}$$

$$\dot{\Gamma} = \frac{d}{dt} \int_{\Sigma_c} \gamma \, dA \tag{4.6}$$

In the above equations ρ denotes the mass density, u_k the displacement component, n_k the components of the unit vector normal to the bounding surface, e the internal energy density per unit mass, q_k the heat conduction vector per unit surface, h the nonmechanical heat source per unit mass and γ the energy required to form a unit of new surface. A dot over a letter denotes ordinary differentiation with respect to time.

For the present discussion it is assumed that the applied loads are time independent and the crack grows slowly in a stable manner. Under these conditions the velocity field $\dot{u}_k$ developed in the continuum is small and the kinetic energy, which is proportional to the square of the velocity at each point (Equation (4.5)), may be ignored. Phenomena in which the kinetic energy term is not negligible and therefore cannot be omitted from the energy balance equation will be studied in Chapter 7, which is devoted to dynamic fracture. Furthermore, it is assumed that there is neither heat flux nor a nonmechanical heat source in the body so that crack growth takes place under isothermal or adiabatic conditions. Under these conditions the thermal energy term in Equation (4.1) which is expressed by Equation (4.3) can be ignored. The rate of change of the internal energy $\dot{E}$ given by Equation (4.4) can be put into the form

$$\dot{E} = \int_R \sigma_{ij}\dot{\epsilon}_{ij}\,dV \tag{4.7}$$

where σ_{ij} and ϵ_{ij} denote the components of stress and strain tensor.

Under the usual assumption of plasticity, the strain increment $d\epsilon_{ij}$ may be split into a recoverable elastic part $d\epsilon_{ij}^e$ and a permanent plastic part $d\epsilon_{ij}^p$, with $d\epsilon_{ij} = d\epsilon_{ij}^e + d\epsilon_{ij}^p$. Thus, Equation (4.7) takes the form

$$\dot{E} = \int_R \sigma_{ij}\dot{\epsilon}_{ij}^e\,dV + \int_{R^p} \sigma_{ij}\dot{\epsilon}_{ij}^p\,dV \tag{4.8}$$

where R^p denotes the plastically deformed region of the body. The first term of Equation (4.8)

$$\dot{U}^e = \int_R \sigma_{ij}\dot{\epsilon}_{ij}^e\,dV \tag{4.9}$$

represents the elastic strain energy rate, while the second term

$$\dot{U}^p = \int_R \sigma_{ij}\dot{\epsilon}^p\,dV \tag{4.10}$$

represents the plastic strain work rate.

For the case considered here all changes with respect to time are caused by changes in crack size, and so we can state

$$\frac{\partial}{\partial t} = \frac{\partial A}{\partial t}\frac{\partial}{\partial A} = \dot{A}\frac{\partial}{\partial A}, \quad \dot{A} \geq 0. \tag{4.11}$$

Equations (4.2) and (4.8) can then take the form

$$\dot{W} = \frac{\partial W}{\partial A}\dot{A} = \left(\int_{\Sigma_T} T_k\frac{\partial u_k}{\partial A}\,d\Sigma + \int_R \rho F_k\frac{\partial u_k}{\partial A}\,dV\right)\dot{A} \tag{4.12}$$

and

$$\dot{E} = \frac{\partial}{\partial A}(U^e + U^p)\dot{A} = \left(\int_R \sigma_{ij}\frac{\partial \epsilon_{ij}^e}{\partial A}\,dV + \int_{R^p} \sigma_{ij}\frac{\partial \epsilon_{ij}^p}{\partial A}\,dV\right)\dot{A}. \tag{4.13}$$

Thus, the energy balance equation (4.1) during quasi-static stable crack growth where dynamic effects and thermal energy are ignored takes the form

$$\frac{\partial W}{\partial A} = \left(\frac{\partial U^e}{\partial A} + \frac{\partial U^p}{\partial A}\right) + \frac{\partial \Gamma}{\partial A}. \tag{4.14}$$

Equation (4.14) indicates that the work rate supplied to the continuum by the external tractions and body forces is equal to the rate of the elastic strain energy and plastic strain work plus the energy dissipated in crack propagation. Equation (4.14) may be put in the form

$$-\frac{\partial \Pi}{\partial A} = \frac{\partial U^p}{\partial A} + \frac{\partial \Gamma}{\partial A} \qquad (4.15)$$

where

$$\Pi = U^e - W \qquad (4.16)$$

is the potential energy of the system. Equation (4.15) shows that the rate of potential energy decrease during crack growth is equal to the rate of energy dissipated in plastic deformation and crack growth. Both forms of energy balance expressed by Equations (4.14) and (4.15) will be used in the sequel.

It is recognized that plastic deformation at the microscopic level can take place along the path of a macrocrack extension, but this cannot concurrently take place with macroplastic deformation that is off to the side of the macrocrack. Under the condition of plane strain, the energy in the small plastic ligament ahead of the crack would have negligible contribution on the release of the elastic energy corresponding to the onset of rapid fracture. It is this sudden creation of a small segment of crack surface leading to global instablity that defines the fracture toughness quantity [1.35] in fracture mechanics. Interpolation of the quantity $\partial U^p/\partial A$ in Equation (4.15) should be clarified in relation to the time when U^p is dissipated and when the crack area A is created.

4.3. Griffith theory

Griffith [1.8, 1.9] approached the problem of fracture of an ideally brittle material by appealing to the minimum potential theorem of elastostatics. Furthermore, he assumed that the energy required to form a unit of new material surface is constant for a given material and environmental conditions. In the following, the general equations of the Griffith theory will be deduced from the previous derivations based on the energy balance and will be applied to two- and three-dimensional problems.

(a) General equations

For an ideally brittle material the energy dissipated in plastic deformation is negligible and can be omitted from Equation (4.14). If γ represents the energy required to form a unit of new material surface, then Equation (4.14) takes the form

$$G = \frac{\partial W}{\partial A} - \frac{\partial U^e}{\partial A} = 2\gamma \qquad (4.17)$$

where the factor 2 appearing on the right-hand side of the equation refers to the two new material surfaces formed during crack growth.

The left hand side of the equation represents the energy available for crack growth, and is given the symbol G in honor of Griffith. Because G is derived from a potential function in a manner analogous to that of a conservative force, it is often referred to as the crack driving force. The right-hand side of Equation (4.17) represents the resistance of the material that must be overcome for crack growth and is a material constant.

Equation (4.17) represents the fracture criterion for crack growth. Two limiting cases, the 'fixed-grips' and 'dead-load' loading, are usually encountered in the literature. In the fixed-grips loading the surface of the continuum on which the loads are applied is assumed to remain stationary during crack growth. By ignoring the work of the body forces, the work performed by the applied loads vanishes and Equation (4.17) takes the form

$$G = -\frac{\partial U^e}{\partial A} = 2\gamma. \tag{4.18}$$

Equation (4.18) indicates that the energy rate for crack growth is supplied by the existing elastic strain energy of the solid. Because of this property the symbol G is usually referred to as the 'elastic strain energy release rate'.

In the dead-load situation the applied loads on the surface of the solid are kept constant during crack growth. For a power law relationship between stress and strain the equation expressing the principle of virtual work takes the form

$$W = \int_{\Sigma_t} T_k u_k \, d\Sigma + \int_R \rho F_k u_k \, dV = \frac{n+1}{n} U^e \tag{4.19}$$

where n is the stress-strain power law coefficient. For a linear elastic material ($n = 1$), Equation (4.19) becomes the Clapeyron theorem of linear elastostatics according to which the work performed by the constant surface tractions and body forces is twice the increase of elastic strain energy. Thus $\partial W/\partial A = 2\partial U^e/\partial A$, and Equation (4.17) takes the form

$$G = \frac{\partial U^e}{\partial A} = 2\gamma. \tag{4.20}$$

Here the energy required for crack growth is supplied by the work performed by the external loads and the elastic strain energy of the solid is increased. The term 'strain energy release rate' for G in this case is physically inappropriate.

From Equations (4.18) and (4.20) it is observed that for either 'fixed-grips' or 'dead-load' loading the magnitude of the elastic strain energy release rate necessary for crack growth is the same. However, the elastic strain energy of the system decreases for 'fixed grips' and increases for 'dead-load' conditions. Equations (4.18) and (4.20) can be put in the form

$$G = -\frac{\partial \Pi}{\partial A} = 2\gamma \tag{4.21}$$

where the potential energy Π is defined from Equation (4.16). Equation (4.21) may be written as

$$\frac{\partial (\Pi + \Gamma)}{\partial A} = 0 \tag{4.22}$$

which, in Griffith's terminology, states that the 'total potential energy' of the system $(\Pi + \Gamma)$ has a stationary value.

The above general equations of the Griffith theory will be applied in the following for the determination of the critical failure load in some characteristic crack problems.

(b) A line crack in an infinite plate

Let us now consider an infinite plate of unit thickness with a line crack of length $2a$ subjected to a uniform stress σ perpendicular and $\epsilon\sigma$ parallel to the crack under conditions of plane strain or generalized plane stress. For this case the change in elastic strain energy due to the presence of a crack, as calculated by Sih and Liebowitz [2.13], is given by

$$U^e = \frac{\pi a^2 \sigma^2}{8\mu}(\kappa + 1). \tag{4.23}$$

For $A = 2a \times 1$, Equation (4.21) gives the critical stress required for unstable crack growth as

$$\sigma_c = \sqrt{\frac{2E\gamma}{\pi a(1-\nu^2)}} \tag{4.24}$$

for plane strain, and

$$\sigma_c = \sqrt{\frac{2E\gamma}{\pi a}} \tag{4.25}$$

for generalized plane stress.

Observe that the stress σ_{cr} is inversely proportional to the square root of half the crack length. This result was verified experimentally by Griffith on glass for a wide range of crack lengths. Equations (4.24) and (4.25) indicate that the stress $\epsilon\sigma$ parallel to the crack has no effect on the critical fracture load.

(c). An elliptical hole in an infinite plate

We now examine an elliptical hole with major and minor semi-axes a and b in an infinite plate of unit thickness subjected to a uniform stress σ perpendicular and $\epsilon\sigma$ parallel to the major axis of the ellipse. The increase in surface energy due to the presence of the hole is

$$\Gamma = 4\gamma a \int_0^{\pi/2} \sqrt{1 - k^2 \sin^2 \beta}\, d\beta = 4\gamma a E(k) \tag{4.26}$$

where $E(k)$ is the complete elliptic integral of the second kind and

$$ak = \sqrt{a^2 - b^2}, \qquad k = \sqrt{1 - k'^2}. \tag{4.27}$$

The change in elastic strain energy due to the presence of the hole calculated by Sih and Liebowitz [2.13] is given by

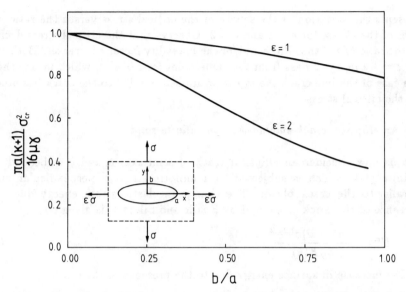

Figure 4.2. Normalized critical stress versus the ratio of the semi-axes of an elliptical hole for the values of the biaxiality coefficient $\epsilon = 1$ and 2 [2.13].

$$U^e = \frac{\pi\sigma^2}{32\mu}(\kappa+1)\big[(1-\epsilon)^2(a+b)^2 + 2(1-\epsilon^2)(a^2-b^2) + $$
$$+ (1+\epsilon)^2(a^2+b^2)\big]. \tag{4.28}$$

It is assumed that the radius of curvature $\rho = b^2/a$ at the apexes of the major axis of the ellipse remains constant during the growth of the hole, i.e.

$$\frac{\partial b}{\partial a} = \frac{b}{2a} + \frac{a}{2b}\left(\frac{\partial \rho}{\partial a}\right) \simeq \frac{b}{2a}. \tag{4.29}$$

Thus, we have

$$\frac{\partial \Gamma}{\partial a} = \frac{2\gamma}{k^2}[(1+k^2)E(k) - k'^2 K(k)] \tag{4.30}$$

where $K(k)$ is the complete elliptic integral of the first kind.

Application of the Griffith criterion, Equation (4.22), yields for the critical stress:

$$\frac{\pi a(\kappa+1)}{16\mu\gamma}\sigma_c^2 =$$
$$\frac{4}{k^2}\left[\frac{(1+k^2)E(k) - k'^2 K(k)}{(1+\epsilon)^2(3-k^2) + 2(1-\epsilon^2)(1+k^2) + (1-\epsilon)^2(1+k')(2+k')}\right]. \tag{4.31}$$

For $b \to 0$, which implies that $k \to 1$ and $k' \to 0$, the right-hand side of Equation (4.31) becomes equal to 1 and Equations (4.24) and (4.25) for a line crack under plane strain and generalized plane stress are obtained. Figure 4.2

presents the variation of the square of the critical stress versus the ratio of the axes of the ellipse for $\epsilon = 1$ and $\epsilon = 2$. Observe that the critical stress decreases as the ratio b/a of the ellipse and/or the biaxiality factor ϵ increases. Both curves for $\epsilon = 1$ and $\epsilon = 2$ pass from the same point for $b/a = 0$, which means that for the case of the line crack the stress $\epsilon\sigma$ applied parallel to the crack has no effect on the critical stress.

(d) An elliptical crack embedded in an infinite solid

Let us now consider an elliptical crack of semi-axes a and b embedded in an infinite plate which is subjected to a uniform stress σ perpendicular and $\epsilon\sigma$ parallel to the crack plane. The change in elastic strain energy due to the presence of the crack calculated by Kassir and Sih [4.1] is given by

$$U^e = \frac{2\pi(1-\nu)ab^2\sigma^2}{3\mu E(k)}. \tag{4.32}$$

The increase in surface energy due to the presence of the crack is

$$\Gamma = 2\pi ab\gamma. \tag{4.33}$$

As in the previous case of the elliptical hole, application of the Griffith criterion is not possible without an assumption regarding the new shape of the crack after its growth. Based on the simplifying assumption of Kassir and Sih [4.1] that the ellipse with semi-axes a and b (given by

$$a = c\cosh\xi_0, \qquad b = c\sinh\xi_0 \tag{4.34}$$

where $c = \sqrt{a^2 - b^2}$) grows into another ellipse with the same foci, the Griffith criterion may be stated as

$$\frac{\partial(U^e - \Gamma)}{\partial\xi} = 0 \tag{4.35}$$

which gives the following expression for the critical stress:

$$\frac{b(1-\nu)}{3\gamma\mu}\sigma_c^2 = \frac{(1+k'^2)E^2(k)}{2(1+k'^2)E(k) - k'^2 K(k)}. \tag{4.36}$$

Equation (4.36) is plotted in Figure 4.3. Observe that the critical stress increases as the ratio a/b ($a/b > 1$) decreases. Its maximum value for the penny-shaped crack is

$$\sigma_c^2 = \frac{\pi\gamma\mu}{a(1-\nu)}. \tag{4.37}$$

(e) Historical remarks and discussion

Griffith [1.8, 1.9], by the energy balance approach, laid down the foundations of fracture mechanics. He resolved the paradox appearing in the Inglis solution

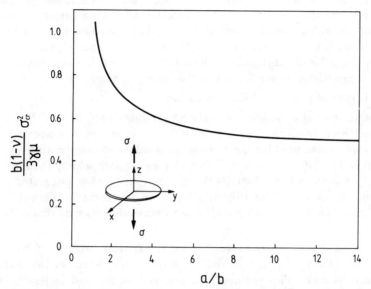

Figure 4.3. Normalized critical stress versus the ratio of the semi-axes of an elliptical crack [2.13].

that infinite stresses develop around a sharp crack and therefore a body with a crack could sustain no applied load. At this point it seems appropriate to discuss a few points appearing in Griffith's two monumental papers.

In his first paper [1.8] Griffith, using Inglis's solution for an elliptical hole in an infinite plate, gave the following expression for the change of strain energy due to the presence of the hole:

$$U^e = \frac{(3-\kappa)(1+\nu)}{4E} \pi \sigma^2 a^2 \cosh 2\alpha_0 \tag{4.38}$$

which, for the case of a crack, becomes

$$U^e = \frac{(3-\kappa)(1+\nu)}{4E} \pi \sigma^2 a^2. \tag{4.39}$$

Using this result he obtained for the critical fracture stress

$$\sigma_{cr} = \sqrt{\frac{2E\gamma}{\nu(1+\nu)\pi a}} \tag{4.40}$$

for plane strain, and

$$\sigma_{cr} = \sqrt{\frac{2E\gamma}{\nu \pi a}} \tag{4.41}$$

for generalized plane strain.

He came to believe, however, that some mistake could be made in the calculation of strain energy, so he added a footnote at the end of his first paper:

It has been found that the method of calculating the strain energy of a cracked plate ... requires correction. The correction affects the numerical values of all quantities calculated from equations ..., but not their order of magnitude. The main argument of the paper is therefore not impaired, since it deals only with the order of magnitude of the results involved, but some reconsideration of the experimental verification of the theory is necessary.

In his second paper [1.9] Griffith quotes:

A solution of this problem was given in a paper read in 1920 but, in the solution there given, the calculation of the strain energy was erroneous, in that the expressions used for the stresses gave values at infinity differing from the postulated uniform stress at infinity by an amount which, though infinitesimal, yet made a finite contribution to the energy when integrated round the infinite boundary. This difficulty has been overcome by slightly modifying the expressions for the stresses, so as to make this contribution to the energy vanish.

He obtained the correct expression for the critical stress given by Equations (4.24) and (4.25), although he did not refer to the details of the derivation of the strain energy. This problem was resolved by Sih and Liebowitz [4.2] who provided the integral expression for obtaining the correct form of strain energy in an infinite body with a cavity of arbitrary shape subjected to general loading conditions at infinity. The energy calculation of the Griffith problem was also provided by Spencer [4.3].

Two points will now be raised in the Griffith theory. First, in calculating the energy release rate it is taken for granted that the crack extends from its tip into the material ahead in a continuum fashion, an assumption that lacks proper justification. Crack growth is rather a stepwise than a continuous process. Second, the concept of potential energy does not apply to a nonconservative system as one involving the process of crack growth. Griffith was aware of this and quotes in this first paper

According to the well-known 'theorem of minimum energy' the equilibrium state of an elastic solid body, deformed by specific surface forces, is such that the potential energy of the whole system is a minimum. The new criterion of rupture is obtained by adding to this theorem the statement that the equilibrium position, if equilibrium is possible, must be one in which rupture of the solid has occured, if the system can pass from the unbroken to the broken condition by a process involving a continuous decrease in potential energy.

Energy rate calculations and related results associated to the Griffith theory can be found in references [4.4–4.10].

4.4. Graphical representation of the energy balance equation

The graphical representation of the various terms appearing in the energy balance equation is useful as it provides a better insight of the variation of the

Crack growth based on energy balance

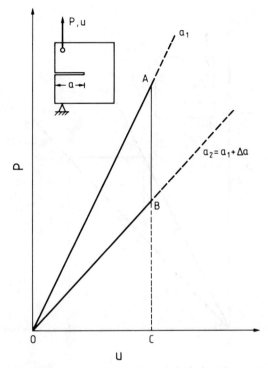

Figure 4.4. Load–displacement response of a cracked plate for propagation of a crack from length a_1 to a_2 under 'fixed grips' conditions along AB.

relevant quantities during crack growth and helps the interpretation of experimental results. The load-displacement response of the body during crack growth, as obtained from a testing machine, is examined separately for the cases of 'fixed-grips', 'dead-load' and the general case of changing both the load and displacement during crack propagation. Finally, the graphical representation in G–a coordinates is introduced.

(a) *'Fixed-grips' loading*

The load–displacement response of a body of unit thickness with a starter crack of length a_1 is represented in Figure 4.4 by the straight line OA. During loading up to the point A elastic strain energy is stored in the body which is represented by the area (OAC). This energy is released when unloading the body. Let us assume that at point A the crack starts to propagate under constant displacement to a new length $a_2 = a_1 + \Delta a$. During crack propagation the load drops from point A to point B lying on the straight line OB which represents the load–displacement response of the body with a longer crack of length a_2. Line OB should lie below line OA since the stiffness of the body decreases with increase of the crack length. The elastic strain energy stored in the body at point B is represented by the area (OBC). If the applied load is removed at point B

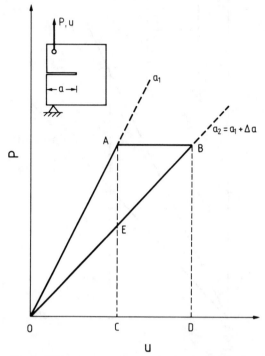

Figure 4.5. Load–displacement response of a cracked plate for propagation of a crack from length a_1 to a_2 under constant load along AB.

the unloading path will follow the line BO. Since the point of application of the load remains fixed during crack growth, no extra work is supplied to the body. The reduction in strain energy during crack growth is represented by the area (OAB). It is that obtained for the elastic energy release rate from Equation (4.18) which is balanced by the material resistance to crack growth 2γ

$$G = \frac{(OAB)}{\Delta a} = 2\gamma. \tag{4.42}$$

(b) *'Dead-load' loading*

The graphical representation of the load–displacement response of a cracked body during crack growth under constant loading is represented in Figure 4.5. The displacement increases from A to B as the crack length increases from a_1 to $a_2 = a_1 + \Delta a$. The energy at the beginning of crack growth is represented by the area (OAC) and at the end by the area (OBD). During crack growth the load P performs work represented by the area $(ABDC)$. The energy supplied to the body for fracture is equal to $(OAC) + (ABDC) - (OBD) = (OAB)$. Equation

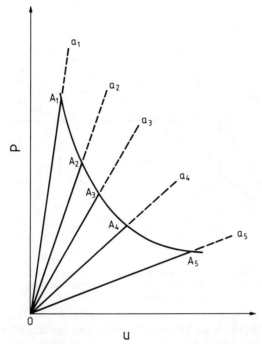

Figure 4.6. Load–displacement response of a cracked plate for propagation of a crack from an initial length a_1 to a final length a_5 under general load–displacement conditions along $A_1 A_2 A_3 A_4 A_5$.

(4.18) takes the form

$$G = \frac{(OAB)}{\Delta a} = 2\gamma. \tag{4.43}$$

Note that the work supplied for crack growth under 'dead-load' loading differs from that necessary for crack growth under 'fixed-grips' loading by the amount ABE which disappears as the crack growth increment Δa tends to zero.

(c) *General load–displacement relation*

Usually, both load and displacement change during crack growth. The load–displacement response mainly depends on the form of specimen and the type of testing machine. In this case no mathematical relation between the crack driving force and the change in elastic strain energy can be found. The load–displacement response during quasi-static growth of a crack of initial length a_1 to a final length a_5 is presented by the curve $A_1 A_2 A_3 A_4 A_5$ in Figure 4.6. Equation (4.42) (or (4.43)), which expresses the crack driving force in terms of segmental areas of the load–displacement curve, still holds for the case of a general relation between load and displacement during crack growth. This equation can be used for the experimental determination of the resistance of the material to crack growth, as was suggested by Gurney and Hunt [4.11]. During

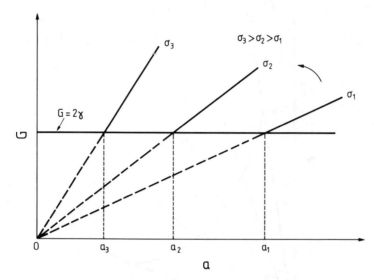

Figure 4.7. Crack driving force G versus crack length a curves for a crack of length $2a$ in an infinite plate subjected to a uniform stress σ perpendicular to the crack axis.

stable crack growth, the load, the displacement and the crack length are recorded simultaneously. This allows the construction of the P–u curve and the drawing of the radial lines OA_i which correspond to different crack lengths. A check of the overall elastic behavior of the specimen is made by removing the applied load and following the unloading lines A_iO which should revert to the origin. When the specimen is reloaded the reloading lines should coincide with the unloading lines. Application of Equation (4.42) allows the determination of γ.

(d) G–a representation

In the previous graphical interpretation, load–displacement coordinates were used, while the crack length and the material resistance to crack growth appeared as parameters. It is sometimes advantageous to use crack driving force/crack growth resistance–crack length coordinates with the load appearing as a parameter. This is shown in Figure 4.7 for the case of a crack of length $2a$ in an infinite plate subjected to a uniform stress σ perpendicular to the crack axis. The crack driving force G obtained from Equations (4.16) or (4.18) and (4.23) is

$$G = \frac{\kappa + 1}{8\mu} \pi a \sigma^2. \tag{4.44}$$

The G–a relation is represented in Figure 4.7 by straight lines for the three different values of the applied stress σ. The intersection of these lines with the constant line $G = 2\gamma$ gives the critical crack length for crack growth. Or, inversely, for a given crack length a_3 the applied stress should be increased to σ_3 for crack growth. For a larger crack length a_2 a lower stress σ_2 is required for crack growth.

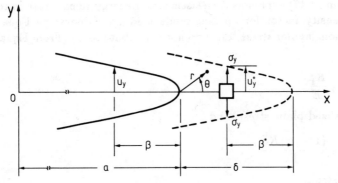

Figure 4.8. Self-similar crack growth.

It is evident that the G–a relation is not necessarily linear but depends on the geometrical configuration of the cracked plate and the loading conditions. The G–a representation of crack growth will be used later in the crack growth resistance curve method.

4.5. Equivalence between strain energy release rate and stress intensity factor

The connection between the strain energy release rate, which is a global quantity, and the stress intensity factor, which expresses the strength of the local elastic stress field in the neighborhood of the crack tip, is very important. Consider the case of an opening mode where the crack extends along its own direction in a self-similar manner. Due to symmetry only normal stresses will be present in elements along the crack direction (Figure 4.8). Assume the crack extends by a length δ. Then the energy released during crack extension is the work performed by the stresses $\sigma_y(\delta - \beta^*, 0)$ acting through the displacements $u_y(\beta, \pi)$. For $\delta \to 0$ the conditions $u_y \to u_y^*$ and $\beta \to \beta^*$ are satisfied, and the work performed at both ends of the crack is

$$G_{\mathrm{I}} = 2 \lim_{\delta \to 0} \frac{1}{\delta} \int_0^\delta \tfrac{1}{2} \sigma_y(\delta - \beta, 0) u_y(\beta, \pi) \, d\beta \tag{4.45}$$

where the subscript I was inserted to denote mode-I loading. Equation (4.45) holds only under 'fixed-grips' or 'dead-load' loading conditions.

Introducing the expressions of σ_y and u_y from Equations (2.76) and (2.78), we obtain for G_{I}

$$G_{\mathrm{I}} = \frac{\kappa + 1}{4\mu} \frac{K_{\mathrm{I}}^2}{2\pi} \int_0^\delta \sqrt{\frac{\beta}{\delta - \beta}} \, d\beta \tag{4.46}$$

or

$$G_{\mathrm{I}} = \frac{\kappa + 1}{8\mu} K_{\mathrm{I}}^2. \tag{4.47}$$

Equation (4.47) expresses the elastic strain energy release rate in terms of the stress intensity factor for opening-mode loading. Observe in Equation (4.45) that the nonsingular stress terms do not contribute to G_I. From Equation (4.47) we get

$$G_I = \frac{K_I^2}{E} \tag{4.48}$$

for generalized plane stress, and

$$G_I = \frac{(1-\nu^2)K_I^2}{E} \tag{4.49}$$

for plane strain.

Equation (4.47) facilitates the determination of G_I when the stress intensity factor K_I is known. Thus, for the problem of a crack in an infinite plate subjected to a uniform uniaxial stress σ perpendicular to the crack, $K_I = \sigma\sqrt{\pi a}$ (Equation (2.77)), and Equation (4.44) is recovered.

The calculation of the strain energy release rate G_{II} for sliding-mode loading is not easy since the crack does not propagate in its own plane but follows a curved path which is not known in advance. This prohibits the analytical computation of G_{II}. Only for the special case when the crack is forced to propagate along its own plane can G_{II} be determined in terms of the stress intensity factor. For this hypothetical situation the shear stresses τ_{xy} have to be released along the segment δ of crack growth and G_{II} takes the form

$$G_{II} = 2 \lim_{\delta \to 0} \frac{1}{\delta} \int_0^\delta \tfrac{1}{2}\tau_{xy}(\delta-\beta,0)u_x(\beta,\pi)\,d\beta \tag{4.50}$$

Introducing Equations (2.91) and (2.92) into Equation (4.50) leads to

$$G_{II} = \frac{\kappa+1}{8\mu}K_{II}^2. \tag{4.51}$$

For out-of-plane shear the direction of crack growth is predetermined and G_{III} is computed as previously by

$$G_{III} = 2 \lim_{\delta \to 0} \frac{1}{\delta} \int_0^\delta \tfrac{1}{2}\tau_{yz}(\delta-\beta,0)w(\beta,\pi)\,d\beta \tag{4.52}$$

which, together with Equation (2.101), gives

$$G_{III} = \frac{K_{III}^2}{2\mu}. \tag{4.53}$$

The calculation of the strain energy release rate for mixed-mode loading is sometimes erroneously made in the literature by adding the values of G for the three modes. This problem will be dealt with later in this chapter.

Equations (4.47), (4.51) and (4.53) establish the equivalence of the strain energy release rate and the stress intensity factor approach in fracture mechanics and form the basis for the critical stress intensity factor fracture criterion.

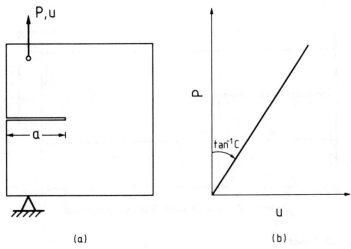

Figure 4.9. Load–displacement response of a cracked plate.

4.6. Compliance

Let us now consider the load–displacement response of a cracked plate of thickness B subjected to a concentrated force P (Figure 4.9(a)). For linear elastic behavior – as long as there is no crack growth – the load–displacement relation is

$$u = CP \tag{4.54}$$

where C is the compliance (reciprocal of stiffness) of the plate. In Figure 4.9(b) the compliance is represented by the tangent of the angle subtending between the load–displacement curve and the P-axis. Analytical expressions of the strain energy release rate expressed by the left-hand side of Equation (4.17) in terms of the compliance are sought. The case of 'fixed-grips' and 'dead-load' conditions during crack growth are considered separately.

(a) 'Fixed-grips' loading

For a constant displacement u during crack growth the applied load does not perform work and the elastic strain energy stored in the plate is

$$U^e = \frac{Pu}{2}. \tag{4.55}$$

Then Equation (4.17) takes the form

$$G = -\tfrac{1}{2} u \frac{dP}{dA} = \frac{1}{2B} \frac{u^2}{C^2} \left(\frac{dC}{da}\right)_u \tag{4.56}$$

where a is the crack length.

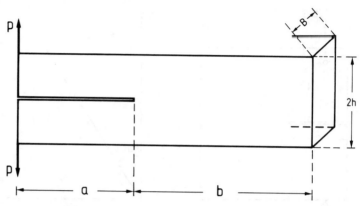

Figure 4.10. Double cantilever beam specimen.

(b) 'Dead-load' loading

For a constant load P the work performed by the load during an infinitesimal crack growth is

$$dW = P\,du \tag{4.57}$$

while the change in elastic strain energy is

$$dU^e = d\left(\frac{Pu}{2}\right) = \frac{P\,du + u\,dP}{2}. \tag{4.58}$$

Then Equation (4.17) becomes

$$G = \tfrac{1}{2}\left(P\frac{du}{dA} - u\frac{dP}{dA}\right) = \frac{1}{2B}P^2\left(\frac{dC}{da}\right)_P. \tag{4.59}$$

Equations (4.56) and (4.59) express the strain energy release rate in terms of the derivative of the compliance of the cracked plate with respect to the crack length for 'fixed-grips' or 'dead-load' loading during crack growth. By combining Equations (4.56) and (4.59) with Equations (4.48) and (4.49) we obtain an opening-mode stress intensity factor of

$$K_I^2 = \frac{EP^2}{2B}\left(\frac{dC}{da}\right)_P = \frac{Eu^2}{2BC^2}\left(\frac{dC}{da}\right)_u \tag{4.60}$$

for generalized plane stress, and

$$K_I^2 = \frac{EP^2}{2(1-\nu^2)B}\left(\frac{dC}{da}\right)_P = \frac{Eu^2}{2(1-\nu^2)BC^2}\left(\frac{dC}{da}\right)_u \tag{4.61}$$

for plane strain.

Equations (4.60) and (4.61) can be used for the analytical or experimental determination of the K_I stress intensity factor.

As an example, consider the case of the double cantilever beam (DCB) specimen shown in Figure 4.10. It is assumed that $a \gg 2h$ and $b \gg 2h$. From

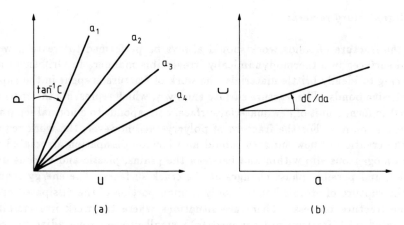

Figure 4.11. (a) Load–displacement response for different lengths and (b) compliance versus crack length.

elementary beam theory we obtain for the relative displacement of the points of application of the loads P under conditions of generalized plane stress

$$u = \frac{2Pa^3}{3EI}, \qquad I = \frac{Bh^3}{12}. \tag{4.62}$$

For plane strain E should be replaced by $E/(1-\nu^2)$. The compliance of the DCB specimen is

$$C = \frac{2a^3}{3EI} \tag{4.63}$$

and Equations (4.60) and (4.61) give for the stress intensity factor K_I for plane stress or plane strain

$$K_I = \sqrt{\frac{12}{h^3}} \frac{Pa}{B}. \tag{4.64}$$

For the experimental determination of the stress intensity factor from Equation (4.60) or (4.61), a series of specimens with different crack lengths are used to calculate the derivative of the compliance with respect to the crack length. This is shown in Figure 4.11.

The accuracy of the experimental determination of K_I depends on the changes in the displacement between loading points remote from the crack to crack extension. This experimental technique provides a quick way of determining K_I in situations where the crack geometry is complicated and the mathematical solution is not readily accessible.

4.7. Critical stress intensity factor fracture criterion

(a) Introductory remarks

For the fracture of solids work should always be performed to create new material surfaces in a thermodynamically irreversible manner. In Griffith's theory referring to ideally brittle materials, the work of fracture is spent in the rupture of cohesive bonds. The fracture surface energy γ, which represents the energy required to form a unit of new material surface, corresponds to a normal separation of atomic planes. For the fracture of polycrystals, however, the work required for the creation of new surfaces should also include dissipation associated with nonhomogeneous slip within and between the grains, plastic and viscous deformation and possible phase changes at the crack surfaces. The energy required for the rupture of atomic bonds is only a small portion of the dissipated energy in the fracture process. There are situations where the work irreversibilities associated with fracture are confined to a small process zone adjacent to the crack surfaces, while the remaining material is deformed elastically. In such case the various work terms associated with fracture may be lumped together in a macroscopic term R (resistance to fracture) which represents the work required for the creation of a unit of new material surface. R may be considered as a material parameter. The plastic zone accompanying the crack tip is very small and the state of affairs around the crack tip can be described by the stress intensity factor.

In the following, a fracture criterion based on the energy balance equation within the framework of the previous discussion is proposed.

(b) Fracture criterion

When the zones of plastic deformation around the crack tip are very small the plastic strain term appearing in the energy balance equation (Equation (4.14)) can be omitted and the work rate supplied to the body for crack growth is represented by the left-hand side of Equation (4.17). In such circumstances, fracture is assumed to occur when the strain energy release rate G, which represents the energy pumped into the fracture zone from the elastic bulk of the solid, becomes equal to the energy required for the creation of a unit area of new material R. The fracture condition is

$$G_\mathrm{I} = G_c = R \tag{4.65}$$

Equation (4.65) is usually expressed in terms of the opening-mode stress intensity factor K_I. By introducing a new material parameter K_c from the equation

$$K_c = \sqrt{\frac{ER}{\beta}}, \tag{4.66}$$

where $\beta = 1$ for plane stress and $\beta = 1 - \nu^2$ for plane strain, and by substitution

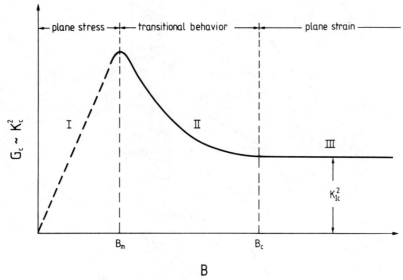

Figure 4.12. Critical fracture toughness G_c (or K_c^2) versus plate thickness B.

of G_I in terms of K_I from Equation (4.48) or (4.49), Equation (4.65) becomes

$$K_I = K_c. \tag{4.67}$$

Equation (4.67) expresses the critical stress intensity factor fracture criterion. The left-hand side of the equation depends on the applied load, the crack length and the geometrical configuration of the cracked plate. Methods for the analytical, numerical or experimental determination of K_I are described in Chapter 2. The right-hand side of the equation is a material parameter and can be determined experimentally. Note that Equation (4.67) was derived from the global energy balance of the continuum, which expresses the law of conservation of energy.

(c) *Variation of K_c with thickness*

Laboratory experiments [4.12–4.14] indicate that K_c varies with the thickness B of the specimen tested. The form of variation of K_c with B is shown in Figure 4.12. Three distinct regions corresponding to 'very thin', 'very thick' and 'intermediate range thickness' specimens can be distinguished. Study of the load–displacement response and the appearance of the fracture surfaces of the specimen are helpful in understanding the mechanisms of fracture in each of the above three regions. The fractures are classified as square or slant according to whether the fracture surface is normal of forms a 45° inclination angle with respect to the direction of the applied tensile load. An analysis of the state of affairs in the three regions of Figure 4.12 follows.

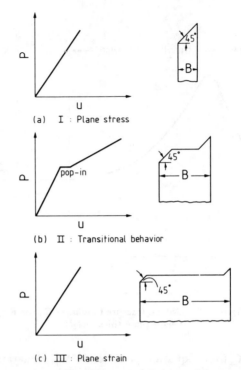

Figure 4.13. Load–displacement response for (a) plane stress, (b) transitional behavior and (c) plane strain.

In region I, corresponding to thin specimens, the critical fracture toughness G_c (which is proportional to K_c^2) increases almost linearly with B up to a maximum value at a critical thickness B_m. The load–displacement response is linear and the fracture surface is completely slant (Figure 4.13(a)). In this case, as explained in Section 3.2, plane stress predominates in the specimen and yielding occurs on planes through thickness at 45° with respect to the specimen flat surfaces. In such circumstances the crack extends mainly under an antiplane shear mode. A model explaining the linear increase of G_c with B based on the theory of dislocations was proposed by Bilby et al. [3.22]. According to this model the relative sliding displacement at the crack tip is

$$S(a) = \frac{4\tau_Y \alpha}{\pi \mu} \ln\left[\sec\left(\frac{\pi q}{2\tau_Y}\right)\right] \tag{4.68}$$

where q is the shear stress in mode-III and τ_Y is the yield stress. Equation (4.68) is completely analogous to Equation (2.74) which expresses the crack-tip opening displacement according to the Dugdale model. For small values of q/τ_Y Equation (4.68) takes the form

$$S(a) = \frac{q^2 \pi \alpha}{2\mu \tau_Y} \tag{4.69}$$

which is analogous to Equation (3.75) of the Dugdale model.

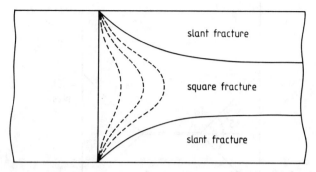

Figure 4.14. Thumbnail crack growth with square and slant fracture.

Assuming that crack extension takes place when $S(a) = \sqrt{2}\,B$, we obtain

$$q = \sqrt{\frac{2\sqrt{2}\,\mu \tau_Y B}{\pi a}}. \tag{4.70}$$

From Equation (4.70) for $\sigma = 2q$ and $\sigma_Y = 2\tau_Y$ where σ is the applied tensile stress, we obtain for the critical value of the strain energy release rate at crack growth

$$G_c = \frac{\sigma^2 \pi a}{E} = \left(\frac{2\sqrt{2}\,\sigma_Y}{1+\nu}\right) B. \tag{4.71}$$

This equation shows that G_c is analogous to B and explains the linear part of the G_c vs B diagram in region I of Figure 4.12.

For very thick specimens (region III) the load–displacement response is linear and the state of stress is predominantly plane strain except a thin layer at the free surfaces where plane stress dominates (see Figure 4.13(c)). The fracture surface is almost completely square with very small slant parts at the free surfaces. A triaxial state of stress is produced in most parts of the specimen which reduces the ductility of the material and fracture takes place at the lowest value of the critical strain energy release rate G_c. For increasing thickness beyond a critical minimum value, B_c, the triaxiality does not change substantially and the fracture toughness remains the same. The critical value of stress intensity factor in region III for plane strain conditions is denoted by K_{Ic} in Figure 4.12 and is independent of the specimen thickness. K_{Ic} is the so-called fracture toughness and represents an important material property. The larger the value of K_{Ic} the larger the resistance of material to crack propagation. Experimental determination of K_{Ic} takes place according to the ASTM specifications described in the next section.

For intermediate values of specimen thickness (region II) the fracture behavior is neither predominantly plane stress nor predominantly plane strain. The thickness is such that the central and edge region under plane strain and plane stress conditions respectively are of comparable size. The fracture toughness in this region changes between the minimum plane strain toughness and the maximum plane stress toughness. In the load–displacement curve (Figure 4.13(b)) at some value of the applied load the crack extends mainly from the center of the

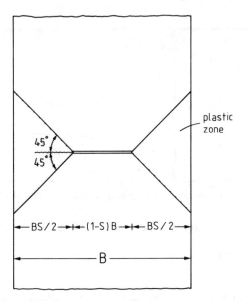

Figure 4.15. Calculation of crack growth resistance according to the model of Krafft et al. [4.13].

thickness of the specimen while the edge regions are plastically deformed. The crack grows in a 'thumbnail' shape (Figure 4.14) under constant or decreasing load while the overall displacement is increased. This behavior is known as 'pop-in'. After crack growth at pop-in the stiffness of the load–displacement curve decreases since it corresponds to a longer crack.

A simplified model was proposed by Krafft et al. [4.13] to explain the decrease of fracture toughness with increase of depth of square fracture. Consider in Figure 4.15 that the square fracture occupies the part $(1-S)B$ of the specimen thickness and that the slant fracture surface is at 45°. The work for plastic deformation $(\mathrm{d}W_p/\mathrm{d}V)$ is assumed constant. If $\mathrm{d}W_f/\mathrm{d}A$ is the work consumed to produce a unit area of flat fracture, the work done for an advance of crack length by $\mathrm{d}a$ is

$$\mathrm{d}W = \left(\frac{\mathrm{d}W_f}{\mathrm{d}A}\right)(1-S)B\,\mathrm{d}a + \left(\frac{\mathrm{d}W_p}{\mathrm{d}V}\right)\frac{B^2 S^2}{2}\,\mathrm{d}a. \tag{4.72}$$

From this equation the strain energy release rate is

$$G = \frac{\mathrm{d}W}{\mathrm{d}A} = \left(\frac{\mathrm{d}W_f}{\mathrm{d}A}\right)(1-S) + \left(\frac{\mathrm{d}W_p}{\mathrm{d}V}\right)\frac{B^2 S^2}{2}. \tag{4.73}$$

By fitting the experimental data to this equation and assuming that the slant fracture has a thickness of 2 mm, Krafft et al. obtained the expression

$$G_c = 20(1-S) + 200S^2. \tag{4.74}$$

Equation (4.74) establishes the dependence of G_c on S. The variation of G_c and $100(1-S)$ versus the thickness B for the experiments of Krafft et al. is shown in Figure 4.16.

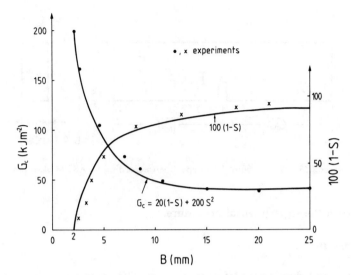

Figure 4.16. Crack growth resistance G_c and percentage square fracture $100(1-S)$ versus plate thickness B according to experiments by Krafft et al [4.13].

Irwin suggested the following semi-empirical equation which relates K_c with the plane strain fracture toughness K_{Ic}:

$$K_c = K_{Ic}\sqrt{1 + \frac{1.4}{B^2}\left(\frac{K_{Ic}}{\sigma_Y}\right)^4} \tag{4.75}$$

where σ_Y is the yield stress.

An analytical model for the explanation of the thickness effect was presented by Sih and Hartranft [4.9]. Based on an approximate elastic three-dimensional theory, they obtained that the critical strain energy release rate is inversely proportional to the plate thickness. The proposed model does not include the effect of plasticity and is not valid for small thicknesses where plastic deformation is not negligible.

4.8. Experimental determination of K_{Ic}

The experimental determination of the plane strain fracture toughness necessitates special requirements to be fullfilled to obtain reproducible values of K_{Ic} under conditions of maximum constraint around the crack tip. Furthermore, the size of the plastic zone accompanying the crack tip must be very small relative to the specimen thickness and the K_I-dominant region. The procedure for measuring K_{Ic} has been standardized by the American Society for Testing and Materials (ASTM) [3.7] to meet these requirements in specimens of minimal possible dimensions that can easily be tested in the laboratory. It is the intent of this section to present the salient points of the ASTM standard test method that would enable the reader to better understand the meaning of K_{Ic} and become

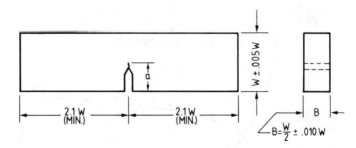

Figure 4.17. Three-point bend specimen according to ASTM standards.

familiar with the experimental procedure.

(a) Test specimens

The type and dimensions of specimens used for measuring K_{Ic} must satisfy the above requirements to ensure that linear elastic behavior occurs over a large stress field relative to the plastic zone and plane strain conditions dominate around the crack tip. According to the ASTM standard, the minimum characteristic specimen dimensions, including the specimen thickness B, the crack length a and the specimen width W, must be fifty times greater than the radius of the plane strain plastic zone at fracture. When the plastic zone size is determined according to the Irwin model (Section 3.5), this condition implies (see Equation (3.61)) that

$$B \geq 2.5 \left(\frac{K_{Ic}}{\sigma_Y} \right)^2 \tag{4.76a}$$

$$a \geq 2.5 \left(\frac{K_{Ic}}{\sigma_Y} \right)^2 \tag{4.76b}$$

$$W \geq 2.5 \left(\frac{K_{Ic}}{\sigma_Y} \right)^2 \tag{4.76c}$$

A variety of precracked test specimens are described in the ASTM Specification E399-81. These include the three-point bend specimen, the compact tension specimen, the arc-shaped specimen and the disk-shaped compact specimen. The geometrical configurations of the most widely used three-point bend specimen and compact tension specimen are shown in Figures 4.17 and 4.18. Several formulas have been proposed for the calculation of stress intensity factor for the standard specimens [4.14–4.16]. According to ASTM standards the following expressions for the computation of K_I proposed by Srawley [4.17] are used:

$$K_I = \frac{PS}{BW^{3/2}} \frac{3 \left(\frac{a}{W} \right)^{1/2} \left[1.99 - \frac{a}{W} \left(1 - \frac{a}{W} \right) \left(2.15 - 3.93 \frac{a}{W} + 2.7 \frac{a^2}{W^2} \right) \right]}{2 \left(1 + 2 \frac{a}{W} \right) \left(1 - \frac{a}{W} \right)^{3/2}} \tag{4.77}$$

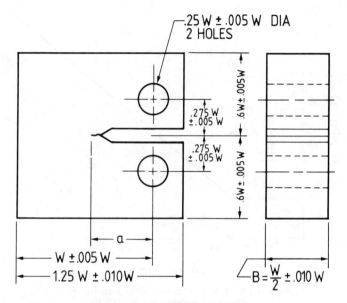

Figure 4.18. Compact tension specimen according to ASTM standards.

for the bend specimen, and

$$K_I = \frac{P}{BW^{1/2}} \frac{\left(2+\frac{a}{W}\right)\left[0.886 + 4.64\frac{a}{W} - 13.32\left(\frac{a}{W}\right)^2 + 14.72\left(\frac{a}{W}\right)^3 - 5.6\left(\frac{a}{W}\right)^4\right]}{\left(1-\frac{a}{W}\right)^{3/2}} \quad (4.78)$$

for the compact tension specimen. The quantities a, W and B are shown in Figures 4.17 and 4.18 and S is the distance between the points of support of beam of Figure 4.17.

Equation (4.77) is accurate to within 0.5 per cent over the entire range of a/W ($a/W < 1$), while Equation (4.78) is accurate to within 0.5 per cent for $0.2 < a/W < 1$. The meaning of symbols entered in Equations (4.77) and (4.78) are explained in Figures 4.17 and 4.18.

(b) Precrack

The precrack introduced in the specimen must simulate the ideal plane crack with zero root radius, as was assumed in the stress intensity factor analysis. The effect of the notch radius ρ on the critical value of the stress intensity factor K_c is shown in Figure 4.19. K_c decreases with decreasing ρ until a limiting radius ρ_c is obtained. Below ρ_c, K_c is approximately constant which shows that a notch with radius smaller than ρ_c can simulate the theoretical crack. The crack front must be normal to the specimen free surfaces and the material around the crack should experience a minimum of damage. To meet these requirements a special technique is used for the construction of the precrack in the specimen.

A chevron starter notch (Figure 4.20) of length $0.45W$ is first machined in the specimen. The notch is then extended by fatigue at a length $0.05W$ beyond the

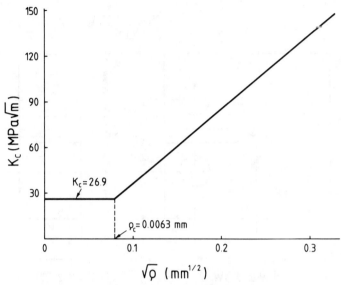

Figure 4.19. Effect of notch radius ρ on the critical stress intensity factor K_c [1.29].

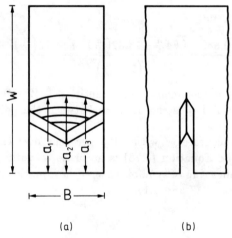

Figure 4.20. Chevron notch.

notch root. The advantage of the chevron notch is that it forces crack initiation in the center so that a straight machined crack front is obtained. If the initial machined notch front was straight it would be difficult to produce a final straight crack front. The crack length a used in the calculations is the average of the crack lengths measured at the center of the crack front and midway between the center and the end of the crack front on each surface ($a = (a_1 + a_2 + a_3)/3$). The surface crack length should not differ from the average length by more than 10 per cent.

In order to ensure that the material around the crack front does not experience

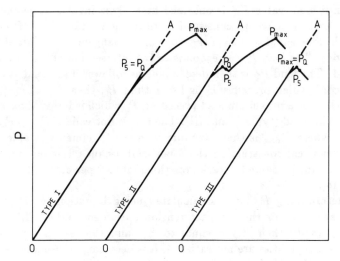

Figure 4.21. Determination of P_Q for three types of load–displacement response according to ASTM standards.

large plastic deformation or damage, and that the fatigue crack is sharp, the fatigue loading should satisfy some requirements. The maximum stress intensity factor to which the specimen is subjected during fatigue must not exceed 60 per cent of K_{Ic} and the last 2.5 per cent of the crack length should be loaded at a maximum K_I such that $K_I/E < 0.002$.

(c) *Experimental procedure*

The precracked standard specimen is loaded to special fixtures recommended by ASTM. The load and the relative displacement of two points located symmetrically on opposite sides of the crack plane are recorded simultaneously during the experiment. The specimen is loaded at a rate such that the rate of increase of stress intensity is within the range 0.55–2.75 MPa.m$^{1/2}$/s. A test record consisting of an autographic plot of the output of the load–sensing transducers versus the output of the displacement gage is obtained. A combination of load–sensing transducer and autographic recorder is selected so that the maximum load can be determined from the test record with an accuracy of 1 per cent. The specimen is tested until it can sustain no further increase of load.

(d) *Interpretation of test record and calculation of K_{Ic}*

For perfectly elastic behavior until fracture the load–displacement curve should be a straight line. Most structural materials, however, present elastoplastic behavior which, combined with some stable crack growth prior to catastrophic fracture, results in nonlinear load–displacement diagrams. The principal types of load–displacement curve usually observed in experiments are shown in Figure 4.21. Type I corresponds to nonlinear behavior, type III to purely linear response and type II reflects the phenomenon of pop-in. For the determination of a valid

K_{Ic}, a conditional value K_Q is obtained first. This involves a geometrical construction on the test record consisting of drawing a secant line OP through the origin with slope equal to 0.95 of the slope of the tangent to the initial linear part of the record. The load P_5 corresponds to the intersection of the secant with the test record. The load P_Q is then determined as follows: if the load at every point on the record which precedes P_5 is lower than P_5 then $P_Q = P_5$ (type I); if, however, there is a maximum load preceding P_5 which is larger than P_5 then P_Q is equal to this load (types II and III). The test is not valid if $P_{\max}/P_Q$ is greater than 1.10, where $P_{\max}$ is the maximum load the specimen was able to sustain. In the geometrical construction the 5 per cent secant offset line represents the change in compliance due to crack growth equal to 2 per cent of the initial crack length [4.18].

After determining P_Q, K_Q is calculated using Equation (4.77) or (4.78) for the bend specimen or the compact tension specimen. When K_Q satisfies the inequalities (4.76) then K_Q is equal to K_{Ic} and the test is a valid K_{Ic} test. When these inequalities are not satisfied, it is necessary to use a larger specimen to determine K_{Ic}. The dimensions of the larger specimen can be estimated on the basis of K_Q.

Results on fracture toughness values for various materials have been published in the literature. Hudson and Seward [4.19, 4.20] presented a compendium of sources of fracture toughness and fatigue crack growth data. Data on K_{Ic} indicate that the fracture toughness depends on temperature and strain rate. Generally speaking, K_{Ic} increases with temperature and decreases with loading rate. For further information and results on K_{Ic} the interested reader should consult references [4.21–4.24].

4.9. Crack stability

In the Griffith energy balance approach of crack growth the critical load is determined from Equation (4.17), which resulted from the conservation of energy in the entire body. Crack growth is considered unstable when the system energy at equilibrium is maximum and stable when it is minimum. A sufficient condition for crack stability is

$$\frac{\partial^2(\Pi + \Gamma)}{\partial A^2} \begin{cases} < 0 : & \text{unstable fracture} \\ > 0 : & \text{stable fracture} \\ = 0 : & \text{neutral equilibrium} \end{cases} \quad (4.79)$$

where the potential energy Π of the system is defined by Equation (4.16).

Two example problems will now be considered with respect to crack stability. The first concerns the case of a line crack in an infinite plate subjected to a uniform stress perpendicular to the crack axis. The potential energy of the system $\Pi = -U^e$, where U^e is given by Equation (4.23) and $\Gamma = 4\gamma a$. The terms Π, Γ and $(\Pi + \Gamma)$ are plotted in Figure 4.22 against half crack length a. Observe that the total potential energy of the system $(\Pi + \Gamma)$ at the critical crack length presents a maximum which corresponds to unstable equilibrium. This result is

Crack growth based on energy balance

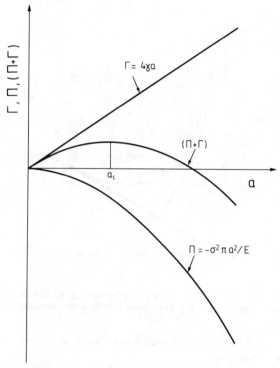

Figure 4.22. Potential energy, Π, surface energy, Γ, and the sum of potential and surface energy, $(\Pi + \Gamma)$, versus crack length a for a line crack in an infinite medium subjected to a uniform stress perpendicular to the crack axis.

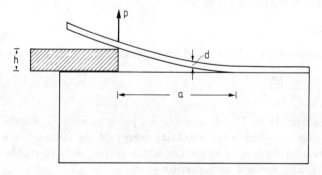

Figure 4.23. Wedge inserted to peel off mica, according to Obreimoff's experiment.

also verified by Equation (4.79).

The second problem concerns the experiment carried out by Obreimoff [4.4] on the cleavage of mica (Figure 4.23). A wedge of thickness h is inserted underneath a flake of mica which is detached from a mica block along a length a. The energy of the system is calculated by considering the mica flake as a cantilever beam with height d built-in at distance a from the point of application of the wedge. During propagation of the crack of length a the force P does not perform work.

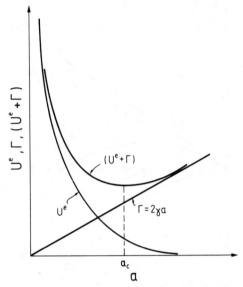

Figure 4.24. Potential energy, Π $(=U^e)$, surface energy, Γ, and the sum of potential and surface energy, $(\Pi+\Gamma)$, versus crack length a for Obreimoff's experiment.

The elastic energy stored in the cantilever beam according to the elementary theory of beam bending is

$$U^e = \frac{Ed^3h^2}{8a^3}. \tag{4.80}$$

The surface energy Γ is given by

$$\Gamma = 2\gamma a \tag{4.81}$$

and from Equation (4.17) the equilibrium crack length a_c is obtained

$$a_c = \left(\frac{2Ed^3h^2}{16\gamma}\right)^{1/4}. \tag{4.82}$$

The quantities $\Pi = U^e$, Γ and $(\Pi + \Gamma)$ are plotted in Figure 4.24 versus the crack length a. The total potential energy of the system $(\Pi + \Gamma)$ presents at critical crack length a_c a minimum which corresponds to stable equilibrium. This result is also verified by Equation (4.79).

The stability condition (4.79) can be put in the form in terms of the crack driving force G, given by Equation (4.21)

$$\frac{\partial(G-R)}{\partial A} \begin{cases} >0: & \text{unstable fracture} \\ <0: & \text{stable fracture} \\ =0: & \text{neutral equilibrium} \end{cases} \tag{4.83}$$

where $R = d\Gamma/dA$. For the case of an ideally brittle material $R = 2\gamma$ =const and the R-term disappears from Equation (4.83). Referring to Equation (4.47), the stability condition of crack growth may be expressed in terms of the stress

Crack growth based on energy balance

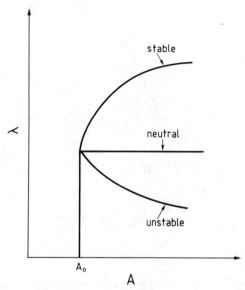

Figure 4.25. Load or displacement factor, λ, versus crack surface, A, for stable, neutral and unstable equilibrium.

intensity factor as follows:

$$\frac{\partial K}{\partial A} \begin{cases} > 0: & \text{unstable fracture} \\ < 0: & \text{stable fracture} \\ = 0: & \text{neutral equilibrium} \end{cases} \qquad (4.84)$$

Consider now the situation when the crack growth is controlled by some load or displacement factor $\lambda(A)$. Then from Equation (4.21) it is deduced for $G = G(\lambda, A)$ that

$$\frac{dG}{dA} = \left(\frac{\partial G}{\partial \lambda}\right)_A \frac{d\lambda}{dA} + \left(\frac{\partial G}{\partial A}\right)_\lambda = 0. \qquad (4.85)$$

Assuming that $(\partial G/\partial \lambda)_A > 0$, which is usually the case, we find that the crack growth is stable when $d\lambda/dA > 0$, unstable when $d\lambda/dA < 0$ and neutral when $d\lambda/dA = 0$. This result is shown in Figure 4.25.

As an example consider the case of the double cantilever beam specimen of Figure 4.10. From Equation (4.56), (4.59) and (4.63) we obtain

$$G = \frac{12P^2a^2}{B^2Eh^3} = \frac{3u^2Eh^3}{16a^4} \qquad (4.86)$$

when the load P or the displacement u are controlled.

Crack growth governed by Equation (4.21) takes place when

$$P = \frac{Bh}{a}\sqrt{\frac{\gamma Eh}{6}} \qquad (4.87)$$

for a controlled load, or

$$u = \frac{a^2}{h}\sqrt{\frac{32\gamma}{3Eh}} \qquad (4.88)$$

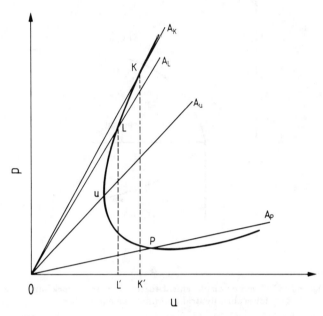

Figure 4.26. Interpretation of crack stability in load–displacement coordinates.

for a controlled displacement.

From Equations (4.83) and (4.86) we find that crack growth is unstable for a controlled load and stable for a controlled displacement. This result is also verified from Equations (4.87) and (4.88) from which it is obtained that $dP/da < 0$ and $du/da > 0$.

In general, both load and displacement change during crack propagation. A general representation of crack stability within the framework of linear elastic fracture mechanics was provided by Clausing [4.25] and Gurney and coworkers [4.11, 4.26–4.28]. Further work on this subject has been done by Mai, Atkins, Chow and coworkers [4.29–4.32]. A graphical interpretation of crack stability in load–displacement (P–u) coordinates is presented in Figure 4.26. The heavy line represents a P–u relation during crack propagation for a given specimen. Two types of prescribed external conditions such as $du/u > 0$ and $dP/P > 0$ are usually considered. The first corresponds to hard testing machines and the second to soft testing machines. In the P–u diagram the condition $du/u > 0$ refers to crack lengths greater than A_u, where u is the point at which the P–u curve has a vertical tangent. Similarly, the condition $dP/P > 0$ refers to crack lengths greater than A_P, where P is the point at which the P–u curve has a horizontal tangent. Consider an increment of crack growth from a crack of area A_K to a new crack of area A_L, with both A_K and A_L smaller than A_u. During crack growth, extra work represented by the area $KLL'K'$ is given to the body in excess of what is required for stable growth. This corresponds to an unstable situation. Thus, on the P–u curve cracks of length smaller than A_u or A_P are unstable when tested in hard or soft machines, respectively. Observe that stable crack growth is more easily achieved in displacement-controlled (hard) than in load-controlled (soft) testing machines.

An analytical expression for crack stability in hard or soft testing machines can easily be obtained. From Equations (4.56) or (4.59) and (4.65) we derive

$$P^2 = \frac{2R}{\frac{d}{dA}\left(\frac{u}{P}\right)} \tag{4.89}$$

which for stability with $dP/P > 0$ (soft machines) gives

$$\frac{1}{R}\frac{dR}{dA} \geq \frac{\frac{d^2}{dA^2}\left(\frac{u}{P}\right)}{\frac{d}{dA}\left(\frac{u}{P}\right)} \tag{4.90}$$

and for stability with $du/u > 0$ (hard machines) gives

$$\frac{1}{R}\frac{dR}{dA} \geq \frac{\frac{d^2}{dA^2}\left(\frac{P}{u}\right)}{\frac{d}{dA}\left(\frac{P}{u}\right)} \tag{4.91}$$

Equations (4.90) and (4.91) were first derived by Gurney and Hunt [4.11]. The right-hand side of inequalities (4.90) and (4.91) depends on the geometry of the specimen and is called the geometry stability factor of the specimen. For example, for the case of the double cantilever beam specimen the compliance $C = u/P$ is given by Equation (4.63) and inequality (4.91) for stability in hard testing machines takes the form

$$\frac{1}{R}\frac{dR}{dA} > -\frac{4}{A} \tag{4.92}$$

which, for constant R, is always satisfied.

Geometry stability factors for a host of testpiece geometries and loading conditions were given by Mai and Atkins [4.32].

4.10. Crack growth resistance curve (R-curve) method

The crack growth resistance curve, or R-curve, method is a one-parameter method for the study of fracture in situations where small, slow, stable crack growth – usually accompanied by inelastic deformation – is observed prior to global instability. The concept was introduced by Irwin and Kies [4.33] who observed that the fracture resistance of thin specimens is represented by a resistance curve rather than a single resistance parameter. A brief description of the method follows.

(a) General remarks

As noted in Section 4.7, the fracture resistance of a material under plane strain conditions with small-scale crack-tip plasticity is described by the critical stress intensity factor K_{Ic}. Under such conditions fracture of the material is sudden and there is either no or very little crack growth before final instability. On the

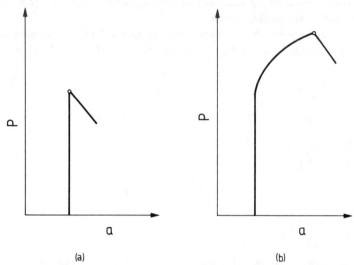

Figure 4.27. Typical load–crack size curves for (a) plane strain and (b) plane stress.

other hand, in thin specimens insufficient material exists to support a triaxial constraint near the crack tip and plane stress dominates. The crack tip plastic enclaves are no longer negligible and final instability is preceded by some slow stable crack growth. As shown in Section 4.7, the fracture resistance depends upon thickness. In such circumstances it was observed experimentally that the fracture resistance increases with increasing crack growth. Typical curves representing the variation of crack size with load under plane strain and plane stress conditions are shown in Figure 4.27.

(b) *R-curve*

The theoretical basis for the R-curve can be provided by the energy balance equation (Equation (4.14)) which applies during stable crack growth. For situations in which the energy dissipated to plastic deformation U^p is not negligible, Equation (4.14) takes the form

$$G = R \qquad (4.93a)$$

with

$$G = \frac{\partial W}{\partial A} - \frac{\partial U^e}{\partial A} \qquad (4.93b)$$

and

$$R = \frac{\partial \Gamma}{\partial A} + \frac{\partial U^p}{\partial A} \qquad (4.93c)$$

R represents the rate of energy dissipation during stable crack growth. It is composed of two parts: the first corresponds to the energy consumed in the creation of new material surfaces and the second refers to the energy dissipated

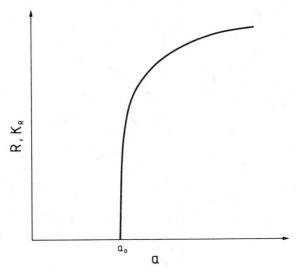

Figure 4.28. R-curve.

in plastic deformation. In situations where the crack-tip zones of plastic irreversibility are relatively small, the two dissipation terms in Equation (4.93c) may be lumped together to form a new material parameter associated with the resistance of the material to fracture. Following crack initiation the plastic zone around the crack tip increases nonlinearly with crack size. Thus, the rate of the energy dissipated to plastic deformation, which constitutes the major part of the dissipation term in Equation (4.93c), increases nonlinearly with crack size. The graphical representation of the variation of R or the critical stress intensity factor K_R plotted against crack extension is called the crack growth resistance curve (R-curve). A typical form of the R-curve is shown in Figure 4.28. The R-curve is considered to be a characteristic of the material for a given thickness, temperature and strain rate, independent of the initial crack size and the geometry of the specimen [4.13, 4.14]. This assumption will be discussed in more detail later.

(c) Determination of the critical load

During stable crack growth, Equation (4.93a) and inequality (4.83) should be satisfied. The strain energy release rate G according to the R-curve method is calculated from Equation (4.47) for 'fixed-grips' or 'dead-load' loading conditions. For example, for the case of a line crack of length $2a$ in an infinite plate subjected to a stress σ perpendicular to the crack, G is given by Equation (4.44). For a crack of length $2a$ in the finite plate of width W, K_I is given from Equation (2.83) and G is calculated from Equation (4.48) under plane stress as

$$G(a,\sigma) = \frac{\sigma^2 W}{E} \tan\left(\frac{\pi a}{W}\right). \tag{4.94}$$

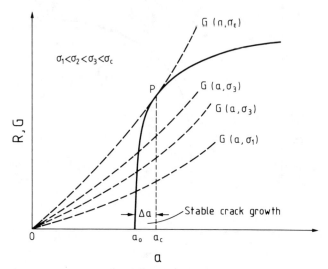

Figure 4.29. R-curve and a family of rising G-curves.

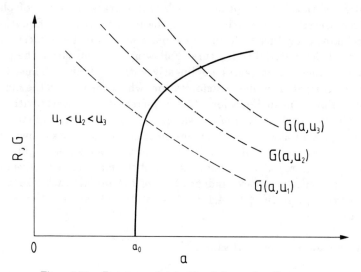

Figure 4.30. R-curve and a family of decreasing G-curves.

In graphical form both parts of Equation (4.93a) are represented in Figure 4.29 in G–a coordinates. The R-curve is displaced at the initial crack length a_0 while the $G(a, \sigma_i)$-curves correspond to different values of the applied stress σ. The points of intersection of the G- and R-curves refer to stable crack growth since Equation (4.93a) and inequality (4.83) are satisfied. Stable crack growth continues up to the point P at which the $G(a, \sigma_c)$-curve that corresponds to the value σ_c of the applied stress is tangent to the R-curve. Beyond point P crack propagation is unstable according to inequality (4.93a). Point P defines the critical stress σ_c and the critical crack length a_c at instability.

For specimen configurations for which K and therefore G decrease with increasing crack length, the graphical representation of the quantities $G(\sigma, a)$ and $R(a - a_0)$ is shown in Figure 4.30. Observe that if R increases monotonically then crack growth is stable for all crack lengths up to a plateau level. In this way the entire R-curve is obtained.

(d) *Experimental determination of the R-curve*

ASTM [4.34] issued a standard for the experimental determination of the R-curve of a material for given thickness, temperature and strain rate. The R-curve is regarded independent of the starting crack length and the specimen configuration. It is a function of crack extension only. Three types of fatigue-precracked standard specimens – namely, the center cracked tension specimen, the compact specimen and the crack line wedge-loaded specimen – are recommended by ASTM. The specimen dimensions are chosen so that the ligament in the plane of the crack is predominantly elastic at all values of the applied load. The first two types of specimen are tested under load control while in the third type the displacement is controlled. The specimens are loaded incrementally to specially designed fixtures, which are described in detail in [4.34]. During the test the load and the crack length are recorded simultaneously. The physical crack length is measured using optical microscopy or the electrical potential method. The effective crack length is obtained by adding to the physical crack length the Irwin plastic zone radius r_y $(= c/2)$ given by Equation (3.59). Calibration formulas are used to obtain the stress intensity factor during stable crack growth. The K–Δa or G–Δa relationship thus obtained constitutes the R-curve. For more details on the R-curve method the interested reader is referred to reference [4.35].

(e) *Discussion of the method*

The R-curve method extends the realm of applicability of the critical stress intensity factor fracture criterion under plane strain conditions presented in Section 4.7 to situations where some amount of stable crack growth, accompanied by non-negligible plastic deformation, precedes fracture instability. It offers a method for assessing fracture toughness in situations of semi-brittle fracture where the plate thickness is not enough to guarantee plane strain conditions around the crack tip. The R-curve describes the changing resistance to fracture with increasing crack size. The slope of the R-curve, known as the tearing modulus, expresses the increase of fracture toughness as the crack length increases prior to instability. It is of utmost importance to clear up some of the basic points of the method.

(i) The assumption that the R-curve is a material parameter for a given thickness, temperature and strain rate independent of initial crack length and specimen configuration is open to question. In general, the energy rate dissipated to plastic deformation (which constitutes the major part of resistance to fracture) depends on the state of stress near the crack tip which is influenced by the specimen geometry and the initial crack length. Thus, R must depend on specimen configuration. Experiments in favor [4.13, 4.14] or against [4.36, 4.37] geometry independence have been published in the literature and this question is still a subject of controversy.

(ii) The way that the strain energy release rate G is calculated from the stress intensity factor through Equation (4.47) holds only under 'fixed-grips' or 'deadload' loading, which is not generally justified during stable crack growth where both load and geometry may vary. Furthermore, Equation (4.47) was derived under the assumption that the plastic zone accompanying the crack tip is small so that the state of affairs can be described by linear elastic analysis. These remarks make questionable not only the applicability of the method to situations of stable crack growth that are generally accompanied by plastic deformation, but also the practice used in the experimental determination of the R-curve.

(iii) A final remark concerns the way the fracture toughness R is defined by Equation (4.93c). Fracture toughness is a parameter that describes the material resistance to crack growth and has to do with the phenomenon of material separation itself. R in Equation (4.93c) is composed of the term $\partial \Gamma/\partial A$ related to the loss of continuity of the material and the term $\partial U^p/\partial A$ which may have a negligible contribution to the energy released at the onset of rapid fracture. This is because U^p is not available at the instant when a new crack surface is created.

From the previous discussion it is clear that the R-curve method uses linear elastic fracture mechanics concepts and analyses in an effort to extend the realm of applicability of the linear analysis to situations of ductile fracture. It can, therefore, be used to describe semi-brittle fracture in situations of small crack-tip plastic deformation and stable crack growth prior to instability. An illuminating discussion of the R-curve method was given by Eftis, Liebowitz and Jones [4.38, 4.39].

(f) Irwin-Orowan theory

In an effort to extend the Griffith theory to situations of semi-brittle fracture, Irwin [4.40] and Orowan [4.41] introduced independently a modification to the Griffith formula (4.24) or (4.25). The Irwin–Orowan theory can easily be described by the energy balance analysis of the R-curve method.

For the case of a crack of length $2a$ in an infinite plate subjected to a stress σ perpendicular to the crack axis, G in Equation (4.93a) is given by Equation (4.44). Putting

$$\frac{\partial \Gamma}{\partial A} = 2\gamma, \qquad \frac{\partial U^p}{\partial A} = 2\gamma_p \tag{4.95}$$

the crack growth resistance R is

$$R = 2(\gamma + \gamma_p) \tag{4.96}$$

and Equation (4.93a) gives for the critical stress

$$\sigma_{\text{cr}} = \sqrt{\frac{2E(\gamma + \gamma_p)}{\pi a}} \tag{4.97}$$

for plane stress, and

$$\sigma_{cr} = \sqrt{\frac{2E(\gamma + \gamma_p)}{(1-\nu^2)\pi a}} \qquad (4.98)$$

for plane strain.

The modified version of Griffith's formula as given by Equations (4.97) and (4.98) should be interpreted with care. The quantity γ_p in Equation (4.95) and γ in Equation (4.20) are defined, respectively, by the rate change of U^p and U^e with reference to the *same crack area* A. Release of the elastic energy U^e and plastic energy U^p, however, may not occur at the same time and location. The difference of γ and γ_p by three orders of magnitude, as claimed by Orowan, cannot be regarded as additive without consideration to time and location. A detailed discussion of γ and γ_p as related to the creation of micro- and macrocrack surface has been given by Sih [1.35].

(g) Critical energy release rate

For the case of fracture extension with plastic deformation where substantial stable crack growth is preceded by crack instability, Equation (4.14) may be written as

$$\tilde{G} = \frac{\partial \Gamma}{\partial A} = \frac{\partial W}{\partial A} - \left(\frac{\partial U^e}{\partial A} + \frac{\partial U^p}{\partial A}\right). \qquad (4.99)$$

The above relation was proposed by Eftis, Liebowitz and Jones [4.38, 4.39] who suggested an empirical estimate of $\tilde{G}$ from the load–displacement diagram of a fracture toughness test. They distinguished two cases. In the first, the nonlinearity of the load–displacement diagram is solely due to extensive plastic deformation, while in the second the nonlinearity is caused by coupled plastic deformation and stable crack growth. In the first case, when the load–displacement test record is described by equation

$$u = \frac{P}{M} + k\left(\frac{P}{M}\right)^n \qquad (4.100)$$

where P is the load, u the displacement, M the specimen compliance and k and n are parameters characterizing the nonlinear part of the curve, the critical fracture toughness $\tilde{G}_c$ is given by

$$\tilde{G}_c = \left[1 + \frac{2nk}{n+1}\left[\frac{P_c}{M(a_0)}\right]^{n-1}\right]\frac{P_c^2}{2B}\frac{d}{da}\left[\frac{1}{M(a_0)}\right]. \qquad (4.101)$$

In Equation (4.101) P_c is the critical load at instability, a_0 is the initial crack length and B is the specimen thickness.

When extensive subcritical crack growth precedes unstable fracture, $\tilde{G}_c$ is given by

$$\tilde{G}_c = \left[1 + \frac{2nk}{n+1}\left[\frac{P_c}{M(a_c)}\right]^{n-1}\right]\left[\frac{M(a_0)}{M(a_c)}\right]\frac{P_c^2}{2B}\frac{d}{da}\left[\frac{1}{M(a_0)}\right]. \qquad (4.102)$$

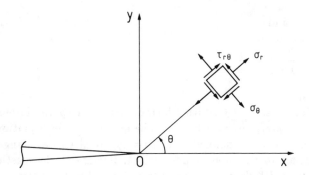

Figure 4.31. Polar stress components in an element around a crack tip.

where a_c represents the critical crack length at instability. The quantities n, k, $M(a_0)$, $M(a_c)$, P_c and a_c are determined by experiment.

The results in Equations (4.101) and (4.102) are subject to clarification of the quantity $\partial U^p/\partial A$ in Equation (4.99), which has already been pointed out when discussing Equations (4.14), (4.15), (4.20) and (4.95).

4.11. Mixed-mode crack propagation

In the energy balance analysis of crack growth it was assumed that the crack extends in a self-similar manner, which occurs only when the applied loads are directed normal to the crack plane. However, nonalignment of loads with respect to crack orientation is a common practice. As a rule, the crack follows a curved path. This section is devoted to the extension of the strain energy release rate fracture criterion to cases of mixed mode loading where the crack growth direction is not known *a priori*. The stress criterion for non-self-similar crack growth is also presented.

(*a*) *The stress criterion*

Although Griffith [1.8, 1.9] formulated the energy balance theory of crack growth he did not abandon the classical concept of fracture related to values in excess of the critical tensile strength in a material element. We quote from reference [1.9]: '... we may infer that the general condition for rupture will be the attainment of a specific tensile stress at the edge of one of the cracks.' Using this concept, and basing his work on the Inglis solution, Griffith calculated the maximum tensile stress at the edge of an elliptical hole in an infinite plate subjected to a biaxial stress field. Griffith's analysis laid down the foundation of what is called today the maximum stress criterion for crack growth under mixed-mode conditions.

The maximum circumferential stress criterion for mixed-mode crack growth was proposed by Erdogan and Sih [4.42]. Consider a crack in a mixed-mode stress field governed by the values of the opening-mode K_I and sliding-mode K_{II} stress intensity factors (Figure 4.31). The singular stress field in the vicinity of the crack tip is expressed by (see Equations (2.20) and (2.22))

$$\sigma_r = \frac{K_I}{\sqrt{2\pi r}}\left(\tfrac{5}{4}\cos\frac{\theta}{2} - \tfrac{1}{4}\cos\frac{3\theta}{2}\right) + \frac{K_{II}}{\sqrt{2\pi r}}\left(-\tfrac{5}{4}\sin\frac{\theta}{2} + \tfrac{3}{4}\sin\frac{3\theta}{2}\right) \quad (4.103a)$$

$$\sigma_\theta = \frac{K_I}{\sqrt{2\pi r}}\left(\tfrac{3}{4}\cos\frac{\theta}{2} + \tfrac{1}{4}\cos\frac{3\theta}{2}\right) + \frac{K_{II}}{\sqrt{2\pi r}}\left(-\tfrac{3}{4}\sin\frac{\theta}{2} - \tfrac{3}{4}\sin\frac{3\theta}{2}\right) \quad (4.103b)$$

$$\tau_{r\theta} = \frac{K_I}{\sqrt{2\pi r}}\left(\tfrac{1}{4}\sin\frac{\theta}{2} + \tfrac{1}{4}\sin\frac{3\theta}{2}\right) + \frac{K_{II}}{\sqrt{2\pi r}}\left(\tfrac{1}{4}\cos\frac{\theta}{2} + \tfrac{3}{4}\cos\frac{3\theta}{2}\right). \quad (4.103c)$$

The assumptions made in the criterion for crack extension in brittle materials may be stated as

(i) The crack extension starts from its tip along the radial direction $\theta = \theta_c$ on which σ_θ becomes maximum.
(ii) Fracture starts when that maximum of σ_θ reaches a critical stress σ_c equal to fracture stress in uniaxial tension.

These hypotheses can be expressed mathematically by the relations

$$\frac{\partial \sigma_\theta}{\partial \theta} = 0; \qquad \frac{\partial^2 \sigma_\theta}{\partial \theta^2} < 0 \quad (4.104a)$$

$$\sigma_\theta(\theta_c) = \sigma_c. \quad (4.104b)$$

Observe that the circumferential stress σ_θ at the direction of crack extension is a principal stress and the shear stress $\tau_{r\theta}$ at that direction vanishes. The crack extension angle θ_c is calculated (see Equation (4.103c)) by

$$K_I\left(\sin\frac{\theta}{2} + \sin\frac{3\theta}{2}\right) + K_{II}\left(\cos\frac{\theta}{2} + 3\cos\frac{3\theta}{2}\right) = 0 \quad (4.105)$$

or

$$K_I \sin\theta + K_{II}(3\cos\theta - 1) = 0. \quad (4.106)$$

For the calculation of the stress σ_θ from Equation (4.103b) a critical distance r_0 measured from the crack tip must be introduced [4.43]. The concept of a core region surrounding the crack tip has been proposed by Sih [4.44]. The idea is that the continuum mechanics solution, as well as experimental measurement, stop at a distance r_0 from the crack tip. r_0 serves as a scale size of analysis at the continuum level. The concept of the core region is further discussed in Chapter 6.

To circumvent the determination of the core region radius r_0 the second hypothesis of the stress criterion is often referred to as follows:

Fracture starts when σ_θ has the same value as in an equivalent opening mode (Figure 4.32), that is

$$\sigma_\theta = \frac{K_{Ic}}{\sqrt{2\pi r}}. \quad (4.107)$$

The fracture condition following from Equations (4.107) and (4.103b) takes the form

$$K_I\left(3\cos\frac{\theta_c}{2} + \cos\frac{3\theta_c}{2}\right) - 3K_{II}\left(\sin\frac{\theta_c}{2} + \sin\frac{3\theta_c}{2}\right) = 4K_{Ic}. \quad (4.108)$$

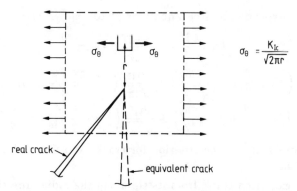

Figure 4.32. Equivalent opening-mode crack model according to the maximum circumferential stress criterion.

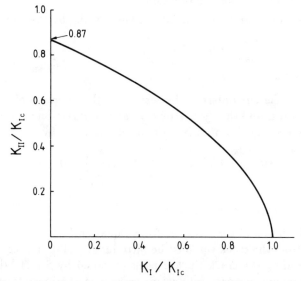

Figure 4.33. Fracture locus for mixed-mode conditions according to the maximum circumferential stress criterion.

For opening-mode loading ($K_I \neq 0$, $K_{II} = 0$), Equations (4.106) and (4.108) yield $\theta_c = 0$, $K_I = K_{Ic}$, while for sliding-mode loading we obtain

$$\theta_c = -\arccos \tfrac{1}{3} = -70.6°, \qquad K_{II} = K_{IIc} = \sqrt{\tfrac{3}{4}}\, K_{Ic}. \tag{4.109}$$

Eliminating θ_c in Equations (4.106) and (4.108) gives the fracture locus in K_I–K_{II} coordinates shown in Figure 4.33. The problem of an inclined crack in a plate subjected to a uniaxial stress field has extensively been used in the literature to check the validity of the maximum circumferential stress criterion. Most of the experimental results obtained regarding the angle of initial crack extension and the critical fracture stress corroborate the theoretical predictions. Other versions of the stress criterion based on the maximum tangential principal

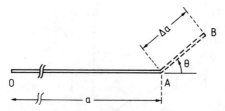

Figure 4.34. Off-axis infinitesimal crack extension.

stress and the maximum tangential strain have been proposed in the literature [4.45–4.48]. Experimental results related to crack propagation under mixed-mode loading can be found in references [4.49–4.51].

The stress criteria used in mixed-mode crack growth are inadequate for a number of reasons. First, the location of fracture may not always be governed by only one of the six independent stress components. The combination may play a role. Next, a stress quantity cannot be used to describe the fracture resistance of a material. An even more serious objection is the contradiction that occurs for the case of a moving crack, where, as we will see in Chapter 7, the normal stress parallel to the crack is greater than the stress perpendicular to the crack. This example shows that the maximum stress criterion is in direct disagreement with experimental observation.

(b) *The strain energy release rate criterion*

The application of the strain energy release rate concept to crack growth under mixed-mode loading is in principle quite simple and straightforward. The crack will run in the direction along which the strain energy release rate $G = G(\theta)$ becomes maximum (Figure 4.34). $G(\theta)$ is determined from Equation (4.21) by

$$G(\theta) = - \lim_{\Delta a \to 0} \frac{\Pi(a + \Delta a) - \Pi(a)}{\Delta a} \tag{4.110}$$

where $\Pi(a)$ and $\Pi(a + \Delta a)$ represent the potential energy of the body prior to and after crack extension. The critical energy release rate at crack growth is

$$G(\theta_c) = \max[G(\theta)] = \lim_{\theta \to \theta_c} \left[- \lim_{\Delta a \to 0} \frac{\Pi(a + \Delta a) - \Pi(a)}{\Delta a} \right]. \tag{4.111}$$

For the analytical calculation of the crack extension angle θ_c a stress analysis of the branched crack problem has first to be performed. A closed form solution for this problem, possessing two different stress singularities at points A and B, is very difficult to obtain even for the most simple geometry, say, a branched crack in an infinite plate under tension. But even if an analytical closed form solution for the branched crack problem were to exist, the proofs for the existence and uniqueness of the strain energy release rate in Equation (4.111) would be a very difficult task for the applied mathematician. The question of whether the limit $\Delta a \to 0$ should precede the limit $\theta \to \theta_c$ or vice versa further aggravates the situation.

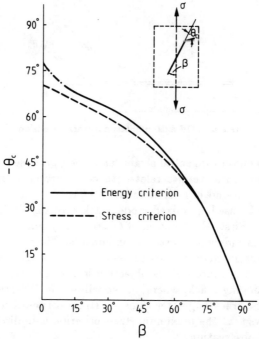

Figure 4.35. Crack extension angle $-\theta_c$ versus crack inclination angle β according to the maximum circumferential stress and the strain energy release rate criterion.

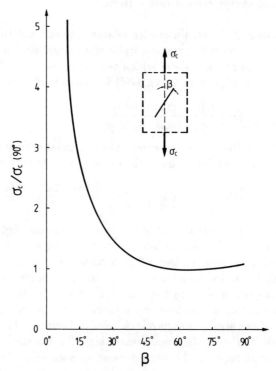

Figure 4.36. Normalized critical stress $\sigma_c/\sigma_c(90°)$ versus crack inclination angle β according to the strain energy release rate criterion.

A numerical solution of the problem was first performed by Palaniswamy and Knauss [4.52, 4.53] for the case of a crack in an infinite plate under tension. Results for the crack growth angle and the critical stress as a function of the crack inclination angle with respect to the applied stress are shown in Figures 4.35 and 4.36. In Figure 4.35 the results of the maximum circumferential stress criterion are also shown. It can be observed that the two theories are in good agreement for $\beta > 60°$.

Because of the inherent difficulties in the numerical analysis and the formidable computation work involved, the application of the strain energy release rate concept to problems of mixed-mode crack growth has not made much progress. For further information on this section the reader should consult references [4.54–4.57].

References

4.1. Kassir, M. K. and Sih, G. C., Griffith's theory of brittle fracture in three dimensions, *International Journal of Engineering Sciences* 5, 899–918 (1967).
4.2. Sih, G, C. and Liebowitz, H., On the Griffith energy criterion for brittle fracture, *International Journal of Solids and Structures* 3, 1–22 (1967).
4.3. Spencer, A. J. M., On the energy of the Griffith crack, *International Journal of Engineering Sciences* 3, 441–449 (1965).
4.4. Obreimoff, J. W., The splitting strength of mica, *Proceedings of the Royal Society of London* A127, 290–297 (1930).
4.5. Bueckner, H. F., The propagation of cracks and the energy of elastic deformation, *Transactions of the American Society of Mechanical Engineers* 80, pp. 1225–1230 (1958).
4.6. Rice, J. R., An examination of the fracture mechanics energy balance from the point of view of continuum mechanics, *Proceedings of the First International Conference on Fracture* (eds T. Yokobori, T. Kawasaki and J. L. Swedlow), Sendai, Japan, Vol. 1, pp. 309–340 (1966).
4.7. Rice, J. R. and Drucker, D. C., Energy changes in stressed bodies due to void and crack growth, *International Journal of Fracture Mechanics* 3, 19–28 (1967).
4.8. Cherepanov, G. P., On crack propagation in solids, *International Journal of Solids and Structures* 5, 863–871 (1969).
4.9. Sih, G. C. and Hartranft, R. J., Variation of strain energy release rate with plate thickness, *International Journal of Fracture* 9, 75–82 (1973).
4.10. Kfouri, A. P., Continuous crack growth or quantized growth steps, *International Journal of Fracture* 15, 23–29 (1979).
4.11. Gurney, C. and Hunt, J., Quasi-static crack growth, *Proceedings of the Royal Society of London* A299, 508–524 (1967).
4.12. Irwin, G. R., Kies, J. A. and Smith, H. L., Fracture strengths relative to onset and arrest of crack propagation, *Proceedings of the American Society for Testing and Materials* 58, 640–657 (1958).
4.13. Krafft, J. M., Sullivan, A. M. and Boyle, R. W., Effect of dimensions on fast fracture instability of notched sheets, *Proceedings of Crack Propagation Symposium*, College of Aeronautics, Cranfield (England), Vol. 1, pp. 8–28 (1961).
4.14. Srawley, J. E. and Brown, W. F. Jr, *Fracture Toughness Testing Methods*, ASTM, STP 381, American Society for Testing and Materials, Philadelphia, pp. 133–198 (1965).
4.15. Brown, W. F. Jr and Srawley, J. E., *Plane Strain Crack Toughness Testing*, ASTM, STP 410, American Society for Testing and Materials, Philadelphia (1966).
4.16. Srawley, J. E., Plane strain fracture toughness, in *Fracture – An Advanced Treatise*, Vol. IV, *Engineering Fracture Design* (ed. H. Liebowitz), Academic Press, pp. 45–68 (1969).
4.17. Srawley, J. E., Wide range stress intensity factor expressions for ASTM E399 standard fracture toughness specimens, *International Journal of Fracture* 12, 475–476 (1976).
4.18. Knott, J. F., *Fundamentals of Fracture Mechanics*, Butterworths, pp. 140–144 (1973).

4.19. Hudson, C. M. and Seward, S. K., A compendium of sources of fracture toughness and fatigue crack growth data for metallic alloys, *International Journal of Fracture* **14**, R151–R184 (1978).

4.20. Hudson, C. M. and Seward, S. K., A compendium of sources of fracture toughness and fatigue crack growth data for metallic alloys–Part II, *International Journal of Fracture* **20**, R59–R117 (1982).

4.21. Wessel, E. T., State of the art of the WOL specimen for K_{Ic} fracture toughness testing, *Engineering Fracture Mechanics* **1**, 77–103 (1968).

4.22. Brown, W. F., Jr, *Review of Developments in Plane Strain Fracture Toughness Testing*, ASTM STP 463, American Society for Testing and Materials, Philadelphia (1970).

4.23. Brown, W. F., Jr and Kaufman, J. G., *Developments in Fracture Mechanics Test Methods Standardization*, ASTM STP 632, American Society for Testing and Materials, Philadelphia (1977).

4.24. Barsom, J. M. and Rolfe, S. T., *Fracture and Fatigue Control in Structures* (2nd edn), Prentice–Hall (1987).

4.25. Clausing, D. P., Crack stability in linear elastic fracture mechanics, *International Journal of Fracture Mechanics* **5**, 211–227 (1969).

4.26. Gurney, C. and Ngan, K. M., Quasistatic crack propagation in nonlinear structures, *Proceedings of the Royal Society of London* **A325**, 207–222 (1971).

4.27. Gurney, C. and Mai, Y. W., Stability of cracking, *Engineering Fracture Mechanics* **4**, 853–863 (1972).

4.28. Gurney, C., Mai, Y. W. and Owen, R. C., Quasi-static crack propagation in materials with high toughness and low yield stress, *Proceedings of the Royal Society of London* **A340**, 213–231 (1974).

4.29. Chow, C. L. and Lam, P. M., Stability conditions in quasi-static crack propagation for constant strain energy release rate, *Journal of Engineering Materials and Technology, Trans. ASME* **96**, 41–48 (1974).

4.30. Chow, C. L. and Lau, K. J., Fracture studies of point-loaded center-cracked plates, *Journal of Strain Analysis* **12**, 286–292 (1977).

4.31. Mai, Y. W., Atkins, A. G. and Caddell, R. M., On the stability of cracking in tapered DCB testpieces, *International Journal of Fracture* **11**, 939–953 (1975).

4.32. Mai, Y. W. and Atkins, A. G., Crack stability in fracture toughness testing, *Journal of Strain Analysis* **15**, 63–74 (1980).

4.33. Irwin, G. R. and Kies, J. A., Critical energy rate analysis of fracture strength, *Welding Research Supplement* **19**, 193–198 (1954).

4.34. *Standard Practice for R-Curve Determination*. ASTM Annual Book of Standards, Part 10, American Society for Testing and Materials, E561–81, pp. 680–699 (1981).

4.35. *Fracture Toughness Evaluation by R-Curve Methods*, ASTM STP 527, American Society for Testing and Materials, Philadelphia, pp. 1–112 (1973).

4.36. Heyer, R. H. and McCabe, D. E., Plane stress fracture toughness testing using a crack-line loaded specimen, *Engineering Fracture Mechanics* **4**, 393–412 (1972).

4.37. Adams, N. J., Influence of configuration on R-curve shape and G_c when plane stress prevails, in *Cracks and Fracture*, ASTM STP 601, American Society for Testing and Materials, Philadelphia, pp. 330–345 (1976).

4.38. Eftis, J. and Liebowitz, H., On fracture toughness evaluation for semi-brittle fracture, *Engineering Fracture Mechanics* **7**, 101–135 (1975).

4.39. Eftis, J., Jones, D. L. and Liebowitz, H., On fracture toughness in the nonlinear range, *Engineering Fracture Mechanics* **7**, 491–503 (1975).

4.40. Irwin, G. R., Fracture dynamics, in *Fracture of Metals*, American Society for Metals, Cleveland, U.S.A., pp. 147–166 (1948).

4.41. Orowan, E., Fracture and strength of solids, in *Reports on Progress in Physics* **XII**, pp. 185–232 (1948).

4.42. Erdogan, F. and Sih, G. C., On the crack extension in plates under plane loading and transverse shear, *Journal of Basic Engineering, Trans. ASME* **85D**, 519–527 (1963).

4.43. Williams, J. G. and Ewing, P. D., Fracture under complex stress – The angled crack problem, *International Journal of Fracture* **10**, 441–446 (1972).

4.44. Sih, G. C., Some basic problems in fracture mechanics and new concepts, *Engineering Fracture Mechanics* **5**, 365–377 (1973).

4.45. Maiti, S. K., Prediction of the path of unstable extension of internal and edge cracks, *Journal of Strain Analysis* **15**, 183–194 (1980).

4.46. Maiti, S. K. and Smith, R. A., Comparison of the criteria for mixed mode brittle fracture based on the preinstability stress-strain field, Part I: Slit and elliptical cracks under uniaxial loading and Part II: Pure shear and uniaxial compressive loading, *International Journal of Fracture* **23**, 281–294 (1983) and **24**, 5–22 (1984).

4.47. Chang, K. J., On the maximum strain criterion – A new approach to the angled crack problem, *Engineering Fracture Mechanics* **14**, 107–124 (1981).

4.48. Chang, K. J., Further studies of the maximum stress criterion on the angled crack problem, *Engineering Fracture Mechanics* **14**, 125–142 (1981).

4.49. Pook, L. P., The effect of crack angle on fracture toughness, *Engineering Fracture Mechanics* **3**, 205–218 (1971).

4.50. Wu, H. C., Yao, R. F. and Yip, M. C., Experimental investigation of the angled elliptic notch problem in tension, *Journal of Applied Mechanics, Trans. ASME* **44**, 455–461 (1977).

4.51. Liu, A. F., Crack growth and failure of aluminum plate under in-plane shear, *Journal of the American Institute of Aeronautics and Astronautics* **12**, 180–185 (1974).

4.52. Palaniswamy, K. and Knauss, W. G., Propagation of a crack under general in-plane tension, *International Journal of Fracture Mechanics* **8**, 114–117 (1972).

4.53. Palaniswamy, K. and Knauss, W. G., On the problem of crack extension in brittle solids under general loading, in *Mechanics Today*, Vol. 4 (ed. S. Nemat-Nasser), Pergamon Press, pp. 87–148 (1978).

4.54. Harrison, N. L., Strain energy release rates for turning cracks, *Fibre Science and Technology* **5**, 197–212 (1972).

4.55. Hussain, M. A., Pu, S. L. and Underwood, J., Strain energy release rate for a crack under combined mode I and II, in *Fracture Analysis*, ASTM STP 560, American Society for Testing and Materials, Philadelphia, pp. 1–28 (1974).

4.56. Nuismer, R. J., An energy release rate criterion for mixed mode fracture, *International Journal of Fracture* **11**, 245–250 (1975).

4.57. Shah, R. C., Fracture under combined modes in 4340 steel, in *Fracture Analysis*, ASTM STP 560, American Society for Testing and Materials, Philadelphia, pp. 29–52 (1974).

5

J-Integral and crack opening displacement fracture criteria

5.1. Introduction

The mathematical formulation of conservation laws applicable in elastostatics in the form of path independent integrals of some functionals of the elastic field over the bounding surface of a closed region originates from the work of Günther [5.1]. Some years earlier Eshelby [5.2] obtained a representation of the 'force on an elastic singularity or inhomogeneity' in the form of a surface integral which, in the absence of such defects, leads to a conservation law in elastostatics. The two-dimensional analogue of the conservation law as a path independent line integral applied to notch problems has been introduced by Rice [5.3, 3.3]. Earlier investigations by Sanders [5.4] and Cherepanov [5.5] are closely related to the work contained in [5.3, 3.3]. Path independent integrals related to energy release rates were discovered by Knowles and Stenberg [5.6] and Eshelby [5.7]. Kröner and Zorski [5.8] have pointed out the importance of including the time variable when discussing path length independent integrals.

The present chapter is devoted to the theoretical foundation of the path independent J-integral introduced into fracture mechanics by Rice and its use as a fracture criterion. The critical value of the opening of the crack faces near the crack tip is also introduced as a fracture criterion.

First, we present the definition of the J-integral in two-dimensional crack problems and its physical interpretation in terms of the rate of change of potential energy with respect to an incremental extension of the crack. The path independent nature of the integral allows the integration path to be taken close or sufficiently far from the crack tip. The J-integral, as was shown in Section 3.7, characterizes the stress and strain fields in an area surrounding the crack tip in the HRR singular solution. These derivations of the integral are strictly valid for a nonlinear elastic material where unloading occurs along the same path as the initial loading. The use of J for plasticity-type materials is supported by appealing to experimentation or to numerical analysis. Experimental methods for the evaluation of the integral are also presented. A failure criterion based on the J-integral is introduced and its applicability and limitations when applied to situations of extensive crack-tip plasticity and/or stable crack growth are discussed. The chapter concludes with a brief presentation of the crack opening displacement fracture criterion.

5.2. Path-independent integrals

Some path-independent surface integrals in the three-dimensional space are first introduced. The solid body to which we are referred to is considered homogeneous elastic, linear or nonlinear and generally anisotropic in a state of static equilibrium under the action of a system of traction T_k. Denote by Σ the bounding surface of the region R occupied by the body and refer all quantities to a fixed Cartesian coordinate system $Ox_1x_2x_3$. To simplify the analysis the assumption of small deformations is invoked. The stress tensor σ_{ij} is obtained from the elastic strain energy w by

$$\sigma_{ij} = \sigma_{ji} = \frac{\partial w}{\partial \epsilon_{ij}}, \qquad w(0) = 0 \tag{5.1}$$

where ϵ_{ij} denotes the strain tensor. The strain energy density w is considered to be continuously differentiable with respect to strain. For elastic behavior Equation (5.1) renders

$$w = \int_0^{\epsilon_{kl}} \sigma_{ij}\, d\epsilon_{ij} \tag{5.2}$$

where the integral is path independent in the strain space.

Equations of equilibrium in the absense of body forces take the form

$$\sigma_{ij,i} = \frac{\partial \sigma_{ij}}{\partial x_i} = 0 \tag{5.3}$$

and the traction vector T_i on surface S is given by

$$T_i = \sigma_{ij} n_i \tag{5.4}$$

where n_i denotes the normal vector.

For small deformation the strain tensor is derived from the displacement field by

$$\epsilon_{ij} = \epsilon_{ji} = \tfrac{1}{2}(u_{i,j} + u_{j,i}). \tag{5.5}$$

Under the above conditions consider the integrals

$$Q_j = \int_\Sigma (w n_j - T_k u_{k,j})\, d\Sigma, \quad j,k = 1,2,3 \tag{5.6}$$

where Σ is every closed surface bounding a region R which is assumed to be free of singularities.

Equation (5.6) can be put in the form

$$Q_j = \int_\Sigma (w n_j - \sigma_{lk} n_l u_{k,j})\, d\Sigma = \int_\Sigma (w \delta_{jl} - \sigma_{lk} u_{k,j}) n_l\, d\Sigma \tag{5.7}$$

and applying Gauss's divergence theorem renders

$$Q_j = \int_R \left(w\delta_{jl} - \sigma_{lk}u_{k,j} \right)_{,l} dV. \tag{5.8}$$

The integrand takes the form

$$\begin{aligned}
\left(w\delta_{jl} - \sigma_{lk}u_{k,j} \right)_{,l} &= w_{,j} - \sigma_{lk,l}u_{k,j} - \sigma_{lk}u_{k,jl} \\
&= \frac{\partial w}{\partial \epsilon_{lk}}\epsilon_{lk,j} - \sigma_{lk}u_{k,lj} = \sigma_{lk}\left(\epsilon_{lk,j} - u_{k,lj}\right) \\
&= \sigma_{lk}\left(\epsilon_{lk} - u_{k,l}\right)_{,j} = \sigma_{kl}\left(\epsilon_{kl} - u_{k,l}\right)_{,j} \\
&= -\sigma_{kl}r_{kl,j} = 0
\end{aligned}$$

In the above derivations Equations (5.1) to (5.5) were used. r_{kl} denotes the nonsymmetrical rotation tensor for small deformation given by

$$r_{ij} = -r_{ji} = \tfrac{1}{2}(u_{i,j} - u_{j,i}). \tag{5.9}$$

Thus, we have

$$Q_j = 0. \tag{5.10}$$

In a similar manner it can be proven that Equation (5.10) holds when the elastic strain energy density w in Equation (5.6) is replaced by the complementary elastic strain energy density Ω given by

$$\Omega = \int_0^{\sigma_{kl}} \epsilon_{ij}\, d\sigma_{ij} \tag{5.11}$$

or

$$\Omega(\sigma_{ij}) = \sigma_{ij}\epsilon_{ij} - w(\epsilon_{ij}), \quad \Omega(0) = 0. \tag{5.12}$$

The integral in Equation (5.11) is path independent in the stress space. The strain is obtained from Ω by

$$\epsilon_{ij} = \frac{\partial \Omega}{\partial \sigma_{ij}} \tag{5.13}$$

in the same way that the stress is obtained from w by Equation (5.1).

5.3. J-integral

(a) *Definition*

For the particular case of the two-dimensional plane elastic problem, consider the integral

$$J = Q_1 = \int_\Gamma \left(w n_1 - T_k u_{k,1} \right) ds \tag{5.14}$$

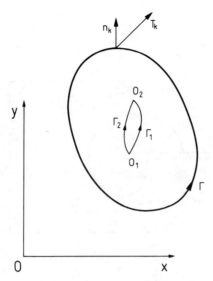

Figure 5.1. Paths Γ_1 and Γ_2 between two points O_1 and O_2 in a continuum.

where Γ is a closed contour bounding a region R (Figure 5.1). With

$$n_l = \frac{dy}{ds} \tag{5.15}$$

Equation (5.14) becomes

$$J = \int_\Gamma w\,dy - T_k \frac{\partial u_k}{\partial x}\,ds \quad (k=1,2) \tag{5.16}$$

and defines the J-integral along a closed contour in the two-dimensional space. From Equation (5.10) it follows that $J = 0$. The zeroing of J along a closed path implies path independence when J is calculated along arbitrary paths connecting any two points O_1 and O_2 (Figure 5.1). We have

$$J_1 = \int_{\Gamma_1} [\ldots] = J_2 = \int_{\Gamma_2} [\ldots]. \tag{5.17}$$

(b) *Application to notches and cracks*

Consider a notch or crack with flat surfaces parallel to the x-axis which may have an arbitrary root radius (Figure 5.2). The J-integral defined from Equation (5.16) is calculated along a path Γ starting from an arbitrary point on the flat part of the lower notch surface and terminating at an arbitrary point on the flat part of the upper surface of the notch. The region R bounded by the closed contour $AB\Gamma_1CD\Gamma_2A$ is free of singularities and the J-integral calculated along $AB\Gamma_1CD\Gamma_2A$ is zero. We have

$$J_{AB\Gamma_1CD\Gamma_2A} = J_{B\Gamma_1C} + J_{CD} + J_{D\Gamma_2A} + J_{AB} = 0. \tag{5.18}$$

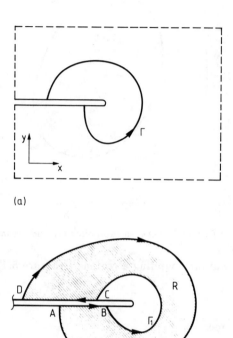

Figure 5.2. (a) Path Γ starting from the lower and ending to the upper face of a notch in a two-dimensional body. The flat notch surfaces are parallel to the x-axis. (b) Paths Γ_1 and Γ_2 around a notch tip.

When parts AB and CD of the notch surfaces that are parallel to the x-axis are traction free, $dy = 0$ and $T_k = 0$ and Equation (5.16) implies that

$$J_{CD} = J_{AB} = 0.$$

Equation (5.18) takes the form

$$J_{B\Gamma_1 C} + J_{D\Gamma_2 A} = 0$$

or

$$J_{B\Gamma_1 C} = J_{A\Gamma_2 D} \tag{5.19}$$

when the contour $A\Gamma_2 D$ is described in a counterclockwise sense. Equation (5.19) establishes the path independence of the J-integral defined by Equation

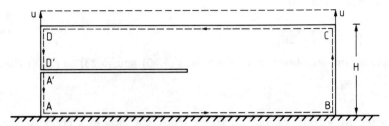

Figure 5.3. An infinite strip with a semi-infinite crack subjected to constant displacement u along its upper face. The J-integral is calculated along the dashed-line path shown.

(5.16) for notch problems.

(c) Example problems

The expressions of the J-integral for some characteristic crack problems will be obtained. Since J is path independent the integration path will be chosen in a convenient way to make the calculations easier.

As a first problem, we consider the case of an infinite strip of height H with a semi-infinite crack (Figure 5.3). The upper surface of the strip is subjected to a constant vertical displacement u. For the determination of the value of the J-integral the path $A'ABCDD'$ extended along the upper and lower surfaces of the strip up to infinity and traversing the strip perpendicularly to the crack is considered. J is calculated from

$$J = J_{A'ABCDD'} = J_{AB} + J_{BC} + J_{CD} + J_{DD'} + J_{A'A}. \tag{5.20}$$

We have: for path AB, CD: $dy = 0$, $u_1 = 0$, $u_2 = u$ =const, implying that

$$J_{AB} = J_{CD} = 0; \tag{5.21}$$

for path DD', $A'A$: the stresses vanish and $\partial u_{1,2}/\partial x = 0$, implying that

$$J_{DD'} = J_{A'A} = 0; \tag{5.22}$$

and for path BC: $\partial u_{1,2}/\partial x = 0$, implying that

$$J_{BC} = \int_0^H w|_{x \to \infty} \, dy. \tag{5.23}$$

For linear elastic material

$$w|_{x \to \infty} = \tfrac{1}{2}\sigma_y \epsilon_y. \tag{5.24}$$

Putting

$$\epsilon_y = \frac{u}{H}, \qquad \sigma_y = nE\epsilon_y = \frac{nEu}{H} \tag{5.25}$$

where

$$n = \frac{1}{1 - \nu^2} \tag{5.26a}$$

for plane stress and

$$n = \frac{1-\nu}{(1+\nu)(1-2\nu)} \tag{5.26b}$$

for plane strain, we obtain from Equations (5.20) and (5.23) to (5.26) that

$$J = \frac{nEu^2}{2H}. \tag{5.27}$$

As a second example problem, let us study the case of a crack in a mixed-mode stress field governed by the values of the three stress intensity factors K_I, K_{II}, K_{III}. For a circular path of radius r encompassing the crack tip, J from Equation (5.16) becomes

$$J = r \int_{-\pi}^{\pi} \left[w(r,\theta) \cos\theta - T_k(r,\theta) \frac{\partial u_k(r,\theta)}{\partial x} \right] d\theta. \tag{5.28}$$

Letting $r \to 0$ it is clear from Equation (5.28) that only the singular terms of the stress field around the crack tip contribute to J. Using the equations of the singular stresses and displacements for the three modes of deformation given in Section 2.5, the following equation for the J-integral is obtained:

$$J = \frac{\beta K_I^2}{E} + \frac{\beta K_{II}^2}{E} + \frac{1+\nu}{E} K_{III}^2 \tag{5.29}$$

where $\beta = 1$ for plane stress and $\beta = 1 - \nu^2$ for plane strain.

The third example concerns the case of the Dugdale model studied in Section 3.6. Referring to Figure 3.19, the integration path is taken along line ABC around the yield strip boundary from the lower side at $x = a$ passing through the tip B of the effective crack. We have

$$J = -\int_{a}^{a+c} \sigma_Y \frac{\partial}{\partial x_1}(u_2^+ - u_2^-) \, dx_1 = \int_0^{\delta} \sigma_Y \, d\delta$$

or

$$J = \sigma_Y \delta \tag{5.30}$$

where δ is the opening of the effective crack at the tip of the physical crack given by Equation (3.74). Equation (5.30) relates the J-integral to δ through the material yield stress σ_Y in uniaxial tension. For small-scale yielding δ is given by Equation (3.75) and the value of J, determined from Equation (5.29) under mode-I loading for plane stress conditions, is obtained.

5.4. Relationship between the J-integral and potential energy

A physical interpretation of the J-integral in terms of the rate of change of potential energy with respect to incremental change of crack size is sought [3.3]. The derivation concerns a linear or nonlinear elastic plane body with a crack of length a subjected to prescribed tractions and displacements along parts of its boundary. Tractions and displacements are assumed to be independent of crack

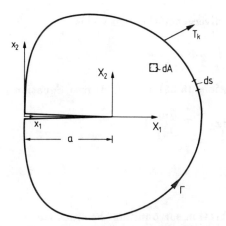

Figure 5.4. A two-dimensional cracked body.

length. The body is referred to a fixed system of Cartesian coordinates $x_1 x_2$ with the x_1-axis parallel to the crack faces (Figure 5.4). It is also assumed that the crack extends in a self-similar manner.

The potential energy $\Pi(a)$ of the body is given by

$$\Pi(a) = \int_A w \, dA - \int_\Gamma T_k u_k \, ds \tag{5.31}$$

where A is the area of the body and Γ its boundary.

Equation (5.31), by differentiation with respect to the crack length a, yields under the previous assumptions

$$\frac{d\Pi}{da} = \int_A \frac{dw}{da} dA - \int_\Gamma T_k \frac{du_k}{da} ds. \tag{5.32}$$

A new coordinate system $X_1 X_2$ attached to the crack tip is introduced:

$$X_1 = x_1 - a; \qquad X_2 = x_2 \tag{5.33a}$$

and

$$\frac{d}{da} = \frac{\partial}{\partial a} + \frac{\partial}{\partial X_1}\frac{\partial X_1}{\partial a} = \frac{\partial}{\partial a} - \frac{\partial}{\partial x_1}. \tag{5.33b}$$

Thus, Equation (5.32) takes the form

$$\frac{d\Pi}{da} = \int_A \left(\frac{\partial w}{\partial a} - \frac{\partial w}{\partial x_1}\right) dA - \int_\Gamma T_k \left(\frac{\partial u_k}{\partial a} - \frac{\partial u_k}{\partial x_1}\right) ds. \tag{5.34}$$

We have

$$\frac{\partial w}{\partial a} = \frac{\partial w}{\partial \epsilon_{ij}}\frac{\partial \epsilon_{ij}}{\partial a} = \sigma_{ij}\frac{\partial \epsilon_{ij}}{\partial a}$$

and by applying the principle of virtual work we obtain

$$\int_A \frac{\partial w}{\partial a} dA = \int_A \sigma_{ij}\frac{\partial \epsilon_{ij}}{\partial a} dA = \int_\Gamma T_k \frac{\partial u_k}{\partial a} ds. \tag{5.35}$$

Furthermore, the divergence theorem yields

$$\int_A \frac{\partial w}{\partial x_1} \, dA = \int_\Gamma w \, dx_2. \tag{5.36}$$

Introducing Equations (5.35) and (5.36) into Equation (5.44) we get

$$-\frac{d\Pi}{da} = \int_\Gamma w \, dx_2 - T_k \frac{\partial u_k}{\partial x_1} \, ds \tag{5.37}$$

or

$$J = -\frac{d\Pi}{da} \tag{5.38}$$

for any path of integration surrounding the crack tip.

Equation (5.38) expresses the J-integral as the rate of decrease of potential energy with respect to the crack length and holds only for self-similar crack growth. From Equations (5.38) and (4.21) we obtain that

$$J = G \tag{5.39}$$

for 'fixed-grips' or 'dead-load' loading during crack growth. For 'dead-load' loading Equation (5.39) holds only for a linear elastic response (see Section 4.3).

For a crack in a mixed-mode stress field the value of the J-integral given by Equation (5.29) represents the elastic strain energy release rate only for self-similar crack growth. However, according to the results of Section 4.11, a crack under mixed-mode loading does not extend along its own plane and therefore the value of energy release rate G given from Equation (5.29) is physically unrealistic.

5.5. J-integral fracture criterion

The J-integral can be viewed as a parameter which characterizes the state of affairs in the region around the crack tip. This argument is supported by the following fundamental properties of J which were proven in the previous sections:

   (i) J is path independent for linear or nonlinear elastic material response;
  (ii) J is equal to $-d\Pi/da$ for linear or nonlinear elastic material response;
 (iii) J is equal to G, and this property holds only for linear elastic material response under 'dead-load' loading during crack growth;
  (iv) J can easily be determined experimentally from the equation $J = -d\Pi/da$;
   (v) J is a measure of the intensity of the stress, strain and displacement fields in the HRR solution valid for nonlinear elastic material response (Equations (3.100) to (3.102));
  (vi) J can be related to the crack-tip opening displacement δ by a simple relation of the form $J = M\sigma_Y \delta$ (for the Dugdale model $M = 1$ (Equation (5.30))).

Based on these characteristic properties J has been suggested [3.14, 5.9] as an attractive candidate for a fracture criterion. Under opening-mode loading the criterion for crack initiation takes the form

$$J = J_c \tag{5.40}$$

where J_c is a material property for a given thickness under specified environmental conditions. Under plane strain conditions the critical value of J, J_{Ic}, is related to the plane strain fracture toughness K_{Ic} by (Equation (5.29))

$$J_{Ic} = \frac{1-\nu^2}{E} K_{Ic}^2. \tag{5.41}$$

The above properties of the J-integral which support its use as a fracture criterion were derived under elastic material response. Attempts have been made to extend the realm of applicability of the J-integral fracture criterion to ductile fracture where extensive plastic deformation and possibly stable crack growth precede fracture instability. Let us now discuss the potentialities and limitations of this approach, which is basically an extension of linear elastic fracture mechanics to account for large-scale inelastic effects.

Strictly speaking, the presence of plastic enclaves nullifies the path independence property of the J-integral. For any closed path surrounding the crack tip and taken entirely within the plastic zone or within the elastic zone, the necessary requirements for path independence (Equations (5.1) and (5.2)) are not satisfied. The stress is not uniquely determined by the strain, and the stress-strain constitutive equations relate strain increments to stresses and stress increments. In an effort to establish path independence for the J-integral the deformation theory of plasticity is invoked. This theory is a nonlinear elasticity theory and no unloading is permitted. Any solution based on the deformation theory of plasticity coincides exactly with a solution based on the flow (incremental) theory of plasticity under proportional loading (the stress components change in fixed proportion to one another). No unloading is permitted at any point of the plastic zone. Although, strictly speaking, the condition of proportional loading is not satisfied in practice it is argued that in a number of stationary crack problems under a single monotonically applied load the loading condition is close to proportionality. In support of this proposition a number of finite element solutions have appeared in the literature [5.10, 5.11].

According to the HRR small-strain solution the J-integral represents the amplitude of the singular stress and strain fields in a manner analogous to the stress intensity factor in the linear elastic case. The singular solution dominates in a region close to the crack tip and outside of the process or core region. Within the core region the continuum mathematical model can no longer adequately describe the physical behavior of the material which is being highly strained and may even be inhomogeneous. If the core region is small compared to the J-dominance region it can be argued that the state of affairs near the crack tip is controlled by the value of the J-integral.

A large amount of effort has been spent in assessing the J-dominance mainly for opening-mode loading under plane strain conditions. In this respect the separation of the crack faces near the tip, the so-called crack opening displacement (COD), plays an important role. According to Equation (3.102) which

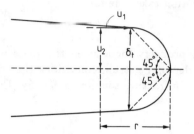

Figure 5.5. Definition of effective crack-tip opening displacement.

expresses the displacement components in the HRR solution, COD, being equal to $2u_2(r, \pi)$, tends to zero with r. Because of this property an effective crack-tip opening displacement δ_t was introduced and defined as the separation where two 45° lines from the deformed crack tip intercept the crack faces ([5.12] and Figure 5.5). We have, from Figure 5.5,

$$r - u_1(\pi) = u_2(\pi) = \tfrac{1}{2}\delta_t. \tag{5.42}$$

Equations (5.42) and (3.102) give for the value of r at the point of intersection

$$r = (\alpha\epsilon_Y)^{1/m}\bigl[\tilde{u}_1(\pi) + \tilde{u}_2(\pi)\bigr]^{(m+1)/m}\frac{J}{\sigma_Y I_n}. \tag{5.43}$$

If we introduce this value of r into Equation (3.102) we obtain for δ_t

$$\delta_t = \frac{d_n J}{\sigma_Y} \tag{5.44a}$$

where

$$d_n = \frac{2(\alpha\epsilon_Y)^{1/m}\bigl[\tilde{u}_1(\pi) + \tilde{u}_2(\pi)\bigr]^{1/m}\tilde{u}_2(\pi)}{I_n}. \tag{5.44b}$$

Equation (5.44a) provides a relationship between the J-integral and the effective crack-tip opening displacement δ_t. Values of d_n for plane stress and plane strain conditions were given by Shih [5.13]. For plane stress and perfectly plastic behavior $(m \to \infty)$ it was found that $d_n = 1$, which coincides with the result of the Dugdale model (Equation (5.30)). For plane strain d_n ranges from about 0.8 for $m \to \infty$ to about 0.1 for $m = 2$ with a very weak dependence on $\alpha\epsilon_Y$. From finite element studies based on finite strain incremental plasticity [5.11, 5.14] it was found that for distances from the crack tip greater than 2 or 3 times δ_t the state of affairs is governed by the small strain singular HRR solution. When R represents the radius of the region where the HRR solution describes the stress and strain fields accurately, the condition for J-dominance is

$$R > 3\delta_t. \tag{5.45}$$

R should also be greater than the radius of the core region encompassing the crack tip. From numerical solutions under small-scale yielding it was deduced that the HRR solution gives a good approximation of the state of affairs up to a distance of about $(1/5 - 1/4)c$ from the crack tip, where c is the extent of the plastic zone ahead of the crack according to the Irwin model (Equations (3.59) and (3.60), Section 3.5). For fully plastic conditions, when the plastic zone extends along the entire uncracked ligament b, R depends on the geometry of the specimen and can be represented as a fraction of b. From finite element computations based on finite strain incremental plasticity it was deduced for the bend specimen that [5.15]

$$R \simeq 0.07b \tag{5.46}$$

for most values of the hardening exponent m including perfect plasticity.

For the center-cracked tensile specimen the J-dominance region is smaller and for intermediate values of the hardening exponent ($m \simeq 0.1$) R can roughly be given by [5.16, 5.17]

$$R =\simeq 0.01b. \tag{5.47}$$

A condition for J-dominance can be established. For intermediate values of the hardening exponent, δ_t can be approximated by

$$\delta_t \simeq 0.6 \frac{J}{\sigma_Y}. \tag{5.48}$$

Combining Equations (5.45)–(5.48) the condition for J-dominance takes the form

$$\frac{b\sigma_Y}{J} > 25 \tag{5.49}$$

for the bend specimen, and

$$\frac{b\sigma_Y}{J} > 175 \tag{5.50}$$

for the center-cracked specimen.

Another point that needs clarification is the interpretation of J as the crack driving force (Section 5.4). This property holds only for linear elastic material response. However, plastic deformation is irreversible, and therefore loading a cracked body and then extending the crack does not give the same result as extending the crack first and then loading the body. Thus J cannot be identified with the energy available for crack extension in the presence of plastic deformation.

5.6. Experimental determination of the J-integral

This section deals with the experimental determination of the J-integral and its critical value J_{Ic}. The multiple-specimen method, the one-specimen method and the standard test method according to the ASTM specifications are briefly presented. Before proceeding to the details of these methods some general equations inherent in the test methods are derived.

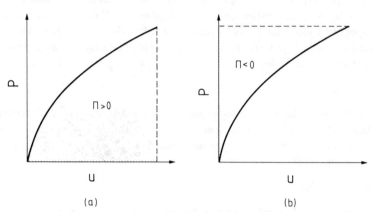

Figure 5.6. Potential energy shown as shaded area for (a) 'fixed-grips' and (b) 'dead-load' conditions.

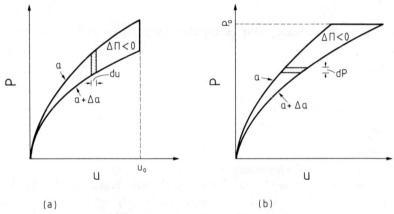

Figure 5.7. Load–displacement curves for crack lengths a and $a + \Delta a$ under (a) 'fixed-grips' and (b) 'dead-load' conditions.

(a) General equations

The experimental determination of J follows from Equation (5.38) according to which J is equal to the rate of decrease of potential energy (defined from Equation (5.31)) with respect to the crack length. Experiments are usually performed under either "fixed-grips" (prescribed displacement) or "dead-load" (prescribed load) conditions. In the load–displacement diagram the potential energy is equal to the area included between the load–displacement curve and the displacement or the load axis for fixed-grips or dead-load conditions, respectively (shaded areas of Figures 5.6(a) and 5.6(b)). Observe that the potential energy is positive for fixed-grips and negative for dead-load conditions.

J-Integral and crack opening displacement fracture criteria

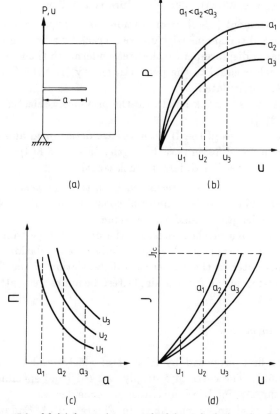

Figure 5.8. Multiple-specimen method for calculating the J-integral.

Consider the load–displacement curves corresponding to the crack lengths a and $a + \Delta a$ for "fixed-grips" or "dead-load" conditions in Figures 5.7(a) and 5.7(b). The area included between the two curves represents the value of $J\Delta a$. We have, from Equation (5.38) and Figures 5.7(a) and 5.7(b) for crack growth under fixed-grips

$$J = -\left(\frac{\partial \Pi}{\partial a}\right)_u = -\int_0^{u_0} \left(\frac{\partial P}{\partial a}\right)_u du \tag{5.51}$$

and under dead-load

$$J = -\left(\frac{\partial \Pi}{\partial a}\right)_P = -\int_0^{P_0} \left(\frac{\partial u}{\partial a}\right)_P dP \tag{5.52}$$

Equations (5.38), (5.51) and (5.52) form the basis for the experimental determination of J.

(b) *Multiple-specimen method*

This method is based on Equation (5.38) and was first introduced by Begley and Landes [5.9]. A number of identically loaded specimens with neighboring crack

lengths is used (Figure 5.8(a)). The procedure is as follows:
  (i) Load–displacement (Γ–u) records under fixed-grips are obtained for several precracked specimens with different crack lengths (Figure 5.8(b)). For given values of displacement u the area underneath the load–displacement record, which is equal to the potential energy Π of the body at that displacement, is calculated.
  (ii) Π is plotted versus crack length for the previously selected displacements (Figure 5.8(c)).
  (iii) The negative slopes of the Π–a curves are determined and plotted versus displacement for different crack lengths (Figure 5.8(d)). Thus the J–u curves are obtained for different crack lengths.

The critical value J_{Ic} of J is determined from the displacement at the onset of crack extension. Since J_{Ic} is a material constant the values of J_{Ic} obtained from different crack lengths should be the same.

The multiple-specimen method presents the disadvantage that several specimens are required to obtain the J versus displacement relation. Furthermore, accuracy problems enter in the numerical differentiation of the Π–a curves. A technique for determining J from a single test becomes very attractive and is described next.

(c) *Single-specimen method*

This method was first proposed by Rice et al. [5.18] and is based on Equation (5.51) or (5.52). The cases of a deeply cracked bend specimen, a compact specimen and a three-point bend specimen are considered.

For the cracked bend specimen shown in Figure 5.9(a), Equation (5.52) becomes

$$J = \int_0^M \left(\frac{\partial \theta}{\partial a}\right)_M dM \qquad (5.53)$$

where M is the applied moment per unit thickness and θ is the angle of relative rotation of the end sections of the specimen. The angle θ can be put in the form

$$\theta = \theta_{nc} + \theta_c \qquad (5.54)$$

where θ_{nc} represents the relative rotation of the uncracked specimen and θ_c is the additional rotation caused by the presence of the crack. It is now assumed that the ligament b is small compared to W so that the rotation θ_c is mainly due to deformation of the ligament. Equation (5.53) takes the form

$$J = \int_0^M \left(\frac{\partial \theta_c}{\partial a}\right)_M dM \qquad (5.55)$$

since θ_{nc} is independent of a. When L is large compared to W it can be assumed that θ_c depends only on M/M_0, where M_0 is the plastic limit moment. We have

$$\theta_c = f\left(\frac{M}{M_0}\right) \qquad (5.56)$$

J-Integral and crack opening displacement fracture criteria

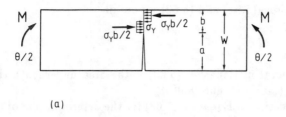

(a)

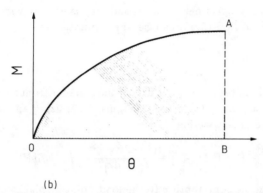

(b)

Figure 5.9. (a) A deeply cracked bend specimen and (b) bending moment M versus angle θ of relative rotation of the specimen end sections.

where

$$M_0 = \frac{\sigma_Y b^2}{4} \tag{5.57}$$

Equation (5.56) gives

$$\left(\frac{\partial \theta_c}{\partial a}\right)_M = -\frac{\partial \theta_c}{2b} = \frac{M}{M_0^2}\frac{dM_0}{db}\frac{\partial f}{\partial(M/M_0)} \tag{5.58}$$

and

$$\left(\frac{\partial \theta_c}{\partial M}\right)_a = \frac{1}{M_0}\frac{\partial f}{\partial(M/M_0)}. \tag{5.59}$$

From equations (5.58) and (5.59) we obtain

$$\left(\frac{\partial \theta_c}{\partial a}\right)_M = \frac{M}{M_0}\frac{dM_0}{db}\left(\frac{\partial \theta_c}{\partial M}\right)_a. \tag{5.60}$$

Substituting Equation (5.60) into Equation (5.55) we obtain

$$J = \frac{1}{M_0}\frac{dM_0}{db}\int_0^{\theta_c} M\,d\theta_c. \tag{5.61}$$

Equation (5.57) renders

$$\frac{1}{M_0}\frac{dM_0}{db} = \frac{2}{b}. \tag{5.62}$$

From Equations (5.61) and (5.62) we obtain

$$J = \frac{2}{b} \int_0^{\theta_c} M \, d\theta_c \qquad (5.63)$$

where the integral in Equation (5.63) is the area underneath the $M = M(\theta_c)$ curve and the θ_c-axis (Figure 5.9(b)).

Direct application of Equation (5.63) for the determination of the J-integral is not possible since in an experiment the total rotation angle θ is measured. The critical value of J, J_{Ic}, can be obtained by determining the area under the M versus angle θ up to the point of crack extension and subtracting the area of a similar but uncracked specimen. The second area is usually very small compared to the first and Equation (5.63) can be approximated as

$$J = \frac{2}{b} \int_0^{\theta} M \, d\theta. \qquad (5.64)$$

Equation (5.64) enables the direct determination of J from a single experiment.

For the case of a compact specimen (Figure 5.10) using analogous derivations as in the bend specimen, we find [5.19]

$$J = \frac{2}{b} \frac{1+\beta}{1+\beta^2} \int_0^{\delta_p} P \, d\delta_p + \frac{2}{b} \frac{\beta(1-2\beta-\beta^2)}{(1+\beta^2)^2} \int_0^P \delta_p \, dP \qquad (5.65)$$

where δ_p is the plastic contribution to the load-point displacement and β is given by

$$\beta = 2\left[\left(\frac{a}{b}\right)^2 + \frac{a}{b} + \tfrac{1}{2}\right]^{1/2} - 2\left(\frac{a}{b} + \tfrac{1}{2}\right). \qquad (5.66)$$

For $a/W > 0.5$ it has been found that the total displacement δ instead of δ_p can be used in Equation (5.65). Furthermore, for deeply cracked specimens $\beta \simeq 0$ and Equation (5.65) becomes

$$J = \frac{2}{b} \int_0^{\delta} P \, d\delta \qquad (5.67)$$

which is analogous to Equation (5.64).

For the deeply cracked three-point bend specimen J is again given from Equation (5.67).

For further information on the results of this section the reader is referred to references [5.20–5.23].

(d) *Standard test method*

ASTM [5.24] issued a standard test method for determining J_{Ic}, the plane strain value of J at initiation of crack growth for metallic materials. The recommended specimens are the three-point bend specimen and the compact specimen that contain deep initial cracks. The specimens are loaded to special fixtures and applied loads and load-point displacements are simultaneously recorded during the test. For a valid J_{Ic} value the crack ligament b and the specimen thickness B must

J-Integral and crack opening displacement fracture criteria

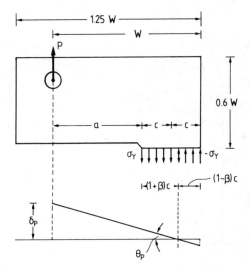

Figure 5.10. A deeply cracked compact specimen at plastic collapse.

be greater than $25 J_{Ic}/\sigma_y$ (see relation (5.49) for conditions of J-dominance). The initial crack length for the three-point bend specimen must be at least $0.5W$ but not greater than $0.75W$, where W is the specimen width. The overall specimen length is $4.5W$ and the specimen thickness is $0.5W$. The geometry of the compact specimen is shown in Figure 5.10 where the initial crack length a is taken to satisfy $0.5W < a < 0.75W$ and the specimen thickness is $0.5W$.

To determine the value of J_{Ic} that corresponds to the onset of slow stable crack propagation the following procedure is followed. The J-integral is determined for the bend specimen from Equation (5.64) and for the compact specimen from Equation (5.65), which can be approximated as

$$J = \frac{2}{b}\frac{1+\beta}{1+\beta^2} \int_0^\delta P \, d\delta \tag{5.68}$$

where β is given from Equation (5.66). J is plotted against physical crack growth Δa_p using at least four data points within specified limits of crack growth (Figure 5.11). A straight line which better fits the experimental points is drawn and the point at which it intersects the blunting line

$$J = 2\sigma_Y \Delta a \tag{5.69}$$

is determined. The value of J which corresponds to the point of intersection is J_{Ic}.

The blunting line approximates the apparent crack advance due to crack-tip blunting in the absence of slow, stable crack tearing. The line is defined on the assumption that, prior to tearing, the crack advance is equal to one half of the crack-tip opening displacement ($\Delta a = 0.5\delta$). Then Equation (5.30) which is based on the Dugdale model results in Equation (5.69).

Two additional offset lines parallel to the blunting line and starting from the

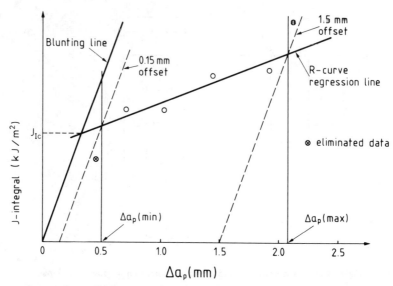

Figure 5.11. Determination of J_{Ic} according to ASTM standards [5.24].

points $\Delta a_p = 0.15$ mm and 1.5 mm are drawn. For a valid test all data should be placed inside the area enclosed by the two parallel offset lines. Data outside these limits are not valid. The valid data points are used to determine the final linear regression line.

The value of J_{Ic} can also be used to obtain an estimate of K_{Ic} from Equation (5.29) which, for opening-mode, takes the form

$$K_{Ic}^2 = J_{Ic} E. \tag{5.70}$$

Equation (5.70) is used in situations where large specimen dimensions are required for a valid K_{Ic} test according to the ASTM specifications (Section 4.8).

5.7. Stable crack growth studied by the J-integral

The crack growth resistance curve method for the study of crack growth under small-scale yielding developed in Section 4.10 has been extended to large-scale yielding using J instead of G or K (Figure 5.12). The J_R-resistance curve is assumed to be a geometry-independent material property for given thickness and environmental conditions. During crack growth the J-integral, which is interpreted as the crack driving force, must be equal to the material resistance to crack growth. Stability of crack growth requires

$$J(P, a) = J_R(\Delta a) \tag{5.71a}$$

$$\frac{\mathrm{d}J(P, a)}{\mathrm{d}a} < \frac{\mathrm{d}J_R(\Delta a)}{\mathrm{d}a} \tag{5.71b}$$

where P represents the applied loading. Crack growth becomes unstable when

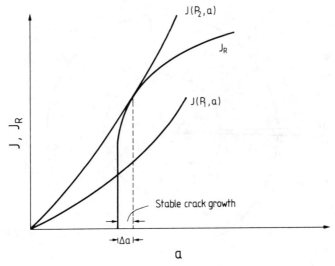

Figure 5.12. Stable crack growth by J-resistance curve analysis.

the inequality (5.71b) is reversed. The methodology developed in Section 4.10 for the study of slow stable crack propagation can equally be applied to the J-resistance curve method.

Paris *et al.* [5.25] introduced the nondimensional tearing moduli

$$T = \frac{E}{\sigma_0^2}\frac{dJ}{da}, \qquad T_R = \frac{E}{\sigma_0^2}\frac{dJ_R}{da} \qquad (5.72)$$

where σ_0 is an appropriate yield stress in tension when the material has strain hardening. σ_0 is usually taken equal to $(\sigma_Y + \sigma_u)/2$ with σ_u being the ultimate stress of the material in tension. The tearing modulus T_R is a material parameter that can reasonably be assumed temperature independent. Using the tearing moduli the stability condition becomes

$$T < T_R. \qquad (5.73)$$

The above analysis, which uses the J-integral for the study of slow stable crack growth, is seriously questioned. Crack growth involves some elastic unloading and, therefore, nonproportional plastic deformation near the crack tip. However, the J-integral is based on deformation plasticity theory which is incapable of adequately modeling both of these aspects of plastic crack propagation. The conditions for J-controlled crack growth have first been studied by Hutchinson and Paris [5.26].

The argument for J-controlled crack growth requires that the region of elastic unloading and nonproportional plastic loading be well contained within the J-dominance zone (Figure 5.13). Let us denote, by R, the characteristic radius of the J-dominance zone. Since the wake of elastic unloading and the region of nonproportional plastic loading are of the order of crack growth Δa, one

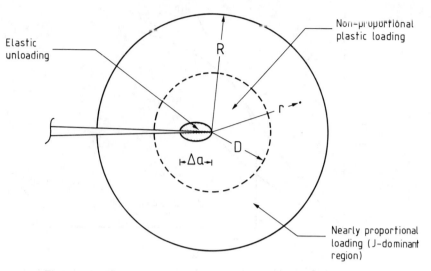

Figure 5.19. Elastic unloading, nonproportional plastic loading and J-dominant region around the tip of a growing crack.

condition for J-controlled growth is

$$\Delta a \ll R. \tag{5.74}$$

Consider next the strain increments due to increments of J and a. Putting Equation (3.101) in the form

$$\epsilon_{ij} = k_n \left(\frac{J}{r}\right)^{n/(n+1)} \tilde{\epsilon}_{ij}(\theta) \tag{5.75}$$

the strain increments are

$$d\epsilon_{ij} = \frac{n}{n+1} k_n \left(\frac{J}{r}\right)^{n/(n+1)} \frac{dJ}{J} \tilde{\epsilon}_{ij}(\theta) - k_n J^{n/(n+1)} \, da \frac{\partial}{\partial x}\left[r^{-n/(n+1)} \tilde{\epsilon}_{ij}(\theta)\right]. \tag{5.76}$$

Using

$$\frac{\partial}{\partial x_1} = \cos\theta \frac{\partial}{\partial r} - \frac{\sin\theta}{r}\frac{\partial}{\partial \theta}$$

Equation (5.76) becomes

$$d\epsilon_{ij} = k_n \left(\frac{J}{r}\right)^{n/(n+1)} \left[\frac{n}{n+1}\frac{dJ}{J}\tilde{\epsilon}_{ij}(\theta) + \frac{da}{r}\tilde{\beta}_{ij}(\theta)\right] \tag{5.77}$$

where

$$\tilde{\beta}_{ij} = \frac{n}{n+1}\cos\theta\,\tilde{\epsilon}_{ij} + \sin\theta\,\frac{\partial \tilde{\epsilon}_{ij}}{\partial \theta}. \tag{5.78}$$

The first term in the brackets of Equation (5.77) corresponds to proportional loading, while the second term corresponds to nonproportional loading. Since $\tilde{\epsilon}_{ij}$ and $\tilde{\beta}_{ij}$ are of comparable magnitude, the first term dominates the second if

$$\frac{dJ}{J} \gg \frac{da}{r}. \tag{5.79}$$

Define a length quantity D as

$$\frac{1}{D} = \frac{dJ}{da}\frac{1}{J} \tag{5.80}$$

where D can be interpreted as the crack growth just beyond initiation associated with a doubling of J above J_{Ic}. If, further,

$$D \ll R \tag{5.81}$$

then there exists an annular region

$$D \ll r < R \tag{5.82}$$

in which the plastic loading is predominantly proportional and the HRR solution dominates.

For a fully yielded specimen

$$D \ll b \tag{5.83}$$

where b is the crack ligament. The relation (5.80) yields

$$\omega = \frac{b}{D} = \frac{b\,dJ}{J\,da} \gg 1. \tag{5.84}$$

Relation (5.84) expresses the requirement for J-controlled crack growth. Finite element analyses and experimental results for crack initiation and stable crack growth have been reported by Shih et al. [5.27]. For further studies related to this subject the reader should consult references [5.28–5.38].

5.8. Mixed-mode crack growth

Attempts have been made to study the problem of crack growth under mixed-mode loading using path-independent line integrals. Consider the vector $\mathbf{Q}$ for the two-dimensional elastic crack problem

$$\mathbf{Q} = Q_1\mathbf{i} + Q_2\mathbf{j} = \mathbf{J} = J_1\mathbf{i} + J_2\mathbf{j} \tag{5.85}$$

where Q_1 and Q_2 are given by Equation (5.6). J_1 is equal to the J-integral according to Equation (5.14). J_2 is expressed by

$$J_2 = \int_\Gamma \left(W n_2 - T_k u_{k,2}\right) ds. \tag{5.86}$$

For a crack in a mixed-mode stress field governed by the values of stress intensity factors K_I and K_{II}, J_1 is given by

$$J_1 = \frac{\kappa + 1}{8\mu}\left(K_I^2 + K_{II}^2\right) \tag{5.87}$$

according to Equation (5.29).

The integral J_2 can be calculated in a manner analogous to J_1, that is,

$$J_2 = -\frac{\kappa + 1}{4\mu} K_I K_{II}. \tag{5.88}$$

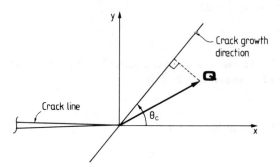

Figure 5.14. Projection of **Q** vector along the direction of crack growth.

Consider the projection $Q(\theta)$ of the vector **Q** along a direction making an angle θ with the x-axis (Figure 5.14). We have

$$J(\theta) = J_1 \cos\theta + J_2 \sin\theta. \tag{5.89}$$

Initiation of crack growth is governed by the following hypotheses:
(i) The crack extends along the radial direction $\theta = \theta_c$ on which $J(\theta)$ becomes maximum.
(ii) Fracture starts when that maximum of $J(\theta)$ reaches the value 2γ, where γ is the energy required to form a unit of new surface in the Grifith theory.

These hypotheses may be expressed mathematically as

$$\frac{\partial J(\theta)}{\partial \theta} = 0, \qquad \frac{\partial^2 J(\theta)}{\partial \theta^2} < 0 \tag{5.90a}$$

$$J(\theta_c) = 2\gamma. \tag{5.90b}$$

Using Equations (5.87) and (5.88) the crack growth direction θ_c is expressed by

$$\theta_c = \tan^{-1}\left(\frac{-2K_I K_{II}}{K_I^2 + K_{II}^2}\right). \tag{5.91}$$

The theory is applied to the case of a crack of length $2a$ in an infinite plate which subtends an angle β with the direction of applied uniform stress σ. The stress intensity factors K_I and K_{II} are given by

$$K_I = \sigma\sqrt{\pi a}\,\sin^2\beta \tag{5.92a}$$

$$K_{II} = \sigma\sqrt{\pi a}\,\sin\beta\cos\beta \tag{5.92b}$$

Equation (5.91) yields for the crack extension angle θ_c

$$\tan\theta_c = -\sin 2\beta \tag{5.93}$$

while from Equation (5.90b) we obtain for the critical stress σ_c

$$\sigma_c = \frac{16\mu\gamma}{(\kappa+1)\pi a}\frac{1}{\sin^2\beta\sqrt{1+\sin^2 2\beta}}. \tag{5.94}$$

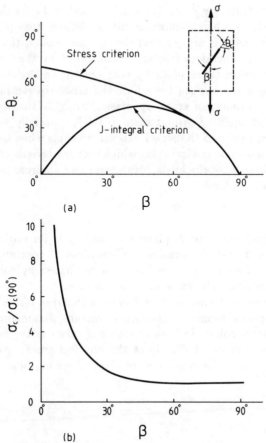

Figure 5.15. (a) Crack extension angle $-\theta_c$ and (b) normalized critical stress $\sigma_c/\sigma_c(90°)$ versus crack inclination angle β according to the J-integral criterion. The results of the maximum circumferential stress criterion for the crack extension are also shown.

The crack extension angle θ_c and the critical stress σ_c are plotted in Figure 5.15 against the crack inclination angle β. The predictions of the maximum circumferential stress theory are also shown in Figure 5.15(a). Observe that the results of the two theories are close to each other for large crack angles, while they deviate substantially for small crack angles. Since the experimental results are very close to the predictions of the maximum stress theory, it can be concluded that the study of mixed-mode crack growth cannot be adequately modeled by the J-approach.

Results related to this section can be found in references [5.39–5.43].

5.9. Crack opening displacement (COD) fracture criterion

(a) Outline of the method

The concept of a critical crack opening displacement as a fracture criterion was introduced independently by Wells [5.44] and Cottrell [5.45] for the study of crack initiation in situations where significant plastic deformation precedes fracture. Under such conditions it is argued that the stresses around the crack tip reach the critical value and therefore fracture is controlled by the amount of plastic strain. Crack extension takes place by void growth and coalescence with the original crack tip, a mechanism for which the crack-tip strain is responsible. A measure of the amount of crack-tip plastic strain is the separation of the crack faces or crack opening displacement (COD), especially very close to the crack tip. It is thus expected that crack extension begins when the crack opening displacement reaches some critical value which is characteristic of the material at a given temperature, plate thickness, strain rate and environmental conditions. The criterion takes the form

$$\delta = \delta_c \tag{5.95}$$

where δ is the crack opening displacement and δ_c is its critical value. It is assumed that δ_c is a material constant independent of specimen configuration and crack length. This assumption has been confirmed by some experiments which indicate that δ has the same value at fracture.

To obtain an analytical expression of Equation (5.95) in terms of applied load, crack length, specimen geometry and other fracture parameters, the Irwin or Dugdale models are invoked. In both models δ is taken as the separation of the faces of the effective crack at the tip of the physical crack. According to the Irwin model δ is given under conditions of plane stress by Equation (3.63). δ_c is expressed by

$$\delta_c = \frac{4}{\pi} \frac{K_c^2}{E\sigma_Y}. \tag{5.96}$$

For the Dugdale model δ is given by Equation (3.74) which, for small values of σ/σ_Y, reduces to Equation (3.75). Under such conditions δ_c is given by

$$\delta_c = \frac{K_c^2}{E\sigma_Y}. \tag{5.97}$$

Equations (3.63) and (3.74) combined with Equation (4.48) yield, respectively,

$$G = \frac{\pi}{4}\sigma_Y \delta \tag{5.98a}$$

$$G = \sigma_Y \delta. \tag{5.98b}$$

Equations (5.98) express the strain energy release rate in terms of the crack opening displacement. From Equations (5.96) to (5.98) it follows that under conditions of small-scale yielding the fracture criteria based on the stress intensity factor, the strain energy release rate and the crack opening displacement are equivalent.

A different definition of crack opening displacement as 'the vertical separation of the points subtended by an angle of 90° from the current crack tip and its relation to the J-integral' was given in Section 5.5.

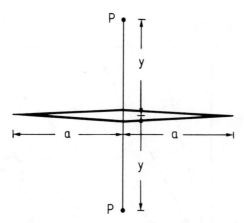

Figure 5.16. Points P at equal distances y from a crack of length 2a.

(b) COD design curve

The objective of the COD design curve is to establish a relationship between the crack opening displacement and the applied load and crack length. Knowledge of the critical crack opening displacement would thus enable determination of the maximum permissible stress or the maximum allowable crack length in a structure. For determining the COD design curve an analytical model should be selected. For the case of the Dugdale model (Section 3.6), Burdekin and Stone [5.46] obtained the following equation which expresses the overall strain ϵ ($\epsilon = u/2y$) of two equidistant points P (Figure 5.16):

$$\frac{\epsilon}{\epsilon_Y} = \frac{2}{\pi}\left[2n\coth^{-1}\left[\frac{1}{n}\sqrt{\frac{k^2+n^2}{1-k^2}}\right] + \right.$$

$$\left. +(1-\nu)\cot^{-1}\sqrt{\frac{k^2+n^2}{1-k^2}} + \nu\cos^{-1}k\right] \quad (5.99a)$$

with

$$n = \frac{a}{y}, \quad k = \cos\left(\frac{\pi\sigma}{2\sigma_Y}\right), \quad \epsilon_Y = \frac{\sigma_Y}{E}. \quad (5.99b)$$

In this equation a is half the crack length, y is the distance of point P from the crack, σ_Y is the yield stress of the material in tension and σ is the applied stress.

Define the dimensionless crack opening displacement Φ by

$$\Phi = \frac{\delta}{2\pi\epsilon_Y a} \quad (5.100)$$

where δ is given for the Dugdale model from Equation (3.74). Eliminating stress σ from Equations (5.99) and (5.100), the design curves shown in Figure 5.17 are obtained. This figure enables the determination of the maximum allowable

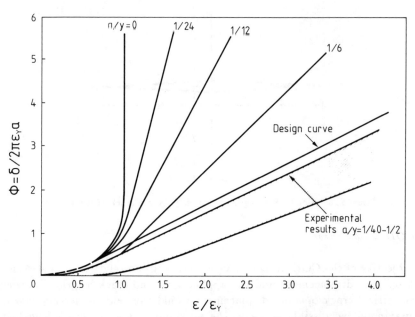

Figure 5.17. Design curves according to crack opening displacement criterion [5.48].

overall strain in a cracked structure at position a/y when the critical crack opening displacement δ_c and the crack length a are known.

Experimental data, however, relating δ_c and the maximum strain at fracture, fell into a single scatter band of Figure 5.17 for a wide range of a/y values. It is thus shown that the design curve based on the Dugdale model is far from reality. An empirical equation was obtained [5.47] to describe the experimental data of Figure 5.17. This equation has the form

$$\Phi = \begin{cases} \left(\frac{\epsilon}{\epsilon_Y}\right)^2, & \left(\frac{\epsilon}{\epsilon_Y}\right) < 0.5 \\ \frac{\epsilon}{\epsilon_Y} - 0.25, & \left(\frac{\epsilon}{\epsilon_Y}\right) > 0.5 \end{cases} \qquad (5.101)$$

For small cracks ($a/W < 0.1$, W being the plate width) and applied stresses below the yield value, Dawes [5.48] argued that

$$\frac{\epsilon}{\epsilon_Y} = \frac{\sigma}{\sigma_Y}. \qquad (5.102)$$

From Equations (5.101) and (5.102) the maximum allowable crack length $a_{\max}$ is

$$a_{\max} = \begin{cases} \dfrac{\delta_c E \sigma_Y}{2\pi\sigma^2}, & \dfrac{\sigma}{\sigma_Y} < 0.5 \\ \dfrac{\delta_c E}{2\pi(\sigma - 0.25\sigma_Y)}, & 0.5 < \dfrac{\sigma}{\sigma_Y} < 1. \end{cases} \qquad (5.103)$$

(c) *Stable crack growth*

The concept of the R-curve method which was introduced in Sections 4.10 and 5.7 for the study of slow stable crack growth in terms of the stress intensity factor and the J-integral has been extended to the crack opening displacement. A rising R-curve is obtained when the COD at the original crack tip δ_f is plotted against crack extension. However, a physical relationship between the opening at the original tip that is left behind as the crack propagates and the fracture toughness of the material at the advancing tip hardly exists. The value of COD at the advancing crack tip δ_a has been introduced to study stable crack growth. It was obtained experimentally that δ_a remains constant during crack propagation at a value less than the initiation value δ_i. For steel it was found that $\delta_i/\delta_a \simeq 4$.

Rice and Sorensen [5.49] studied the stress and deformation fields at growing plane strain crack tips in elastic-perfectly plastic solids. From an asymptotic analysis it was obtained that a crack-tip stress state similar to that of the classical Prandtl field exists. Rice et al. [5.50] found it necessary to include a zone of elastic unloading between the centered fan region and the trailing constant stress plastic region. The following expression for the rate of opening displacement δ at distance r from the growing tip was obtained

$$\dot{\delta} = \alpha \frac{\dot{J}}{\sigma_Y} + \beta \left(\frac{\sigma_Y}{E}\right) \dot{a} \ln \left(\frac{R}{r}\right), \quad r \to 0 \tag{5.104}$$

where $\beta = 5.08$ for a Poisson's ratio $\nu = 0.3$; α is a dimensionless term whose value is approximately equal to $1/d_n$, where d_n is defined by Equation (5.44); and R is about 15 to 30 per cent larger than the maximum extent of the plastic zone. The dot denotes differentiation with respect to time.

When the crack grows so that a increases continuously with J, it is obtained from the asymptotic integration of Equation (5.104):

$$\frac{E\delta}{\beta \sigma_0 R} = \left[\frac{\alpha}{\beta} T + \ln \left(\frac{eR}{r}\right)\right] \frac{r}{R} \tag{5.105}$$

where e is the base of the natural logarithm and T is the tearing modulus given by Equation (5.72). When the critical crack opening displacement condition $\delta = \delta_c$ at $r = r_m$ is introduced into Equation (5.105) it takes the form

$$\frac{\delta_c}{r_m} = \frac{\alpha}{\sigma_Y} \frac{\mathrm{d}J}{\mathrm{d}a} + \beta \frac{\sigma_Y}{E} \ln \left(\frac{eR}{r_m}\right). \tag{5.106}$$

Since R is taken as a function of J, Equation (5.106) can be regarded as a first-order differential equation which determines the form of variation of J with a during crack growth.

For further information on the results of this section the reader is referred to references [5.51–5.53].

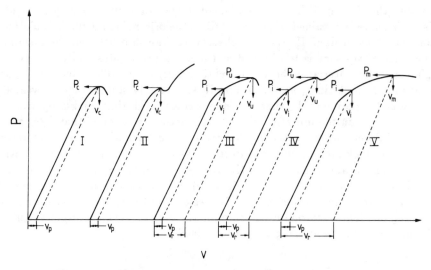

Figure 5.18. Different types of load–clip gauge displacement records according to British standards.

(d) Standard COD test

Determination of the critical crack opening displacement is the subject of the British Standard BS 5762 [5.54]. We use the edge-notched three-point bend specimen which has been described in Section 4.8 for the determination of the fracture toughness K_{Ic}. The specimen thickness B is taken about equal to the application thickness and the beam width W is twice the thickness ($W = 2B$). The specimen is fatigue precracked as in the K_{Ic} standard test with the exception that a straight starter notch is recommended rather than a chevron notch.

The load versus crack mouth displacement is recorded from the experiment. Clip gages are usually installed at a distance z from the specimen surface. The load–displacement records fall into the five categories shown in Figure 5.18. Four categories of crack-tip opening displacement are defined in relation to Figure 5.18: δ_c at the onset of unstable crack growth (case I) or pop-in (case II) when no stable crack growth is observed. δ_u at the onset of unstable crack growth (case III) or pop-in (case IV) when stable crack growth takes place prior to instability. δ_i at the commencement of stable crack growth (cases III, IV and V) and δ_m at the maximum load P_m (case V) when it is preceded by stable crack growth.

The critical crack-tip opening displacement δ (δ_i, δ_c, δ_u or δ_m) is determined from the test record by

$$\delta = \delta_e + \delta_p \qquad (5.107a)$$

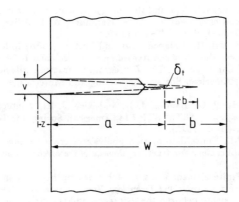

Figure 5.19. Definition of δ_t and its relation to v.

where

$$\delta_e = \frac{K_I^2(1-\nu^2)}{2\sigma_Y E}, \qquad \delta_p = \frac{v_p rb}{rb + a + z} \qquad (5.107b)$$

in which v_p is the plastic component of the measured displacement v (Figure 5.18) and the quantities r, b, a and z are shown in Figure 5.19.

In Equation (5.107a) the crack opening displacement δ is equal to elastic δ_e plus the plastic δ_p contribution. The elastic part δ_e is calculated from Equation (3.75) of the Dugdale model which is modified for plane strain and a plastic constraint factor equal to 2. The plastic part δ_p is obtained by assuming that the crack ligament $b = W - a$ acts as a plastic hinge with a rotation point at a distance rb from the crack tip. From experiments it was found that the value of the rotation factor r lies between 0.33 and 0.48. A nominal value of 0.4 is used for the standard test. Experimental evidence in China [5.55, 5.56] indicates that a value of $r = 0.45$ is more appropriate.

For further information on the standard COD test consult references [5.57, 5.58].

References

5.1. Günther, W., Über einige Randintegrale der Elastomechanik, *Abhandlungen der Braunschweigischen Wissenschaftlichen Gesellschaft* **14**, 53–72 (1962).
5.2. Eshelby, J. D., The continuum theory of lattice defects, in *Solid State Physics* (eds F. Seitz and D. Turnbull), Vol. 3, Academic Press, New York, pp. 79–144 (1956).
5.3. Rice, J. R., A path independent integral and the approximate analysis of strain concentration by notches and cracks, *Journal of Applied Mechanics, Trans. ASME* **35**, 379–386 (1968).
5.4. Sanders, J. L. Jr, On the Griffith–Irwin theory, *Journal of Applied Mechanics, Trans. ASME* **27**, 352–353 (1960).
5.5. Cherepanov, C. P., Crack propagation for continuous media, *Journal of Applied Mathematics and Mechanics (PMM)* **31**, 503–512 (1967).
5.6. Knowles, J. K. and Stenberg, E., On a class of conservation laws in linearized and finite elastostatics, *Archives for Rational Mechanics and Analysis* **44**, 187–211 (1972).

5.7. Eshelby, J. D., Energy relations and the energy-momentum tensor in continuum mechanics, in *Inelastic Behavior of Solids* (eds M. F. Kanninen, W. F. Adler, A. R. Rosenfield and J. I. Jaffee), McGraw-Hill, pp. 77–115 (1970).

5.8. Kröner, E. and Zorski, H., Balance Laws and Surface (Path) Independent Integrals for Bodies with Cracks, *Theoretical and Applied Fracture Mechanics* 1, 249–256 (1984).

5.9. Begley, J. A. and Landes, J. D., The J integral as a fracture criterion, *Fracture Toughness*, ASTM STP 514, American Society for Testing and Materials, Philadelphia, pp. 1–23 and 24–39 (1972).

5.10. Bucci, R. J., Paris, P. C., Landes, J. D. and Rice, J. R., J-integral estimation procedures, *Fracture Toughness*, ASTM STP 514, American Society for Testing and Materials, Philadelphia, pp. 40–69 (1972).

5.11. McMeeking, R. M., Finite deformation analysis of crack-tip opening in elastic–plastic materials and implications for fracture, *Journal of the Mechanics and Physics of Solids* 25, 357–381 (1977).

5.12. Tracey, D. M., Finite element solutions for crack-tip behavior in small-scale yielding, *Journal of Engineering Materials and Technology, Trans. ASME* 98, 146–151 (1976).

5.13. Shih, C. F., Relationship between the J-integral and the crack opening displacement for stationary and extending cracks, *Journal of the Mechanics and Physics of Solids* 29, 305–326 (1981).

5.14. Rice, J. R., McMeeking, R. M., Parks, D. M. and Sorensen, E. P., Recent finite element studies in plasticity and fracture mechanics, *Computer Methods in Applied Mechanics and Engineering*, 17/18, 411–442 (1979).

5.15. Shih, C. F. and German, M. D., Requirements for a one parameter characterization of crack tip fields by the HRR singularity, *International Journal of Fracture* 17, 27–43 (1981).

5.16. McMeeking, R. M. and Parks, D. M., On criteria for J-dominance of crack tip fields in large scale yielding, in *Elastic–Plastic Fracture*, ASTM STP 668, American Society for Testing and Materials, Philadelphia, pp. 175–194 (1979).

5.17. Hutchinson, J. W., Fundamentals of the phenomenological theory of nonlinear fracture mechanics, *Journal of Applied Mechanics, Trans. ASME* 50, 1042–1051 (1983).

5.18. Rice, J. P., Paris, P. C. and Merkle, J. G., Some further results of J-integral analysis and estimates, in *Progress in Flaw Growth and Fracture Toughness Testing*, ASTM STP 536, American Society for Testing and Materials, Philadelphia, pp. 213–245 (1973).

5.19. Merkle, J. G. and Corten, H. T., A J-integral analysis for the compact specimen, considering axial force as well as bending effects, *Journal of Pressure Vessel Technology* 96, 286–292 (1974).

5.20. Landes, J. D., Walker, H. and Clarke, G. A., Evaluation of estimation procedures used in J-integral testing, in *Elastic–Plastic Fracture*, ASTM STP 668, American Society for Testing and Materials, Philadelphia, pp. 266–287 (1979).

5.21. McMeeking, R. M., Estimates of the J-integral for elastic–plastic specimens in large scale yielding, *Journal of Engineering Materials and Technology, Trans. ASME* 106, 278–284 (1984).

5.22. Adams, N. J. I. and Munro, H. G., A single test method for evaluation of the J-integral as a fracture parameter, *Engineering Fracture Mechanics* 6, 119–132 (1974).

5.23. Sönnerlind, H. and Kaiser, S., The J-integral for a SEN specimen under nonproportionally applied bending and tension, *Engineering Fracture Mechanics* 24, 637–646 (1986).

5.24. Standard test method for J_{Ic}, a measure of fracture toughness, *ASTM Annual Book of Standards*, Part 10, E813–81, American Society for Testing and Materials, Philadelphia, pp. 822–840 (1982).

5.25. Paris, P. C., Tada, H., Zahoor, A. and Ernst, H., The theory of instability of the tearing mode of elastic–plastic crack growth, in *Elastic–Plastic Fracture*, ASTM STP 668, American Society for Testing and Materials, Philadelphia, pp. 5–36 and 251–265 (1979).

5.26. Hutchinson, J. W. and Paris, P. C., Stability analysis of J-controlled crack growth, in *Elastic–Plastic Fracture*, ASTM STP 668, American Society for Testing and Materials, Philadelphia, pp. 37–64 (1979).

5.27. Shih, C. F., deLorenzi, H. G. and Andrews, W. R., Studies on crack initiation and stable crack growth, in *Elastic–Plastic Fracture*, ASTM STP 668, American Society for Testing and Materials, Philadelphia, pp. 65–120 (1979).

5.28. Tada H., Paris, P. C. and Gamble, R. M., A stability analysis of circumferential cracks for reactor piping systems, In *Fracture Mechanics – Twelfth Conference*, ASTM STP 700, American Society for Testing and Materials, Philadelphia, pp. 296–313 (1980).

5.29. Bamford, W. H. and Bush, A. J., Fracture behavior of stainless steel, in *Elastic–Plastic Fracture*, ASTM STP 668, American Society for Testing and Materials, Philadelphia, pp.

553–577 (1979).
5.30. Popelar, C. H., Pan, J. and Kanninen, M. F., A tearing instability analysis for strain hardening materials, in *Fracture Mechanics - Fifteenth Symposium*, ASTM STP 833, American Society for Testing and Materials, Philadelphia, pp. 699–720 (1984).
5.31. Pan, J., Ahmad, J., Kanninen, M. F. and Popelar, C. H., Application of a tearing instability analysis for strain hardening materials to a circumferentially cracked pipe in bending, in *Fracture Mechanics - Fifteenth Symposium*, ASTM STP 833, American Society for Testing and Materials, Philadelphia, pp. 721–745 (1984).
5.32. Ernst, H. A., Paris, P. C. and Landes, J. D., Estimations on J-integral and tearing modulus T from a single specimen test record, in *Fracture Mechanics - Thirteenth Conference*, ASTM STP 743, American Society for Testing and Materials, Philadelphia, pp. 476–502 (1981).
5.33. Joyce, J. A. and Vassilaros, M. G., An experimental evaluation of tearing instability using the compact specimen, in *Fracture Mechanics - Thirteenth Conference*, ASTM STP 743, American Society for Testing and Materials, Philadelphia, pp. 525–542 (1981).
5.34. Joyce, J. A., Instability testing of compact and pipe specimens utilizing a test system made compliant by computer control, in *Elastic-Plastic Fracture - Second Symposium*, Vol. 2, ASTM STP 803, American Society for Testing and Materials, Philadelphia, pp. 439–463 (1983).
5.35. Landes, J. D. and McCabe, D. E., Load history effects on the J_R-curve, in *Elastic-Plastic Fracture - Second Symposium*, Vol. 2, ASTM STP 803, American Society for Testing and Materials, Philadelphia, pp. 723–738 (1983).
5.36. Ernst, H. A., Some salient features on the tearing instability theory, in *Elastic-Plastic Fracture - Second Symposium*, Vol. 2, ASTM STP 803, American Society for Testing and Materials, Philadelphia, pp. 133–155 (1983).
5.37. *Post-yield Fracture Mechanics* (2nd edn), (eds D. G. H. Latzko, E. E. Turner, J. D. Landes, D. E. McCabe and T. K. Hellen), Elsevier Applied Science Publishers (1984).
5.38. Cherepanov, G. P., *Mechanics of Brittle Fracture* (translated from Russian), McGraw-Hill, pp. 266–270 (1979).
5.39. Carlson, A. J., Path independent integrals in fracture mechanics and their relation to variational principles, in *Prospects of Fracture Mechanics* (eds G. C. Sih, H. C. Van Elst and D. Broek), Noordhoff Int. Publ., pp. 139–158 (1974).
5.40. Kishimoto, K., Aoki, S. and Sakata, M., On the path independent integral-$\hat{J}$, *Engineering Fracture Mechanics* **13**, 841–850 (1980).
5.41. Ahmad, J., Barnes, C. R. and Kanninen, M. F., An elasto-plastic finite-element investigation of crack initiation under mixed-mode static and dynamic loading, in *Elastic-Plastic Fracture - Second Symposium*, Vol. I, ASTM STP 803, American Society for Testing and Materials, Philadelphia, pp. 214–239 (1983).
5.42. Budiansky, B. and Rice, J. R., Conservation laws and energy-release rates, *Journal of Applied Mechanics, Trans. ASME* **40**, 201–203 (1973).
5.43. Shih, C. F., Small-scale yielding analysis of mixed mode plane-strain crack problems, in *Fracture Analysis*, ASTM STP 560, American Society for Testing and Materials, Philadelphia, pp. 187–210 (1974).
5.44. Wells, A. A., Unstable crack propagation in metals: cleavage and fast fracture, *Proceedings of the Crack Propagation Symposium*, College of Aeronautics, Cranfield, Vol. 1, pp. 210–230 (1961).
5.45. Cottrell, A. H., Theoretical aspects of radiation damage and brittle fracture in steel pressure vessels, *Iron Steel Institute Special Report* No. 69, pp. 281–296 (1961).
5.46. Burdekin, F. M. and Stone, D. E. W., The crack opening displacement approach to fracture mechanics in yielding materials, *Journal of Strain Analysis* **1**, 145–153 (1966).
5.47. Dawes, M. G., Fracture control in high yield strength weldments, *Welding Journal Research Supplement* **53**, 369S–379S (1974).
5.48. Dawes, M. G., The COD design curve, in *Advances in Elasto-Plastic Fracture Mechanics* (ed. L. H. Larsson), Applied Science Publishers, pp. 279–300 (1980).
5.49. Rice, J. R. and Sorensen, E. P., Continuing crack-tip deformation and fracture for plane-strain crack growth in elastic-plastic solids, *Journal of the Mechanics and Physics of Solids* **26**, 163–186 (1978).
5.50. Rice, J. R., Drugan, W. J. and Sham, T-L., Elastic-plastic analysis of growing cracks, in *Fracture Mechanics - Twelfth Conference*, ASTM STP 700, American Society for Testing and Materials, Philadelphia, pp. 189–221 (1980).
5.51. Sorensen, E. P., A numerical investigation of plane strain stable crack growth under small-

scale yielding conditions, in *Elastic-Plastic Fracture*, ASTM STP 668, American Society for Testing and Materials, Philadelphia, pp. 151–174 (1979).
5.52. Dean, R. H. and Hutchinson, J. W., Quasi-static steady crack growth in small-scale yielding, in *Fracture Mechanics - Twelfth Conference*, ASTM STP 700, American Society for Testing and Materials, Philadelphia, pp. 383–405 (1980).
5.53. Turner, C. E., Stable crack growth and resistance curves, in *Developments in Fracture Mechanics-1* (ed. G. G. Chell), Elsevier Applied Science Publishers, pp. 107–144 (1979).
5.54. BS 5762, Methods for crack opening displacement (COD) testing, British Standards Institution, London (1979).
5.55. Xiao, Y.-G. and Huang, G.-H., On the compatibility between J-integral and crack opening displacement, *Engineering Fracture Mechanics* 16, 83–94 (1982).
5.56. Wu, Sh.-X., Plastic rotation factor and J–COD relationship of three-point bend specimen, *Engineering Fracture Mechanics* 18, 83–95 (1983).
5.57. Landes, J. D. and McCabe, D. E., Experimental methods for post-yield fracture toughness determinations, in *Post-yield Fracture Mechanics* (2nd edn), (eds D. G. H. Latzko, C. E. Turner, J. D. Landes, D. E. McCabe and T. K. Hellen), Elsevier Applied Science Publishers, pp. 223–284 (1984).
5.58. Garwood, S. J., The measurement of critical values of crack tip opening displacement (CTOD) and J on parent steels and weldments for use in fracture assessments, in *Elastic-Plastic Fracture Mechanics* (ed. L. H. Larsson), Reidel Publ. Co., Dordrecht, pp. 85–115 (1985).

6

Strain energy density failure criterion

6.1. Introduction

The selection of failure criteria for predicting the allowable load of structural components has historically been one of the problematic areas in design. The conventional approach often tends to penalize the structure in weight and size, if not economically. In the absence of a viable design technology, a structure could fail prematurely owing to overdesign in one location and underdesign in another without being known. There is no way to define 'safety' unless confidence can be placed in predicting the mode of failure by including the combined effect of loading rates, geometry and material.

Generally speaking, a component can fail by yielding and/or fracture. Although the theory of linear elastic fracture mechanics (LEFM) has proved to be a useful tool for designing against the onset of catastrophic or brittle fracture, the discipline is no longer valid when plastic deformation occurs or the failure process is path dependent. The current approach is to switch from one criterion to another when the material behavior and/or failure mode change. Such a practice is undesirable because it is prone to inconsistencies and relies too heavily on test results to identify the failure mode, a procedure that is no doubt costly and often unreliable. The discipline of fracture mechanics obviously should not be confined within the framework of LEFM; it is generally concerned with the failure of materials initiating from defects at the different scale levels. The formation of macrocracks as a result of the coalescence of microcracks and/or voids, for example, is a case in point. From the viewpoint of engineering application, it is not sufficient simply to identify failure modes phenomenologically but the results should be assessed quantitatively. Otherwise, no predictive capability could be established.

The usefulness of a failure criterion should be tested by its consistency and ability to explain the physical phenomena in general. A common procedure for testing a criterion is to compare experimental data with theoretical predictions obtained from the criterion. This approach, however, is not adequate, as sometimes the differences between the results are usually so small that they fall within the scatter of measuring experimental quantities. Many of the criteria presented in the previous chapters, such as critical stress intensity factor, energy release rate, J-integral, crack opening displacement, maximum normal stress, etc., have had some limited success in analyzing the simpler crack problems, but are found to be inadequate and often invalid for more complex situations. Unverified assumptions that were thought to yield conservative results are frequently invalidated when applied to situations in practice.

More specifically, one of the shortcomings of the above-mentioned fracture criteria is their restriction to symmetry between the applied load and crack plane and self-similar crack growth. In other words, the direction and shape of crack growth must be known *a priori*. Such an idealization is seldom encountered in service and must be regarded as the exception rather than the rule. In addition to the treatment of curved crack path, the resistance of the material to fracture should also be included in the theory without additional assumptions. This property is manifested through the fracture toughness parameter which is intended to be characteristic of the material regardless of whether the material deforms elastically and/or plastically. The ability of a criterion to treat mixed-mode fracture cannot be overemphasized. Mixed-mode fracture can occur in the plane of the crack specimen when load and crack are not symmetrically aligned or in the thickness direction when ductile fracture modes are present such as cup-and-cone failure or the development of shear lips near the specimen surfaces. The mixed-mode crack growth problem cannot be dismissed as being academic. It is real and must be accounted for in machine and structural design.

Although the energy release rate approach has had reasonable success in predicting brittle fracture, it failed to recognize the importance of slow stable crack growth prior to the onset of catastrophic fracture. The J-integral criterion suffers serious limitations in addressing the problem of crack growth. First of all, the criterion is restricted to plane problems because contour integration is a two-dimensional concept. The value of J is path independent only for a perfectly elastic material (linear or nonlinear) and self-similar crack growth. This occurs when the load is perpendicular to the crack and the crack border maintains a constant curvature during growth. The extension of J to include stable crack growth involves the assumption of a constant slope of J versus crack length curve, which is contradictory to experimental results. The crack opening displacement criterion relies mostly on measurements which are dependent on material, temperature, strain rate and triaxiality of stress which is affected by specimen geometry and size. The crack opening displacement is, therefore, test specimen sensitive.

A common characteristic feature of the above criteria in the analysis of the problem of ductile fracture is that they attempt to circumvent the real problem and to extend the concepts of linear elastic fracture mechanics in situations of extensive crack-tip plastic deformation and a substantial amount of subcritical crack growth. One of the major shortcomings of the conventional failure criteria is that they failed to separate the fracture energy from other forms of energy dissipation. This is why the critical strain energy release rate (G_{Ic}), stress intensity factor (K_{Ic}), J-integral (J_c), crack opening displacement (δ_c), etc., are all sensitive to change in loading and specimen geometry and size. In this regard, they can hardly be claimed as fracture toughness, even less as material constants.

The strain energy density (SED) criterion was proposed in the early 1970s by Sih [6.1–6.3] in an attempt to circumvent many of the difficulties that could not be overcome by other criteria. The combined effect of specimen size, complex geometries and loadings are the cause of this difficulty. It fundamentally represents a departure from the conventionally accepted criteria mentioned previously. This new criterion does not rely on the existence of an initial defect such as a crack in the solid. It can account for failure of the material by fracture and/or yielding. This permits a consistent treatment of the entire failure process starting from the early stage of fracture initiation to the final separation of the material. The validity of the strain energy density criterion has been established at the Institute of Fracture and Solid Mechanics of Lehigh University by applying the theory to a host of nontrivial problems of fundamental importance. They include two- and three-dimensional crack problems, cracked nonhomogeneous and composite materials, plates and shells with cracks, dynamic crack problems, failure initiating from notches, ductile fracture involving the prediction of crack initiation, slow stable crack growth and final separation, fatigue crack growth, etc. Many of the results have been published in the introductory chapters of the seven volumes of the series *Mechanics of Fracture* edited by Sih [6.4-6.10]. Furthermore, a number of international conferences [6.11, 6.12] have been devoted to the study of the SED criterion.

The fundamental quantity in the SED criterion is the strain energy dW/dV contained in a unit volume of material at a given instance of time. The quantity dW/dV serves as a useful failure criterion and has successfully been applied in the solution of a host of engineering problems of major interest. However, when the theory is used for irreversible material behavior the quantity dW/dV is not sufficient for the determination of stress and/or strain fields and must be applied in conjunction with the theory of plasticity that frequently leads to incompatibility of stress and failure analysis. For this reason the surface energy density dW/dA was incorporated into the theory and was related to dW/dV through the rate of change of volume with surface area dV/dA. This provides a one-to-one correspondence between the uniaxial and multiaxial stress state and allows the simultaneous analysis of stress and failure in systems, including energy dissipation.

In the following the underlying principles of the SED theory using the volume strain energy, dW/dV, are presented. The theory is then applied to a number of problems of brittle and ductile fracture.

6.2. Volume strain energy density

The original concept of the SED theory was based on the idea that a continuum may be viewed as an assembly of small building blocks, each of which contains a unit volume of material and can store a finite amount of energy at a given instance of time. The energy per unit volume will be referred to as the (volume) strain energy density function, dW/dV, and is expected to vary from one location to another.

The strain energy density function can be computed from

$$\frac{dW}{dV} = \int_0^{\epsilon_{ij}} \sigma_{ij}\,d\epsilon_{ij} + f(\Delta T, \Delta C) \tag{6.1}$$

where σ_{ij} and ϵ_{ij} are the stress and strain components and ΔT and ΔC are the changes in temperature and moisture concentration, respectively.

Equation (6.1) shows that a material element can contain energy even when the stresses are zero. Thus, any failure criteria based on stress or strain alone are obviously limited in application.

Historically the energy quantity has been used for the description of failure of a material element by yielding. Two separate theories have been developed. They are the total energy or Beltrami–Haigh theory and the distortional energy or Hubert–von Mises–Hencky theory. According to these theories, failure in a material by yielding occurs when the total or the distortional strain energy per unit volume absorbed by the material equals the energy per unit volume stored in the material loaded in uniaxial tension at yield. This quantity corresponds to the limiting strain energy and is regarded as a material constant. Extensive experimental evidence is available on the use of the strain energy quantity to describe failure by yielding [6.13, 6.14].

For linear elastic material behavior the strain energy density function dW/dV can be written as

$$\frac{dW}{dV} = \frac{1}{2E}(\sigma_x^2 + \sigma_y^2 + \sigma_z^2) - \frac{\nu}{E}(\sigma_x\sigma_y + \sigma_y\sigma_z + \sigma_z\sigma_x) +$$
$$+ \frac{1}{2\mu}(\tau_{xy}^2 + \tau_{yz}^2 + \tau_{zx}^2) \tag{6.2}$$

where σ_x, σ_y, σ_z, τ_{xy}, τ_{yz} and τ_{zx} are the stress components, E the Young modulus, ν the Poisson ratio and μ the shear modulus of elasticity such that $E = 2\mu(1+\nu)$.

For the plane problems of elasticity the quantity dW/dV takes the form

$$\frac{dW}{dV} = \frac{1}{4\mu}\left[\frac{\kappa+1}{4}(\sigma_x + \sigma_y)^2 - 2(\sigma_x\sigma_y - \tau_{xy}^2)\right] \tag{6.3}$$

where $\kappa = 3 - 4\nu$ for plane strain and $\kappa = (3-\nu)/(1+\nu)$ for generalized plane stress.

The strain energy per unit volume, dW/dV, can be further decomposed into two parts:

$$\frac{dW}{dV} = \left(\frac{dW}{dV}\right)_d + \left(\frac{dW}{dV}\right)_v \tag{6.4}$$

in which

$$\left(\frac{dW}{dV}\right)_d = \frac{1+\nu}{6E}\left[(\sigma_x - \sigma_y)^2 + (\sigma_y - \sigma_z)^2 + (\sigma_z - \sigma_x)^2 + \right.$$
$$\left. + 6(\tau_{xy}^2 + \tau_{yz}^2 + \tau_{zx}^2)\right] \tag{6.5}$$

represents the distortional strain energy per unit volume corresponding to the deviatoric stress tensor that is associated with distortion of an element undergoing no volume change.

The quantity

$$\left(\frac{dW}{dV}\right)_v = \frac{1-2\nu}{6E}(\sigma_x + \sigma_y + \sigma_z)^2 \tag{6.6}$$

represents the part of the strain energy per unit volume associated with volume change and no shape change.

Keep in mind that the separation of dW/dV into its dilatational and distortional components invokes linearity and superposition of stress and strain which is mathematically unacceptable if the process is nonlinear and irreversible. The von Mises criterion of yielding considers only the distortional component and neglects the portion of the strain energy density due to volume change which cannot be justified in general.

By means of the plane strain condition

$$\sigma_3 = \nu(\sigma_1 + \sigma_2) \tag{6.7}$$

where σ_1, σ_2, σ_3 are the principal stresses, the ratio $(dW/dV)_v/(dW/dV)_d$ obtained from Equations (6.5) and (6.6) takes the form

$$\frac{(dW/dV)_v}{(dW/dV)_d} = \frac{(1-2\nu)(1+\nu)[(\sigma_1/\sigma_2)+1]^2}{[(\sigma_1/\sigma_2)-1]^2 + [(1-\nu)-\nu(\sigma_1/\sigma_2)]^2 + [(1-\nu)(\sigma_1/\sigma_2)-\nu]^2} \tag{6.8}$$

in terms of the principal stresses.

Figure 6.1 gives the variations of the ratio $(dW/dV)_v/(dW/dV)_d$ with the ratio of the principal stresses σ_1/σ_2 for $\nu = 0, 0.1, 0.2, 0.3$, and 0.4. It is observed from this plot that the greatest volume change takes place for $\sigma_1 = \sigma_2$ corresponding to a two-dimensional hydrostatic stress state. For most metals with ν ranging from 0.2 to 0.3, $(dW/dV)_v/(dW/dV)_d$ varies from 4 to 6.5. From the relative magnitudes of $(dW/dV)_v$ and $(dW/dV)_d$, it is seen that both quantities should be taken into account when considering the failure of material elements either by yielding and/or fracture.

Failure of material elements in a solid is caused by permanent deformation or fracture which can be related to shape change (distortion) and volume change (dilatation). In general, a material element is subjected to both distortion and dilatation and the corresponding energies for linear elastic material response can be computed from Equations (6.5) and (6.6). Realistic modeling of material failure requires knowledge of damage at both the microscopic and macroscopic scale level. X-ray examination of the problem of brittle fracture of a tensile crack in a low carbon steel revealed a thin layer of highly distorted material along the fracture surface. Consider the case of a macrocrack in a tensile stress field (Figure 6.2). The elastic zone ahead of the crack contains shear planes and the plastic zone off the plane of the macrocrack contains cleavage planes. Refer-

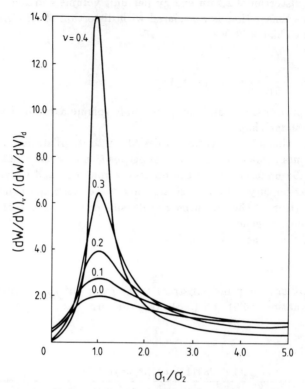

Figure 6.1. Variation of the ratio of the dilatational and distortional strain energy density $(dW/dV)_v/(dW/dV)_d$ versus the ratio σ_1/σ_2 of the principal stresses for a state of plane strain. The Poisson ratio ν takes the values 0, 0.1, 0.2, 0.3 an 0.4.

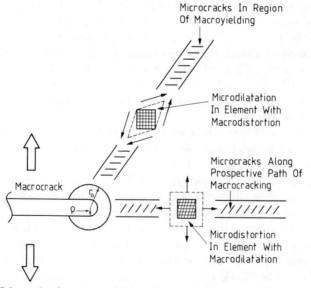

Figure 6.2. Schematic of macro- and micro-damage in the region ahead of the crack tip.

ring to the macroelement ahead of the crack from the continuum mechanics solution of the stress problem it is deduced that the principal stresses σ_1 and σ_2 are equal. From Equation (6.8) it is obtained that for a Poisson's ratio $\nu = 0.3$, $(dW/dV)_v$ is 6.5 times larger than $(dW/dV)_d$. Even though the macroelement ahead of the crack fractures due to the dilatational component of the strain energy density, the distortional component is not negligible and is responsible for the creation of slip planes or microcracks. Similarly, maximum distortion of the macroelement off the axis of the macrocrack takes place while micro cleavage planes appear perpendicular to the direction of tension. From stress analysis it is obtained that in this element the distortional component of the strain energy density becomes maximum, while the dilatation component is not negligible and is responsible for the creation of the cleavage planes.

From the above arguments it appears that, for a complete description of material damage, it is mandatory to consider both components, the dilatational and the distortional, of strain energy density. They both play a role in the material damage process. Microyielding may attribute to macrofracture and microfracture to macroyielding. Thus, the two processes of yielding and fracture are inseparable. They are unique features of material damage and should be treated simultaneously by a single failure criterion. Thus, the von Mises yield condition is incomplete as it considers only the distortional component of the strain energy density and cannot explain the creation of the microcracks in the plastic zone.

Material damage at the macroscopic and microscopic scale levels can be connected by referring to the proportion of $(dW/dV)_v$ and $(dW/dV)_d$ within the macroelement. For the macrocrack under tension, the model of Figure 6.2 suggests macrofracture to coincide with the direction where $(dW/dV)_v > (dW/dV)_d$ and macroyielding with the direction where $(dW/dV)_d > (dW/dV)_v$. These directions are determined by appealing to physical hypotheses.

6.3. Basic hypotheses

The strain energy density criterion provides a complete description of material damage by including both the distortional and dilatational effects. In this respect, and following the previous arguments, both components of strain energy density should be included. Both distortion and dilatation vary in proportion, depending on the load history and location due to nonuniformity in stress or energy fields. Their contributions to distortion or dilatation of a macroelement are weighted automatically by taking the stationary values of the total strain energy density with respect to appropriate space variables referenced from the site of failure initiation. The relative local minimum of dW/dV corresponds to large volume change and is identified with the region dominated by macrodilatation leading to fracture, while the relative local maximum of dW/dV corresponds to large shape change and is identified with the region dominated by macrodistortion leading to yielding.

The strain energy density criterion may be stated in terms of three basic hypotheses, and applies to all materials (reversible or irreversible), loading types (monotonic, cyclic or fatigue) and structure geometries with or without initial defects. The hypotheses are independent of material type or restrictions introduced by constitutive equations. The strain energy density function dW/dV near the crack tip or any other possible failure site such as a re-entrant corner, inclusion, void, etc., decays with distance r. The strain energy density function dW/dV will be used in the form

$$\frac{dW}{dV} = \frac{S}{r} \tag{6.9}$$

where S is the strain energy density factor and r the radial distance measured from the site of possible failure initiation. The singular dependency $1/r$ is a fundamental character of the Newtonian potential and is independent of the constitutive relation.

At this point it should be emphasized that failure criteria based on the amplitude of the asymptotic stresses (stress intensity factor for crack problems) are restrictive since the singular character of the stresses depends on the constitutive relation of the material. Furthermore, the order of singularity for each stress component may be different, as in the case of plate bending (Equations (2.215)) or in finite deformation theory. A characteristic example is the case of a surface crack where the singularity at the intersection of the crack border and the free surface differs from that of $r^{-1/2}$ in the interior. This puts the concept of energy release rate or stress intensity factor on shaky ground as they both rely on the existence of the $r^{-1/2}$ stress singularity. The r^{-1} singular character of the strain energy density for crack problems applies to all materials, ranging from perfectly linear and elastic to nonlinear viscous and plastic and to all embedded and surface crack shapes. The r^{-1} singular nature of dW/dV does not change along the crack border, including the point on the free surface.

The three hypotheses of the strain energy density criterion are:

Hypothesis (1) The location of fracture coincides with the location of relative minimum strain energy density, $(dW/dV)_{\min}$, and yielding with relative maximum strain energy density, $(dW/dV)_{\max}$.

Hypothesis (2) Failure by fracture or yielding occurs when $(dW/dV)_{\min}$ or $(dW/dV)_{\max}$ reach their respective critical values.

Hypothesis (3) The amount of incremental growth $r_1, r_2, \ldots, r_j, \ldots, r_c$ is governed by

$$\left(\frac{dW}{dV}\right)_c = \frac{S_1}{r_1} = \frac{S_2}{r_2} = \ldots = \frac{S_j}{r_j} = \ldots = \frac{S_c}{r_c}. \tag{6.10}$$

There is unstable fracture or yielding when the critical ligament size r_c is reached.

The above hypotheses of the SED criterion will be used in the sequel for the solution of a host of problems of fundamental importance.

6.4. Two-dimensional linear elastic crack problems

This section deals with the general problem of crack extension in a mixed-mode stress field governed by the values of the stress intensity factors K_I and K_{II}.

The fundamental quantity is the strain energy density factor S which is the amplitude of the energy field that possesses an r^{-1}-type of singularity. The situation of brittle crack growth where crack initation coincides with final instability is considered. Unlike the stress intensity factor K, which is a measure of the local stress amplitude, the strain energy density factor S is also direction sensitive. The difference between K and S is analogous to the difference between a scalar and a vector quantity. The factor S, defined as $r(dW/dV)$ in Equation (6.9), represents the local energy release for a segment of crack growth r. There is unstable crack growth when the critical ligament size r_c is reached.

For crack growth in a two-dimensional stress field the first hypothesis of the SED criterion can be expressed mathematically by the relations

$$\frac{\partial S}{\partial \theta} = 0, \qquad \frac{\partial^2 S}{\partial \theta^2} > 0 \tag{6.11}$$

where θ is the polar angle. Crack initiation occurs when

$$S(\theta_c) = S_c \tag{6.12}$$

where θ_c is defined by (6.11) and S_c is the critical value of the strain energy density factor which is a material constant. S_c represents the fracture toughness of the material.

The stress components for the problem of a crack in a two-dimensional mixed-mode stress field governed by the values of the opening-mode, K_I, and sliding-mode, K_{II}, stress intensity factors are given by Equations (4.103). Introducing these equations into Equation (6.3), the following quadratic form for the strain energy density factor S is obtained.

$$S = a_{11}k_I^2 + 2a_{12}k_I k_{II} + a_{22}k_{II}^2 \tag{6.13}$$

where the coefficients a_{ij} $(i,j = 1, 2)$ are given by

$$\begin{aligned}
16\mu a_{11} &= (1 + \cos\theta)(\kappa - \cos\theta) \\
16\mu a_{12} &= \sin\theta[2\cos\theta - (\kappa - 1)] \\
16\mu a_{22} &= (\kappa + 1)(1 - \cos\theta) + (1 + \cos\theta)(3\cos\theta - 1).
\end{aligned} \tag{6.14}$$

In these equations $k_j = K_j/\sqrt{\pi}$ $(j =$I, II$)$ and $\kappa = 3 - 4\nu$ or $\kappa = (3-\nu)/(1+\nu)$ for plane strain or plane stress conditions respectively.

Substituting S from Equation (6.13) into relations (6.11) we obtain

$$[2\cos\theta - (\kappa - 1)\sin\theta]k_I^2 + 2[2\cos 2\theta - (\kappa - 1)\cos\theta]k_I k_{II} +$$
$$+ [(\kappa - 1 - 6\cos\theta)\sin\theta]k_{II}^2 = 0 \tag{6.15a}$$
$$[2\cos 2\theta - (\kappa - 1)\cos\theta]k_I^2 + 2[(\kappa - 1)\sin\theta - 4\sin 2\theta]k_I k_{II} +$$
$$+ [(\kappa - 1)\cos\theta) - 6\cos 2\theta]k_{II}^2 > 0. \tag{6.15b}$$

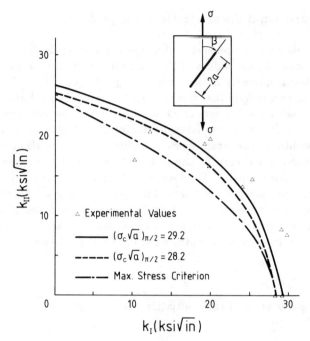

Figure 6.3. Mixed-mode fracture criterion for cracks under tension.

Relations (6.15a) and (6.15b) represent the general formulas of the strain energy density criterion for the case of a crack in a two-dimensional stress field under mixed-mode loading conditions. For a particular problem under consideration with known values of the k_I, k_{II} stress intensity factors introducing these values into Equation (6.15a), we obtain the values of the crack extension angle θ_c as the roots of this equation which satisfy inequality (6.15b). Substituting these roots, θ_c, into Equation (6.13) the minimum values $S_{\min}$ of the strain energy density factor are obtained. When $S_{\min}$ is equated to the critical strain energy density factor S_c, which is a material constant, the critical values of the applied loads corresponding to the onset of rapid crack propagation are obtained.

Eliminating the angle θ_c in Equations (6.15a) and (6.14), the crack growth condition expressed by Equations (6.12) and (6.13) defines the fracture locus in the k_I–k_{II} plane. It is shown in Figure 6.3 for aluminum alloys with $4.8(\mu S_c)^{1/2} =$ 28.2 kip/in$^{5/2}$ and 29.2 kip/in$^{5/2}$ and tensile applied loads. The third curve represents the prediction based on the maximum stress criterion (Section 4.11 (a)). Experimental results obtained from the uniaxial tension plate with an inclined crack are presented. Observe that the SED criterion is closer to the experimental results. The k_I–k_{II} locus for compressive applied stresses for glass is shown in Figure 6.4 together with experimental results. The k_I–k_{II} curve for compression is basically different from that in tension. The curve does not intersect the k_I-axis which implies the obvious fact that a crack under mode-I does not extend in compression.

The usefulness and versatility of the SED criterion for determining the allow-

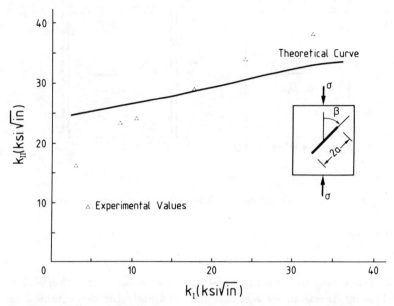

Figure 6.4. Mixed-mode fracture criterion for cracks under compression.

able load corresponding to a variety of engineering problems of practical interest has been demonstrated in the book *Problems of Mixed Mode Crack Propagation* by the author [6.15]. In the next section of this chapter the SED criterion is used for determining the angle of initial crack extension and the corresponding critical load for the case of an inclined crack in an infinite plate under uniform uniaxial stress.

6.5. Uniaxial extension of an inclined crack

Consider a central crack of length $2a$ in an infinite plate subjected to a uniform uniaxial stress σ at infinity where the axis of the crack makes an angle β with the direction of stress σ. Mixed-mode conditions predominate in the vicinity of the crack tip, and the values of the k_I, k_II stress intensity factors are given by [2.27]:

$$k_\text{I} = \sigma a^{1/2} \sin^2 \beta, \qquad k_\text{II} = \sigma a^{1/2} \sin \beta \cos \beta. \tag{6.16}$$

Substituting these values into Equation (6.13) the following equation is obtained for the strain energy density factor S:

$$S = \sigma^2 a (a_{11} \sin^2 \beta + 2a_{12} \sin \beta \cos \beta + a_{22} \cos^2 \beta) \sin^2 \beta \tag{6.17}$$

where the coefficients a_{ij} are given by Equations (6.14).

Equation (6.15a) for the calculation of the angle θ_c of initial crack extension

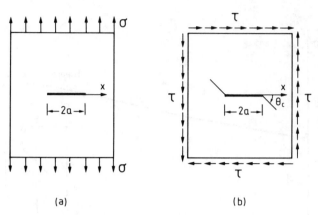

Figure 6.5. (a) Opening-mode and (b) sliding-mode crack extension.

takes the form:

$$(\kappa - 1)\sin(\theta_c - 2\beta) - 2\sin[2(\theta_c - \beta)] - \sin 2\theta_c = 0, \quad \beta \neq 0. \tag{6.18}$$

Before proceeding to the general case of a crack of any inclination with respect to the loading direction, we will consider separately the two common cases of opening-mode and sliding-mode crack extension.

(a) *Opening-mode crack extension*

This case corresponds to the trivial Griffith crack configuration consisting of an infinite body with a central crack of length $2a$ subjected to a uniform uniaxial stress σ at infinity (Figure 6.5(a)). Because of load symmetry the crack propagates in its own plane. Let us suppose that this is an unknown problem and analyze it through the use of the strain energy density theory. Inserting the values of the stress intensity factors k_I, k_{II}:

$$k_I = \sigma a^{1/2}, \qquad k_{II} = 0 \tag{6.19}$$

into Equation (6.13) the following equation is obtained:

$$S = \frac{\sigma^2 a}{16\mu}(1 + \cos\theta)(\kappa - \cos\theta). \tag{6.20}$$

Furthermore, Equation (6.15a), which gives the stationary values of S, takes the form

$$[2\cos\theta - (\kappa - 1)]\sin\theta = 0 \tag{6.21}$$

while inequality (6.15b) becomes

$$2\cos 2\theta - (\kappa - 1)\cos\theta > 0. \tag{6.22}$$

Equation (6.21) is satisfied when $\theta_c = 0$ or $\theta_c = \arccos[(\kappa - 1)/2]$. The second root θ_c does not satisfy inequality (6.22) because for the elastic constant κ is $1 \leq \kappa \leq 3$, and thus it is disregarded. Hence, the minimum value of S corresponds to an initial crack extension angle $\theta_c = 0$, which means that the crack extends in its own plane. The plane $(\theta_c = 0)$ corresponds to the direction of maximum potential energy, a position of unstable equilibrium. For $\theta_c = 0$, Equation (6.20) gives the minimum value $S_{\min}$ of S

$$S_{\min} = \frac{(\kappa - 1)\sigma^2 a}{8\mu}. \qquad (6.23)$$

Equating $S_{\min}$ with S_c, which is a material constant, we obtain the following expression for the critical stress σ_c corresponding to the onset of crack extension:

$$\sigma_c a^{1/2} = \left(\frac{8\mu S_c}{\kappa - 1}\right)^{1/2}. \qquad (6.24)$$

The value of the stress intensity factor given by Equation (6.19) corresponds to an infinite plate. For the general case of a mode-I crack with stress intensity factor K_I, Equation (6.23) takes the form when it is referred to the critical state of unstable crack extension under plane strain conditions

$$S_c = \frac{(1 + \nu)(1 - 2\nu)K_{Ic}^2}{2\pi E}. \qquad (6.25)$$

Equation (6.25) relates the critical strain energy density factor, S_c, with the critical stress intensity factor, K_{Ic}, which can be determined by the methods described in Section 4.8. S_c is a material constant and can account for the characterization of the fracture toughness of the material. Values of S_c, together with the ultimate stress and the critical stress intensity factor K_{Ic} for some common metals and alloys reported by Sih and Macdonald [6.16], are given in Table 6.1.

(b) *Sliding-mode crack extension*

This case corresponds to an infinite body containing a central crack of length $2a$ and subjected to a uniform shear stress τ at infinity (Figure 6.5). The k_I, k_{II} stress intensity factors are given by:

$$k_I = 0, \qquad k_{II} = \tau a^{1/2} \qquad (6.26)$$

and Equation (6.13) gives

$$S = \frac{\tau^2 a}{16\mu}[(\kappa + 1)(1 - \cos\theta) + (1 + \cos\theta)(3\cos\theta - 1)]. \qquad (6.27)$$

Working as in the previous case, we obtain that the angle of crack extension θ_c is given by

$$\theta_c = \arccos\left(\frac{\kappa - 1}{6}\right). \qquad (6.28)$$

Note that θ_c is a function of Poisson's ratio ν. Table 6.2 shows the values of the predicted fracture angle θ_c for ν ranging from 0 to 0.5 for plane strain conditions $(\kappa = 3 - 4\nu)$.

Introducing the value of the angle of initial crack extension θ_c into Equation (6.27) and equating the resulting value of $S_{\min}$ to S_c the following expression for

Table 6.1. Fracture toughness values

Material	Ultimate strength σ_u (ksi)	Critical stress-intensity factor K_{Ic} (ksi$\sqrt{\text{in}}$)	Critical strain-energy-density factor S (lb/in)[a]
A517F Steel (AM)	120	170	95.8
AISI 4130 Steel (AM)	170	100	33.2
AISI 4340 Steel (VAR)	300	40	5.3
AISI 4340 Steel (VAR)	280	40	5.3
AISI 4340 Steel (VAR)	260	45	6.7
AISI 4340 Steel (VAR)	240	60	11.9
AISI 4340 Steel (VAR)	220	75	18.7
300M Steel (VAR)	300	40	5.3
300M Steel (VAR)	280	40	5.3
300M Steel (VAR)	260	45	6.7
300M Steel (VAR)	240	60	11.9
300M Steel (VAR)	220	75	18.7
D6AC Steel (VAR)	240	40–90	5.3–26.9
H-11 Steel (VAR)	320	30	3.0
H-11 Steel (VAR)	300	40	5.3
H-11 Steel (VAR)	280	45	6.7
12Ni-5Cr-3Mo Steel (VAR)	190	220	160.5
18Ni (300) Maraging Steel (VAR)	290	50	8.3
18Ni (250) Maraging Steel (VAR)	260	85	24.0
18Ni (200) Maraging Steel (VAR)	210	120	47.8
18Ni (180) Maraging Steel (VAR)	195	160	84.9
9Ni-4Co-0.3C Steel (VAR)	260	60	11.9
Al 2014-T651	70	23	3.6
Al 2024-T851	65	23	3.6
Al 2219-T851	66	33	7.4
Al 2618-T651	64	32	7.0
Al 7001-T75	90	25	4.2
Al 7075-T651	83	26	4.6
Al 7079-T651	78	29	5.7
Al 7178-T651	83	24	3.9

Table 6.2. Fracture angle $-\theta_0$ under pure shear and plane strain conditions

ν	0	0.1	0.2	0.3	0.4	0.5
$-\theta_0$	70.5°	74.5°	78.5°	82.3°	86.2°	90.0°

[a] The calculation of S_c was based on $E = 30 \times 10^6$ psi, $\nu = 1/4$ for steel and $E = 10.6 \times 10^6$ psi. $\nu = 1/3$ for aluminum.

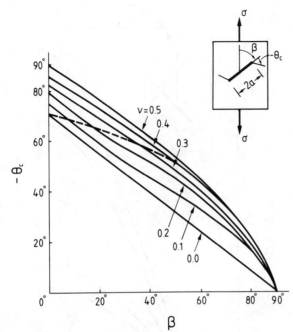

Figure 6.6. Variation of the crack extension angle $-\theta_c$ versus the crack inclination angle β under plane strain conditions for tensile applied loads.

the critical shear stress τ_c is derived:

$$\tau_c a^{1/2} = \left(\frac{192\mu S_c}{-\kappa^2 + 14\kappa - 1}\right)^{1/2}. \tag{6.29}$$

(c) *Inclined crack; tensile loads*

Solution of Equation (6.18) with the crack inclination angle β ranging from 0 to 90° and various values of the material constant κ provides the stationary value of the strain energy density factor. From the resulting values of the angle θ, those which satisfy inequality (6.15b) provide the angles of initial crack extension θ_c. Figure 6.6 displays the variation of $-\theta_c$ versus the crack angle β for $\nu = 0$, 0.1, 0.2, 0.3, 0.4 and 0.5 under plane strain conditions. Results for plane stress conditions can be obtained by replacing ν with $\nu/(1+\nu)$. In the same figure the dashed curve represents the results obtained by the maximum stress criterion (Section 4.11(a)). It is observed that these results agree with those based on Equation (6.18) for large values of β and represent a lower bound for small values of β. In general it can be taken as an average curve. It is worth noting that the crack extension angle θ_c is always negative for uniaxial tensile loads. The results of Figure 6.6 are in good agreement with experimental results obtained from plexiglas plates with a central crack [6.3, 6.4].

Introducing into Equation (6.17) the values of the crack extension angle θ_c as they are displayed in Figure 6.6, the minimum values of the strain energy density factor S are obtained. Equating these values of S with the critical

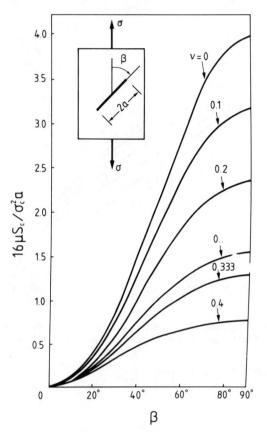

Figure 6.7. Variation of the quantity $16\mu S_c/\sigma_c^2 a$ versus the crack angle β under plane strain conditions for tensile applied loads.

strain energy density factor S_c, which is a material constant, the values of the critical tensile stress σ_c for crack propagation are obtained. The variation of the quantity $16\mu S_c/\sigma_c^2 a$ versus the crack inclination angle β for $\nu = 0, 0.1, 0.2, 0.3, 0.333$ and 0.4 is shown in Figure 6.7 for plane strain conditions prevailing in the plate. It is observed that the quantity $16\mu S_c/\sigma_c^2 a$ increases with the crack angle β, reaching a maximum for opening-mode crack extension. Furthermore, $16\mu S_c/\sigma_c^2 a$ increases as Poisson's ratio ν of the plate decreases. Since S_c is a material constant, the above statements imply that the quantity $\sigma_c^2 a$ decreases as the crack angle β increases, while quite the contrary happens with respect to Poisson's ratio ν. Thus, the lowest value of the applied stress σ_c that will initiate crack propagation occurs at $\beta = \pi/2$ for a material with a low Poisson ratio.

(d) *Inclined crack; compressive loads*

In addition to the negative roots of Equation (6.18) which correspond to uniaxial tensile loads, there exists another set of solutions for positive angles θ_c. Physi-

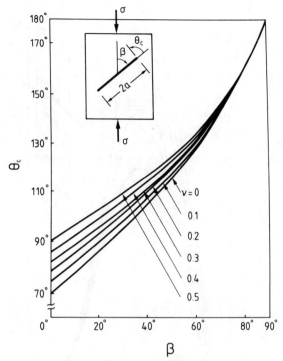

Figure 6.8. Variation of the crack extension angle θ_c versus the crack inclination angle β under plane strain conditions for compressive applied loads.

cally the positive crack extension angles correspond to the case when the inclined crack is under uniaxial compression. Since S depends on σ^2, Equations (6.17) and (6.18) contain both the solutions of uniaxial tension $+\sigma$ and compression $-\sigma$. At this point it should be emphasized that for the case of compression it is assumed that no overlapping between the crack tips takes place.

Figure 6.8 presents the variation of the positive crack extension angle θ_c for compressive applied loads versus the crack inclination angle β for various values of Poisson's ratio ν. It is observed that contrary to tensile loading where the crack tends to become horizontal, the crack path under uniaxial compression extends towards the direction of loading. Such a phenomenon has indeed been observed by Hoek and Bieniawski [6.17] who made tests on a number of glass plates with inclined cracks under uniaxial compression. Unfortunately, they did not report the angle of initial crack extension and therefore a comparison of their experimental results with the theoretical results of Figure 6.8 is not possible.

Following the same procedure as used in the case of tensile loading, the stationary values of the strain energy density factor are obtained. Figure 6.9 presents the variation of the quantity $16\mu S_c/\sigma_c^2 a$ versus the crack inclination angle β for plane strain conditions. From Figure 6.9 we observe that the quantity $16\mu S_c/\sigma_c^2 a$ reaches a maximum in the interval $0 < \beta < 90°$ which depends on the value of Poisson's ratio ν. It is also observed that the critical stress σ_c increases and tends

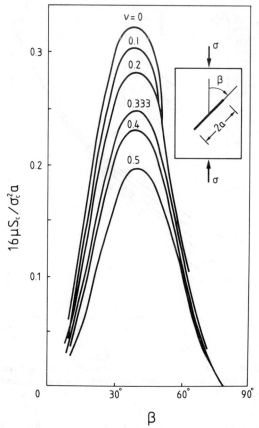

Figure 6.9. Variation of the quantity $16\mu S_c/\sigma_c^2 a$ versus the crack angle β under plane strain conditions for compressive applied loads.

to infinity as the crack becomes parallel ($\beta = 0$) or perpendicular ($\beta = 90°$) to the direction of loading. This result corresponds to the physical observation that a crack in a plate parallel or perpendicular to the direction of a compressive applied stress has no influence on the fracture behavior of the plate.

For the solution of other two-dimensional crack and inclusion problems by the SED criterion, refer to [6.18–6.32].

6.6. Three-dimensional linear elastic crack problems

(a) Introductory remarks

In this section the strain energy density criterion is used to determine the growth characteristics of arbitrary cracks embedded in linear elastic three-dimensional bodies. The local stress field along the crack front was studied in Section 2.9. This can be expressed in terms of three stress intensity factors, all of which are independent of the local coordinates, depending only on the crack geometry, the form of loading and the location of the point along the crack border. The case of an elliptical crack embedded in a solid loaded by a uniform inclined stress with respect to the crack plane is studied. This case is analogous to the two-dimensional problem of an inclined crack under uniform stress, which was considered in the previous section. The critical stress and the new shape of the crack after propagation are determined as a function of the crack shape, the orientation of the applied stress and the Poisson ratio of the material. The solution of this fundamental problem is representative of the application of the strain energy density criterion to linear elastic three-dimensional crack problems.

(b) Basic equations

As was pointed out in Section 2.9 and referring to Figure 2.26, the local stress field at a point Q_0 defined by the spherical coordinates r_0, θ_0, ϕ_0 and not lying in the normal plane is again given by Equations (2.199) when r is substituted by $r_0 \cos \phi_0$. Therefore, the expression for the strain energy density dW/dV referred to each point along the crack periphery will be given by the same expression as in the two-dimensional case with the addition of an extra term corresponding to the mode-III stress intensity factor. Thus, we find from Equations (6.13) and (6.14) for plane strain conditions that

$$\frac{dW}{dV} = \frac{S}{r_0 \cos \phi_0} \tag{6.30}$$

where

$$S = a_{11}k_I^2 + 2a_{12}k_I k_{II} + a_{22}k_{II}^2 + a_{33}k_{III}^2. \tag{6.31}$$

The coefficients a_{ij} $(i,j = 1,2,3)$ are exactly the same as in Equation (6.14) with plane strain conditions and are given by

$$\begin{aligned} 16\mu a_{11} &= (3 - 4\nu - \cos\theta)(1 + \cos\theta) \\ 16\mu a_{12} &= 2\sin\theta(\cos\theta - 1 + 2\nu) \\ 16\mu a_{22} &= 4(1 - \nu)(1 - \cos\theta) + (3\cos\theta - 1)(1 + \cos\theta) \\ 16\mu a_{33} &= 4. \end{aligned} \tag{6.32}$$

The stress intensity factors k_j $(j = I, II, III)$ are related to K_j $(j = I, II, III)$ given in Equations (2.200) by $k_j = K_j/\sqrt{\pi}$.

According to the first hypothesis of the strain energy density criterion (Section 6.3, page 202) crack propagation from each point of the crack front is in the direction of the point that has the minimum value of strain energy density factor as compared with other points of a spherically shaped core of radius r_0. Thus, a line can be drawn from each point of the crack to the point of the sphere with the minimum value of S. The aggregate of these lines defines the new crack surface.

It is readily apparent from (6.30) that the minimum value of dW/dV for the case under consideration occurs when

$$(\phi_0)_c = 0 \tag{6.33}$$

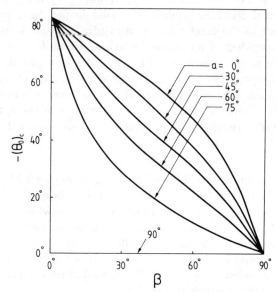

Figure 6.10. Variation of the fracture angle $-(\theta_0)_c$ versus the load inclination angle β for various positions (α) along the crack front. The applied stress is tensile and $b/a = 0.5$, $\nu = 0.33$ and $\omega = 0$.

that is, the crack always propagates from each point of its front in the normal plane. The angle $(\theta_0)_c$ defining the position of the line in the normal plane that gives the new crack surface after propagation can be determined from Equation (6.11).

Having determined the angles $(\phi_0)_c = 0$ and $(\theta_0)_c$ that give the direction of crack propagation from each point of the crack front, the corresponding minimum values $S_{\min} = S\big((\phi_0)_c, (\theta_0)_c\big)$ of S can be obtained. According to the second hypothesis of the strain energy density criterion, crack propagation starts when $S_{\min}$ reaches the critical value S_c tolerable by the material.

(c) *Results*

The values of the strain energy density factor S can be obtained as a function of the spherical angles ϕ and θ, the loading angles β and ω (Figure 2.27), the crack diameter ratio b/a, the position along the crack border α and the quantity $\sigma\sqrt{b}$. Figure 6.10 presents the variation of the fracture angle $(-\theta_0)_c$ versus the angle β for various positions along the crack front when a tensile stress is applied on the xz-plane $(\omega = 0)$, $b/a = 0.5$ and $\nu = 0.33$. Observe that when the applied stress is perpendicular to the crack plane $(\beta = 90°)$, the crack grows from all points along its front in its own plane. Furthermore, the ends of the minor axis $(\alpha = 90°)$ grow in the plane of the crack. The tendency of the fracture angles $(-\theta_0)_c$ to decrease as the point on the crack border moves from the major to the minor axis of the crack is also apparent.

The value of the critical applied stress σ_c for crack extension can be determined

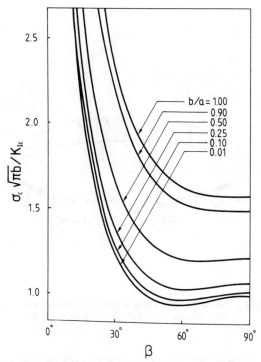

Figure 6.11. Variation of the normalized critical fracture stress $\sigma_c\sqrt{\pi b}/K_{Ic}$ versus the load inclination angle β for various values of the diameter ratio b/a of the crack. The applied stress is tensile and $\nu = 0.33$ and $\omega = 0$.

by equating the values of $S_{\min}$ with the critical value of the strain energy density factor, S_c (which is a material constant). Figure 6.11 presents the variation of the nondimensional quantity $\sigma_c\sqrt{\pi b}/K_{Ic}$ versus the load inclination angle β for various values of the diameter ratio a/b of the ellipse under tensile applied loads. K_{Ic} is the critical value of the strain energy density factor connected with S_c through relation (6.25). Poisson's ratio ν takes the value 0.33, and $\omega = 0$. Crack extension initiates from the end of the minor axis of the ellipse. Note from Figure 6.11 that the critical stress decreases as the ellipse becomes increasingly slender (decreasing the value of b/a).

As previously determined, crack extension always takes place from each point along its front on the normal plane. Thus, a line drawn from each point on the crack front in the normal plane at an angle $(\theta_0)_c$ with respect to the crack plane indicates the directions in which the points of the crack front would move after propagation. The third hypothesis of the strain energy density criterion which states that $S_{\min}/r_0$ remains constant along the crack front is used to determine the new fracture surface during slow stable crack propagation. Thus, the values of r_0 at all points of the crack periphery are calculated from the previously obtained minimum values $S_{\min}$ while specifying the length r_0 of the crack extension at a particular point of the crack. The resulting initial segment

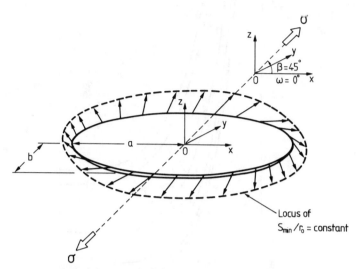

Figure 6.12. Initial fracture surface of an elliptical crack ($b/a = 0.5$) under tension. $\beta = 45°$, $\omega = 0$ and $\nu = 0.33$.

of the new crack surface around an elliptical crack of $b/a = 0.5$ and $\beta = 45°$ with the applied stress in the xz-plane ($\omega = 0$) is shown in Figure 6.12 when the stress is tensile.

For further information on the use of the strain energy density theory for the solution of three-dimensional crack problems refer to references [6.5, 6.33, 6.34].

6.7. Bending of cracked plates

(a) Introductory remarks

In this section the strain energy density criterion is used to analyze the brittle fracture behavior of cracked bent plates. The local stress field along the crack front as determined in Section 2.10 can be expressed in terms of three stress intensity factors. They are independent of the local coordinates and depend only on the crack geometry and the form of loading. The basic problem of a plate with an inclined crack under the direction of the applied uniaxial moment is studied. The equations of the strain energy density criterion are established and used to determine the crack extension angle and the critical value of the applied moment for unstable crack growth.

(b) Basic equations

The local stress field along the crack front, as determined in Section 2.10, is expressed by Equations (2.217) in terms of the stress intensity factors K_I, K_II

and K_{III}. Introducing the values of stress components σ_x, σ_y, τ_{xy}, τ_{xz} and τ_{yz} from Equations (2.217) into Equations (6.3) and (6.9) we obtain for the strain energy density factor S the form

$$S = A_{11}K_I^2 + 2A_{12}K_I K_{II} + A_{22}K_{II}^2 + A_{33}K_{III}^2 \qquad (6.34)$$

where the coefficients A_{ij} $(i,j = 1,2,3)$ are given by

$$\begin{aligned} A_{11} &= \frac{1}{8E}(1+\cos\theta)[3-\nu-(1+\nu)\cos\theta] \\ A_{12} &= \frac{1}{4E}\sin\theta[(1+\nu)\cos\theta - (1-\nu)] \\ A_{22} &= \frac{1}{8E}[4(1-\cos\theta) + (1+\nu)(3\cos\theta - 1)(1+\cos\theta)] \\ A_{33} &= \frac{1+\nu}{2E}. \end{aligned} \qquad (6.35)$$

Introducing the values of the stress intensity factors K_I, K_{II} and K_{III} given by Equations (2.219) into Equation (6.35) we obtain the following expression for the strain energy density factor S:

$$S = \frac{2a}{E}\left[\frac{3\Phi(1)M\sin\beta}{h^2}\right]^2\left\{\left(\frac{z}{h}\right)^2 F(\theta,\beta) + \left[1 - \left(\frac{2z}{h}\right)^2\right]^2 G(\beta)\right\} \qquad (6.36)$$

with the functions $F(\theta,\beta)$ and $G(\beta)$ defined as

$$\begin{aligned} F(\theta,\beta) &= B_{11}\sin^2\beta + 2\lambda B_{12}\sin\beta\cos\beta + \lambda^2 B_{22}\cos^2\beta \\ G(\beta) &= \frac{5}{8(1+\nu)}\gamma^2\cos^2\beta \end{aligned} \qquad (6.37)$$

and the coefficients B_{ij} and λ and γ given by

$$\begin{aligned} B_{ij} &= 8EA_{ij} \quad (i,j=1,2,3) \\ \lambda &= \frac{\Psi(1)}{\Phi(1)}, \quad \gamma = \frac{\Omega(1)}{\Phi(1)}. \end{aligned} \qquad (6.38)$$

The quantities $\Phi(1)$, $\Psi(1)$ and $\Omega(1)$ entering the above equations have been defined in Section 2.10.

One of the basic assumptions of the strain energy density criterion is the requirement of the stationary value of the strain energy density factor S (Equation (6.11)). This is expressed by

$$\begin{aligned} (1+\nu)\{2(1-3\lambda^2)\sin 2\theta_c - (1-\lambda)(1-3\lambda)\sin[2(\theta_c+\beta)]- \\ -(1+\lambda)(1+3\lambda)\sin[2(\theta_c-\beta)]\} - 2(1-\nu)[2(1-\lambda^2)\sin\theta_c - \\ -(1-\lambda)^2\sin(\theta_c+2\beta) - (1+\lambda)^2\sin(\theta_c-2\beta)] = 0. \end{aligned} \qquad (6.39)$$

Solution of Equation (6.39) under the restrictions imposed by inequality (6.11) gives the values of the crack extension angle θ_c. Of the two values of θ_c which meet these requirements, one is negative, corresponding to the tension side of the plate, and the other is positive, corresponding to the compression side. Introducing the previously obtained values of the crack extension angle θ_c into Equation (6.36), the minimum values $S_{\min}$ of strain energy density factor are determined.

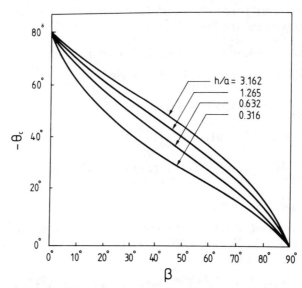

Figure 6.13. Variations of fracture angle with crack angle for $\nu = 0.3$.

When $S_{\min}$ is equated to the critical value S_c of S, which is a material constant, the critical values M_c of the applied moment M for crack extension are obtained. Thus, the following equation is derived

$$\frac{M_c}{(Eh^3 S_c)^{1/2}} = \frac{\sqrt{h/a}}{3\sqrt{2}\,\Phi(1)\sin\beta\{(z/h)^2 F_{\min}(\theta_0,\beta) + [1-(2z/h)^2]G(\beta)\}^{1/2}} \quad (6.40)$$

where $F_{\min}(\theta_0, \beta)$ is the minimum value of the function $F(\theta, \beta)$ which is the only quantity in relation (6.36) dependent on θ. Relation (6.40) gives the normalized critical moment $M_c/(Eh^3 S_c)^{1/2}$ as a function of the crack inclination angle β, the ratio of the plate thickness to the half-crack length h/a, the Poisson ratio ν and the ratio z/h of the distance z of the particular layer considered to the plate thickness.

(c) *Results*

By comparing the critical values of the applied moment for crack extension from different layers of the plate, the critical moment M_c is always found to be smaller at the tensile surface layer. Thus, fracture of the plate always takes place from this layer. Figure 6.13 presents the variation of the angle $-\theta_c$ versus the crack inclination angle for $\nu = 0.3$ and various values of the ratio h/a. Observe that for $\beta = 90°$ the crack extension angle θ_c is equal to zero. In other words, when the direction of the plane of the applied moment is perpendicular to the crack, the crack propagates in its own plane. Figure 6.13 also shows that $|\theta_c|$ increases as h/a also increases. Figure 6.14 establishes the dependence of the critical moment on the ratio h/a for $\nu = 0.3$. Note that M_c decreases with the crack inclination angle (similar to results obtained for the inclined crack under

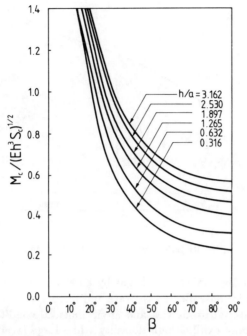

Figure 6.14. Variation of the normalized critical moment versus crack inclination angle for various values of h/a, with $\nu = 0.3$.

extensional loading, Section 6.5, page 209), while it increases with h/a. Thus, when the thickness of the plate increases, the value of the critical moment causing fracture also increases.

For further information on this section refer to [6.6].

6.8. Ductile fracture

(a) Introductory remarks

The term ductile fracture is generally used to indicate failure where unstable crack propagation is preceded by plastic deformation. (Ductile indicates the presence of stable deformation, while fracture designates load instability associated with the sudden creation of a macrocrack surface.) A typical characteristic feature of ductile fracture is that the crack grows slowly at first prior to the onset of unstable crack propagation. The crack growth process can be separated into the phases of crack initiation, and stable and unstable crack growth. Generally speaking, all fracture processes may be regarded as transitions from stable to unstable crack propagation. When the amount of stable crack growth is small it is usually assumed that onset of crack initiation coincides with crack instability. The phenomenon of ductile fracture is associated with a nonlinear load versus deformation relation which is attributed to plastic deformation and slow stable crack growth. These two effects take place simultaneously and it is not possible

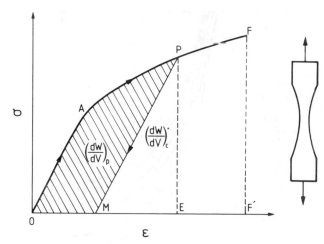

Figure 6.15. Dissipated, $(dW/dV)_p$, and available, $(dW/dV)_c^*$, energy of a tension specimen.

to measure their individual contributions by known experimental methods.

Ductile fracture is controlled by the rate and history of loading, specimen size and geometry, material properties and environmental conditions. Thus, small metal specimens may exhibit high ductility while large structures of the same material can behave in a brittle fashion. Furthermore, a substantial amount of subcritical crack growth may result in a specimen made of a brittle material when it is slowly loaded. On the contrary, brittle fractures may occur in structures made of a ductile material when the load is applied suddenly or when they are subjected to low temperatures. The development of plastic zones during crack growth in ductile fracture which corresponds to material damage at a microscopic scale level reduces the amount of available energy for macrocrack instability.

The strain energy density theory will be used in the following to address the entire history of crack growth including the phases of initiation, stable and unstable crack growth. The theory has successfully been applied to the solution of a number of problems related to ductile fracture. For an in-depth study of such problems the interested reader is referred to references [6.35–6.52].

(b) Energy dissipation

During the process of crack growth not all the energy is consumed by the creation of new macroscopic surfaces. Dissipation of energy also takes place at the microscopic scale level which is referred to as *yielding* in continuum mechanics. Even though yielding prevails in a direction at some angle to the crack propagation direction (Figure 6.2) the amount of energy consumed at the microscopic level along the path of crack growth is not negligible and should be taken into account.

Referring to the true stress–true strain diagram of the material in tension (Figure 6.15), and considering a stress above the yield stress σ_Y, the unloading path will follow the line PM which is almost parallel to the direction OA of the elastic portion of the diagram. If now reloading of the specimen is resumed the new stress–strain curve will be along the line MPF. During the unloading and reloading procedure the amount of energy dissipated is represented by the area $OAPM = (dW/dV)_p$. Thus, for a stress level σ the available strain energy $(dW/dV)_c^*$ for crack growth is represented by the area $MPFF'$, or

$$\left(\frac{dW}{dV}\right)_c^* = \left(\frac{dW}{dV}\right)_c - \left(\frac{dW}{dV}\right)_p. \tag{6.41}$$

This equation gives the total energy per unit volume required for failure of a material element.

Thus, Equation (6.10), which expresses the third hypothesis of the strain energy density theory, should be modified as

$$\left(\frac{dW}{dV}\right)_c^* = \frac{S_1}{r_1} = \frac{S_2}{r_2} = \cdots \frac{S_j}{r_j} = \frac{S_c^*}{r_c^*} \quad \text{or} \quad \frac{S_0^*}{r_0^*} = \text{const.} \tag{6.42}$$

The material damage process increases monotonically up to global instability when

$$\begin{aligned} S_1 < S_2 < \cdots < S_j < \cdots < S_c \\ r_1 < r_2 < \cdots < r_j < \cdots < r_c \end{aligned} \tag{6.43}$$

and comes to rest when

$$\begin{aligned} S_1 > S_2 > \cdots > S_j > \cdots > S_c \\ r_1 > r_2 > \cdots > r_j > \cdots > r_0 \end{aligned} \tag{6.44}$$

where r_0 is the radius of the fracture core region.

(c) Resistance curves

During stable crack growth the relationship between load and crack length increment is usually nonlinear and changes with specimen size and loading rate. The nonlinear data offer little or no insight into the combined effect of those parameters influencing material damage. The objective of constructing crack growth resistance curves is to seek a parameter that, when plotted against crack growth, yields straight line relations. Thus, results other than those obtained experimentally can be obtained by interpolation. In this way, the behavior of large structural components can be predicted from small specimen data, or long-time behavior of material can be estimated from short-time results.

Extensive results [6.37, 6.41, 6.43, 6.53–6.56] on the study of slow stable crack growth by the strain energy density theory revealed that the rate of change of the strain energy density factor S with crack length a, dS/da, is constant. Typical forms of the 'S versus a' curve shown in Figures 6.16(a) to 6.16(c) express the effect of specimen size, loading rate and material properties on crack growth. For a fixed material and loading rate Figure 6.16(a) shows that all S–a curves are parallel lines and move to the left as specimen size is increased. Thus, for a fixed value of S_c the amount of subcritical crack growth decreases with increasing specimen size. In this way, smaller specimens are seen to behave in a more ductile manner than larger specimens, which show less subcritical damage prior to failure – a result that is well known in practice. The effect of loading

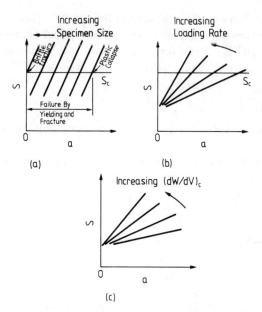

Figure 6.16. Schematic of resistance curves for changes in (a) specimen size, (b) loading rate and (c) $(dW/dV)_c$.

loading rate for a fixed material and specimen size is presented in Figure 6.16(b), which shows that the S–a curves rotate in a counterclockwise sense about a common point as the loading rate is increased. This means that for a given material with a constant value of S_c the critical crack growth length decreases when the rate of loading is increased. Thus, lower loading rates promote slow stable crack instability. Finally, Figure 6.16(c) shows the results of $(dW/dV)_c$ on crack growth. As $(dW/dV)_c$ increases, the S–a lines rotate in a counter-clockwise sense. Thus, materials with higher values of $(dW/dV)_c$ fail with less crack growth than materials with smaller values of $(dW/dV)_c$.

The above results establish the combined effect of specimen size, loading rate and material type on crack growth. They allow the determination of a failure mode that can change from brittle to ductile, involving stable slow crack growth. Results for cases other than those analyzed can be obtained by extrapolation of the data that involve rotation and translation of the S–a lines. Such information is essential for design against ductile fracture.

(d) Development of crack profile

Equation (6.10), expressing the third hypothesis of the strain energy density theory, can be used to determine crack profiles in three dimensions during slow stable crack growth. The procedure is illustrated in Figure 6.17, which refers to a crack in a tensile plate specimen. Six points along one-half of the plate thickness, due to symmetry, are considered. The variation of the strain energy density function dW/dV versus distance r ahead of the crack front is shown in Figure 6.17(b) for the six points. The intersection of the line $(dW/dV)_c =$

Strain energy density failure criterion

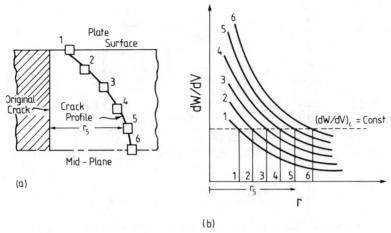

Figure 6.17. Development of crack growth profile.

const with curves $dW/dV - r$ for the six points determines the values of r_j ($j = 1, 2, 3, 4, 5, 6$) which define the crack profile. The same procedure can be repeated to describe the crack growth profile during slow stable crack growth.

This procedure for constructing crack profiles has been applied to the case of a straight crack in a tensile plate specimen [6.36, 6.41]. Larger intervals of crack growth have been obtained for material elements near the plate midsection than for those near the plate boundaries. The influence of plate thickness, material type and loading step on the shape and size of crack growth front during stable crack growth was established. These results explain the well-known crack tunneling effect and verify experimental observation. Crack growth profiles for embedded elliptical, semi-elliptical and quarter-elliptical surface cracks have been determined in references [6.57, 6.58].

6.9. Failure initiation in bodies without pre-existing cracks

The strain energy density theory has been employed so far to analyze the brittle and ductile fracture of solids with pre-existing cracks. This section deals with the application of the theory to problems in which no initial cracks are assumed to exist.

Consider a continuum solid (Figure 6.18) and let $O_j x_j y_j$ be the local coordinate systems attached to the points O_j ($j = 1, 2, \ldots, n$). According to the strain energy density theory the direction of possible failure initiation by fracture or yielding at each point is determined by the minimum and maximum values of the strain energy density function, dW/dV, along the circumference of a circle centered at that point. The location of failure initiation in the continuum coincides with the point at which the maximum of the local minima or maxima of dW/dV occurs. For fracture or yielding initiation, the maximum of the local minima or maxima of dW/dV, respectively, should be equated to the critical value of dW/dV. This value is equal to the area underneath the true stress–true

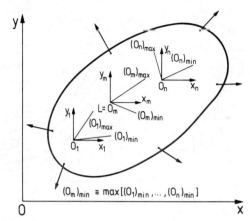

Figure 6.18. Local minima and maxima of dW/dV in a continuum.

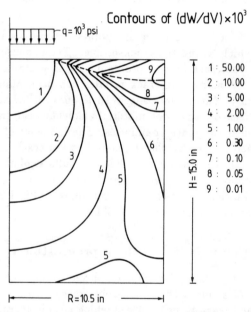

Figure 6.19. Strain energy density contours (continuous lines) and fracture trajectory (dotted line) for a circular cylindrical body pressed by a circular cylindrical punch.

strain diagram of the material in tension up to the point of yield or fracture for yielding or fracture, respectively. This justifies the fact that yielding always precedes fracture.

Having determined the point of fracture initiation, the subsequent fracture path can easily be defined for rapid brittle fracture. In this case failure occurs suddenly and it does not seem unreasonable to assume that the fracture path can be determined from the conditions before failure initiation takes place. In such cases the material does not have sufficient time to redistribute the stresses during the fracture process and the fracture path can be predetermined with high accuracy. The fracture trajectories can be determined from the calculation of the fracture angles for a wide range of radius vectors from the point of fracture initiation.

The above concept has been applied to a number of problems, including notched plates [6.8, 6.59], beams [6.60] and contact problems [6.61]. The case of a circular cylindrical body pressed by a circular cylindrical punch is considered as an example. Figure 6.19 presents the strain energy density contours $(dW/dV) \times 10^3$ for the half-cross-section of a body which has a height of $H = 15$ in. and a radius of its cross-section $R = 10.5$ in. The material of the body has a modulus of elasticity $E = 30 \times 10^6$ lb/in^2 and Poisson's ratio $\nu = 0.3$. Failure initiation of the cylinder starts from the outer end of the applied stress. The subsequent fracture path shown by dashed line in the figure is obtained by the minimum values of dW/dV along various radius vectors centered at the point of failure initiation. Observe that the resulting cone crack has a curved front.

6.10. Other criteria based on energy density

Numerous investigators have attempted to modify the strain energy density criterion on the grounds that a better prediction of the physical events could be achieved [6.62–6.65]. Their argument is based on the decomposition of the energy quantity to its distortional and dilatational components and the use of one of them as the fundamental quantity for the study of crack growth. For a thorough discussion of these and other criteria as related to the strain energy density criterion the reader is referred to [6.66–6.70]. In the following the maximum dilatational strain energy density criterion, as it appeared in [6.65], will be discussed.

This criterion assumes that the crack extends along the direction of maximum dilatational strain energy density $(dW/dV)_v^{max}$ which is calculated along the circumference of a circular core area surrounding the crack tip. For the case of a crack in a mixed-mode stress field governed by the values of the opening-mode, k_I, and sliding-mode, k_{II}, stress intensity factors the dilatational part, $(dW/dV)_v$, of the strain energy density quantity calculated from Equation (6.6) in conjunction with Equations (4.103) takes the form

$$r\left(\frac{dW}{dV}\right)_v = b_{11}k_I^2 + 2b_{12}k_I k_{II} + b_{22}k_{II}^2 \tag{6.45}$$

with

$$\begin{aligned} 12\mu b_{11} &= (1-2\nu)(1+\nu)(1+\cos\theta) \\ 12\mu b_{12} &= -(1-2\nu)(1+\nu)\sin\theta \\ 12\mu b_{22} &= (1-2\nu)(1+\nu)(1-\cos\theta). \end{aligned} \tag{6.46}$$

According to the maximum dilatational strain energy density assumption, the angle θ_c of an initial crack extension can be obtained as

$$\frac{\partial}{\partial\theta}\left[\left(\frac{dW}{dV}\right)_v\right] = 0, \qquad \frac{\partial^2}{\partial\theta^2}\left[\left(\frac{dW}{dV}\right)_v\right] < 0. \tag{6.47}$$

Equation (6.47) gives

$$\sin \theta k_I^2 + 2 \cos \theta k_I k_{II} - \sin \theta k_{II}^2 = 0. \tag{6.48}$$

Let us apply Equation (6.48) to the case of an inclined crack in an infinite plate. Introducing the values of k_I and k_{II} from Equation (6.16) into Equation (6.48), the crack extension angle θ_c is found to be

$$\theta_c = 2\beta - \pi. \tag{6.49}$$

The resulting angles θ_c expressed by the above equation are unrealistic and are far beyond any experimental observation. Thus, for $0 < \beta < \pi/2$ Equation (6.49) predicts angles θ_c in the range $-\pi < \theta_c < 0$, while from experimental observations the angle θ_c varies in the interval $-\pi/2 < \theta_c < 0$. Furthermore, crack extension angles θ_c in the range $-\pi/2 < \theta_c < 0$ have been predicted by most failure criteria (maximum circumferential stress, strain energy density, maximum strain, etc.) used for mixed-mode crack growth predictions.

In conclusion, it should be stated that the soundness of a failure criterion should be judged on its general ability to explain a wide variety of physical phenomena. Furthermore, a failure criterion cannot be responsible for incorrect interpretation of analytical and/or experimental data. Thus a valid criterion should be free from contradictions or inconsistencies when applied to physical problems. In this respect, the maximum dilatational strain energy density criterion leads to predictions that are contrary to physical observations.

More recent advances on the strain energy density theory involve the inclusion of change in the local strain rate on crack extension [6.71]. Data banks for stress–strain relationships with different strain rates are used so that the response of the local elements ahead of the crack can be derived individually according to the load history. Interaction of surface and volume energy [6.72, 6.73] has also been discussed in connection with material damage at the microscopic and macroscopic level. Scaling of size/time/temperature was considered to be an important factor for describing the progressive damage of solids under nonequilibrium conditions.

References

6.1. Sih, G. C., Some basic problems in fracture mechanics and new concepts, *Engineering Fracture Mechanics* 5, 365–377 (1973).
6.2. Sih, G. C., Energy-density concept in fracture mechanics, *Engineering Fracture Mechanics* 5, 1037–1040 (1973).
6.3. Sih, G. C., Strain-energy-density factor applied to mixed mode crack problems, *International Journal of Fracture* 10, 305–321 (1974).
6.4. Sih, G. C., A special theory of crack propagation: methods of analysis and solutions of crack problems, in *Mechanics of Fracture*, Vol. 1 (ed. G. C. Sih), Noordhoff Int. Publ., The Netherlands, pp. XXI-XLV (1973).
6.5. Sih, G. C., A three-dimensional strain energy density factor theory of crack propagation, in *Mechanics of Fracture*, Vol. 2, (M. K. Kassir and G. C. Sih), Noordhoff Int. Publ., The Netherlands, pp. XV–LIII (1975).
6.6. Sih, G. C., Strain energy density theory applied to plate bending problems, in *Mechanics*

of *Fracture*, Vol. 3 (ed. G. C. Sih), Noordhoff Int. Publ., The Netherlands, pp. XVII–XLVIII (1977).
6.7. Sih, G. C., Dynamic crack problems – strain energy density fracture theory, in *Mechanics of Fracture*, Vol. 4, (ed. G. C. Sih), Noordhoff Int. Publ., The Netherlands, pp. XVII–XLVII (1977).
6.8. Sih, G. C., Strain energy density and surface layer energy for blunt cracks or notches, in *Mechanics of Fracture*, Vol. 5 (ed. G. C. Sih), Noordhoff Int. Publ., The Netherlands, pp. XIII-CX (1978).
6.9. Sih, G. C., Failure of composites as predicted by the strain energy density theory, in *Mechanics of Fracture*, Vol. 6 (G. C. Sih and E. P. Chen), Martinus Nijhoff Publ. pp. XV–LXXXI (1981).
6.10. Sih, G. C., Experimental fracture mechanics: strain energy density criterion, in *Mechanics of Fracture*, Vol. 7 (ed. G. C. Sih), Martinus Nijhoff Publ., pp. XVII–LVI (1981).
6.11. Sih, G. C., Czoboly, E. and Gillemot, F., *Absorbed Specific Energy and/or Strain Energy Density Criterion*, Martinus Nijhoff Publ. (1982).
6.12. Sih, G. C. and Gdoutos, E. E., *Mechanics and Physics of Energy Density*, Kluwer Academic Publ. (1990).
6.13. Haigh, B. P., *The Strain Energy Function and the Elastic Limit*, British Association of Advancement of Sciences, pp. 486–495 (1919).
6.14. Nadai, A., *Theory of Flow and Fracture of Solids*, McGraw-Hill, New York (1950).
6.15. Gdoutos, E. E., *Problems of Mixed Mode Crack Propagation*, Martinus Nijhoff Publ. (1984).
6.16. Sih, G. C. and Macdonald, B., Fracture mechanics applied to engineering problems-strain energy density fracture criterion, *Engineering Fracture Mechanics* 6, 361–386 (1974).
6.17. Hoek, E. and Bieniawski, Z. T., Fracture propagation mechanics in hard rock, *Technical Report – Rock Mechanics Division*, South African Council for Scientific and Industrial Research (1965).
6.18. Gdoutos, E. E., Finite width effects on the crack extension angle, *International Journal of Fracture* 15, R111–R114 (1979).
6.19. Gdoutos, E. E., The influence of specimen's geometry on the crack extension angle, *Engineering Fracture Mechanics* 13, 79–84 (1980).
6.20. Gdoutos, E. E., Fracture phenomena in a cracked plate subjected to a concentrated load, *Engineering Fracture Mechanics* 14, 323–335 (1981).
6.21. Gdoutos, E. E., Fracture of cracked plates under localised moments, *International Journal of Mechanical Sciences* 23, 121–128 (1981).
6.22. Gdoutos, E. E., Fracture of plates with circular cracks, *Materialprüfung* 22, 83–86 (1980).
6.23. Gdoutos, E. E., Brittle fracture of a plate with two cracks emanating from a circular hole under inclined tension, *Materialprüfung* 23, 194–196 (1981).
6.24. Gdoutos, E. E., Failure of a plate with a circular hole resulting from an array of surface cracks, *Engineering Fracture Mechanics* 15, 457–467 (1981).
6.25. Gdoutos, E. E., Fracture initiation from singular points of rigid inclusions, *Journal of Applied Mechanics, Trans. ASME* 47, 971–973 (1980).
6.26. Gdoutos, E. E., Fracture of composites with rigid inclusions having cuspidal points, *Proceedings of the International Conference on Analytical and Experimental Fracture Mechanics, Rome, Italy, June 23-27, 1980* (eds G. C. Sih and M. Mirabile), Sijthoff and Noordhoff, pp. 943–958 (1981).
6.27. Gdoutos, E. E., Fracture phenomena in composites with rigid inclusions, *Proceedings of the US-Greece Seminar on Mixed Mode Crack Propagation, Athens, Greece, August 18-22, 1980* (eds G. C. Sih and P. S. Theocaris), Sijthoff and Noordhoff, pp. 109–122 (1981).
6.28. Gdoutos, E. E., Failure of a composite with a rigid fiber inclusion, *Acta Mechanica* 39, 251–262 (1981).
6.29. Gdoutos, E. E., Failure of a bimaterial plate with a crack at an arbitrary angle to the interface, *Fibre Science and Technology* 15, 27–40 (1981).
6.30. Gdoutos, E. E., Fracture of aluminum-epoxy layered composites containing cracks, *Journal of Strain Analysis* 17, pp. 75–78 (1982).
6.31. Gdoutos, E. E., Interaction effects between a crack and a circular inclusion, *Fibre Science and Technology* 15, 173–185 (1981).
6.32. Gdoutos, E. E., Interaction between two equal skew-parallel cracks, *Journal of Strain Analysis* 15, 127–136 (1980).
6.33. Sih, G. C. and Cha, B. C. K., A fracture criterion for three-dimensional crack problems, *Engineering Fracture Mechanics* 6, 699–723 (1974).
6.34. Hartranft, R. J. and Sih, G. C., Stress singularity for a crack with an arbitrary crack

front, *Engineering Fracture Mechanics* 9, 705–718 (1977).
6.35. Sih, G. C., Fracture toughness concept, in *Properties Related to Fracture Toughness*, ASTM STP 605, American Society for Testing and Materials, Philadelphia, pp. 3–15 (1976).
6.36. Sih, G. C. and Kiefer, B. V., Nonlinear response of solids due to crack growth and plastic deformation, in *Nonlinear and Dynamic Fracture Mechanics* (eds N. Perrone and S. W. Atluri), The American Society of Mechanical Engineers, AMD, Vol. 35, pp. 136–156 (1979).
6.37. Sih, G. C. and Madenci, E., Crack growth resistance characterized by strain energy density function, *Engineering Fracture Mechanics* 18, 1159–1171 (1983).
6.38. Sih, G. C., The analytical aspects of macrofracture mechanics, in *Analytical and Experimental Fracture Mechanics* (eds G. C. Sih and M. Mirabile), Sijthoff and Noordhoff Int. Publ., The Netherlands, pp. 3–15 (1981).
6.39. Sih, G. C., Mechanics of crack growth: geometrical size effect in fracture, in *Fracture Mechanics in Engineering Applications* (eds G. C. Sih and S. R. Valluri), Sijthoff and Noordhoff Int. Publ., The Netherlands, pp. 3–29 (1979).
6.40. Sih, G. C. and Madenci, E., Prediction of failure in weldments, Part I: Smooth joint; and Part II: Joint with initial notch and crack, *Theoretical and Applied Fracture Mechanics* 3, 23–29 and 31–40 (1985).
6.41. Sih, G. C. and Chen, C., Non-self-similar crack growth in elastic-plastic finite thickness plate, *Theoretical and Applied Fracture Mechanics* 3, 125–139 (1985).
6.42. Sih, G. C. and Tzou, D. Y., Crack-extension resistance of polycarbonate material, *Theoretical and Applied Fracture Mechanics* 2, 229–234 (1984).
6.43. Carpinteri, A. and Sih, G. C., Damage accumulation and crack growth in bilinear materials with softening: Application of energy density theory, *Theoretical and Applied Fracture Mechanics* 1, 145–159 (1984).
6.44. Gdoutos, E. E., Stable growth of a central crack, *Theoretical and Applied Fracture Mechanics* 1, 139–144 (1984).
6.45. Gdoutos, E. E. and Sih, G. C., Crack growth characteristics influenced by load time record, *Theoretical and Applied Fracture Mechanics* 2, 91–103 (1984).
6.46. Gdoutos, E. E., Path dependence of stable crack growth, in *Advances in Fracture Research, Proceedings of the Sixth International Conference on Fracture* (eds S. R. Valluri et al.), Pergamon Press, Vol. 2, pp. 1021–1028 (1984).
6.47. Gdoutos, E. E., Stable growth of a crack interacting with a circular inclusion, *Theoretical and Applied Fracture Mechanics* 3, 141–150 (1985).
6.48. Gdoutos E. E. and Papakaliatakis, G., The effect of load biaxiality on crack growth in non-linear materials, *Theoretical and Applied Fracture Mechanics* 5, 133–140 (1986).
6.49. Gdoutos, E. E. and Papakaliatakis, G., The influence of plate geometry and material properties on crack growth, *Engineering Fracture Mechanics* 25, 141–156 (1986).
6.50. Gdoutos, E. E., Mixed mode crack growth predictions, *Engineering Fracture Mechanics* 28, 211–221 (1987).
6.51. Gdoutos, E. E. and Papakaliatakis, G., Dependence of crack growth initiation on the form of the stress-strain diagram of the material in tension, *Engineering Fracture Mechanics* 34, 143–151 (1981).
6.52. Sih, G. C., The state of affairs near the crack tip, in *Modeling Problems in Crack Tip Mechanics* (ed. J. T. Pindera), Martinus Nijhoff Publ., pp. 65–90 (1984).
6.53. Sih, G. C. and Tzou D. Y., Mechanics of nonlinear crack growth: effects of specimen size and loading step, *Modeling Problems in Crack Tip Mechanics* (ed. J. T. Pindera), Martinus Nijhoff Publ., The Netherlands, pp. 155–169 (1984).
6.54. Carpinteri, Andrea, Crack growth resistance in non-perfect plasticity: Isotropic versus kinematic hardening, *Theoretical and Applied Fracture Mechanics* 4, 117–122 (1985).
6.55. Sih, G. C. and Chao, C. K., Size effect of cylindrical specimens with fatigue cracks, *Theoretical and Applied Fracture Mechanics* 1, 239–247 (1984).
6.56. Sih, G. C. and Chao, C. K., Influence of load amplitude and uniaxial tensile properties on fatigue crack growth, *Theoretical and Applied Fracture Mechanics* 2, 247–257 (1984).
6.57. Sih, G. C. and Kiefer, B. V., Stable growth of surface cracks, *Journal of Engineering Mechanics Division, ASCE* 106, 245–253 (1980).
6.58. Gdoutos, E. E. and Hatzitrifon, N., Growth of three-dimensional cracks in finite-thickness plates, *Engineering Fracture Mechanics* 26, 883–895 (1987).
6.59. Kipp, M. E. and Sih, G. C., The strain energy density failure criterion applied to notched elastic solids, *International Journal of Solids and Structures* 11, 153–173 (1975).
6.60. Gdoutos, E. E., Continuum aspects of crack initiation, *Application of Fracture Mechanics to*

Materials and Structures (eds G. C. Sih, E. Sommer and W. Dahl), Martinus Nijhoff Publ., pp. 237–249 (1984).
6.61. Gdoutos, E. E. and Drakos, G., Crack initiation and growth in contact problems, *Theoretical and Applied Fracture Mechanics* 3, 227–232 (1985).
6.62. Jayatilaka, A. de S., Jenkins, I. J. and Prasad, S. V., Determination of crack growth in a mixed mode loading system, in *Advances in Research on the Strength and Fracture of Materials* (ed. D. M. R. Taplin), Pergamon Press, Vol. 3A, pp. 15–23 (1978).
6.63. Theocaris, P. S. and Andrianopoulos, N. P., A modified strain-energy density criterion applied to crack propagation, *Journal of Applied Mechanics, Trans. ASME* 49, 81–86 (1982).
6.64. Theocaris, P. S., A higher-order approximation to the T-criterion of fracture in biaxial fields, *Engineering Fracture Mechanics* 19, 975–991 (1984).
6.65. Chow, C. L. and Xu, Jilin, Mixed mode ductile fracture using the strain energy density criterion, *International Journal of Fracture* 28, 17–28 (1985).
6.66. Sih, G. C. and Gdoutos, E. E., Discussion of paper [6.63], *Journal of Applied Mechanics, Trans. ASME* 49, 678–679 (1982).
6.67. Sih, G. C., Moyer, E. T. Jr., and Gdoutos, E. E., Discussion of the paper 'The Mises elastic-plastic boundary as the core region in fracture criteria', *Engineering Fracture Mechanics* 18, 731–733 (1983).
6.68. Sih, G. C. and Tzou, D. Y., Discussion of the paper 'Criteria for brittle fracture in biaxial tension', *Engineering Fracture Mechanics* 21, 977–981 (1985).
6.69. Gdoutos, E. E., Discussion of the papers 'Comparison of the criteria for mixed mode brittle fracture based on the preinstability stress-strain field', Part I: Slit and elliptical cracks under uniaxial tensile loading and Part II: Pure shear and uniaxial compressive loading, *International Journal of Fracture* 27, R23–R29 (1985).
6.70. Gdoutos, E. E., Discussions of paper [6.65], *International Journal of Fracture* 30, R53–R58 (1986) and 33, R71–R72 (1987).
6.71. Sih, G. C. and Tzou D. Y., Plastic deformation and crack growth behavior, in *Series on Fatigue and Fracture*, Volume III, *Plasticity and Failure Behavior* (eds G. C. Sih, A. J. Ishlinsky and S. T. Mileiko), pp. 91–114, Kluwer Academic Publ. (1990)
6.72. Sih, G. C. and Chao, C. K., Scaling of size/time/temperature – Part 1: Progressive damage in uniaxial tensile specimen, *Theoretical and Applied Fracture Mechanics*, 12, 93–108 (1989).
6.73. Sih, G. C. and Chao C. K., Scaling of size/time/temperature – Part 2: Progressive damage in uniaxial compressive specimen, *Theoretical and Applied Fracture Mechanics* 12, 109–119, (1989).

7

Dynamic fracture

7.1. Introduction

The analysis of crack systems considered so far concerned only quasi-static situations in which the kinetic energy is relatively insignificant in comparison with the other energy terms and can be omitted. The crack was assumed either to be stationary or to grow in a controlled stable manner and the applied loads varied quite slowly. The present chapter is devoted entirely to dynamically loaded stationary or growing cracks. In such cases rapid motions are generated in the medium and inertia effects become of significant importance.

Elastodynamic analysis of crack problems indicates that stresses and displacements caused by dynamic loading can differ greatly from those associated with corresponding static loading. It is often found that at some locations in the body the dynamic stresses are higher than the corresponding static stresses. This result may be explained by the interaction of the emitted elastic waves with the crack faces and other characteristic boundaries of the body. Furthermore, the mechanical properties of the material depend markedly on the time for which the applied loading is maintained in the solid. For example, in most metals both the yield and ultimate strength increase with the rate of loading. Dynamic loads give rise to high stress levels near cracks and fracture takes place so rapidly that there is insufficient time for yielding to develop. Energy is therefore released within a short time, resulting in rapid crack propagation which explains the experimental observation that dynamic loads generally promote brittle fractures.

Broadly speaking, problems of dynamic fracture mechanics may be classified into two main categories. The first concerns the situation where a crack reaches a point of instability and moves rapidly, generally, under slowly varying applied loading. Motion of the crack leads to a sudden unloading along the crack path. The second category of dynamic problems arises when a body with a stationary crack is subjected to a rapidly varying load – for example, an impact or impulsive load. Problems of interest concern initiation of rapid crack growth, crack speed, crack branching and crack arrest.

A dynamic crack problem may be stated in its most general form as follows: A solid body with an initial crack is subjected to a time-dependent loading. We are seeking the conditions of crack initiation, growth and arrest but, even in its

Dynamic fracture

most general form, the formulation of the problem is not an easy task. Study is usually restricted to plane symmetric or axisymmetric problems in which the crack area can be characterized by a length parameter a and the crack path is known beforehand. Solution of the problem requires determination of the three displacement components and the crack length as a function of time. The three equations of motion, coupled with a fracture criterion, provide four equations for the determination of the above four unknown quantities.

From the historical point of view the first work on fracture under dynamic loading was performed by J. Hopkinson [7.1]. He measured the strength of steel wires subjected to a falling weight and explained the results in terms of elastic waves propagating along the wire. The next major investigation was carried out by the son of J. Hopkinson, B. Hopkinson, [7.2, 7.3] who detonated explosive charges in contact with metal plates. He demonstrated the effect of 'spalling' or 'scabbing' which results when a compressive pulse is reflected at the opposite free face of a plate as a tensile pulse. After the work of B. Hopkinson very little research had been done until the Second World War. Following the development of fracture mechanics a number of studies related to dynamic crack growth have appeared. Among the pioneering investigations the work performed by Mott [7.4], Schardin et al. [7.5], Kerkhof [7.6, 7.7], Yoffe [7.8] and Wells and Post [7.9] should be mentioned. Since then substantial progress in the field has been made and a vast number of publications have appeared in the literature. For a thorough in-depth study of dynamic fracture mechanics problems the reader is referred to the book by Sih [7.10] and to the review articles by Erdogan [7.11], Achenbach [7.12] and Freund [7.13].

The present chapter presents in a concise form the basic concepts and the salient points of dynamic fracture mechanics. The theory advanced by Mott for the prediction of the speed of a moving crack is first described. The theory, besides its limitations, constitutes the first attempt to include the kinetic energy term into the Griffith energy balance equation. Further extensions and improvements of Mott's results are also outlined. The stress field around a crack moving at constant velocity is then described. The mathematical treatment is simplified by defining a set of coordinate axes attached to the moving crack tip and by solving the elastodynamic equations in the moving coordinate system. Attention is focused on the stress field in the vicinity of the crack tip which presents an inverse square root singularity and is expressed in terms of the dynamic stress intensity factor in a manner analogous to the static case. The concept of the strain energy release rate is then introduced and it is related to the stress intensity factor. The transient response of cracks to impact and the scattering of elastic waves about stationary and moving cracks, based on the stress intensity factor as in the static case, is formulated. The problems of crack branching and crack arrest are also treated. The chapter concludes with a description of the main methods for the experimental study of dynamic crack problems.

7.2. Mott's model

The theory proposed by Mott [7.4] constitutes the first attempt for a quantitative prediction of the speed of a rapidly moving crack. Mott extended Griffith's theory by adding a kinetic energy term to the expression of the total energy of the system and sought the configuration which keeps this total energy constant. The problem considered by Mott was the propagation of a central crack in an infinite plate subjected to a uniform time-independent uniaxial stress σ perpendicular to the plane of the crack. The following key assumptions were made:

(i) The stress and displacement fields for the dynamic problem are the same as those for the static problem with the same crack length.
(ii) The crack is travelling at a constant velocity.
(iii) The crack speed is small compared to the shear wave velocity in the body.

To obtain an expression for the kinetic energy term K given by Equation (4.5) the displacement field in the solid must be known. Mott derived an expression for K on dimensional grounds. The components of the velocity at a given point in the body due to a rapidly propagating crack may be written as

$$\dot{u} = \frac{du}{dt} = \frac{\partial u}{\partial a}\frac{da}{dt} = \dot{a}\frac{\partial u}{\partial a} \tag{7.1a}$$

$$\dot{v} = \frac{dv}{dt} = \frac{\partial v}{\partial a}\frac{da}{dt} = \dot{a}\frac{\partial v}{\partial a} \tag{7.1b}$$

where the dot symbol denotes differentiation with respect to time.

The kinetic energy from Equation (4.5) takes the form

$$K = \tfrac{1}{2}\rho\dot{a}^2 \iint_R \left[\left(\frac{\partial u}{\partial a}\right)^2 + \left(\frac{\partial v}{\partial a}\right)^2\right] dx\,dy \tag{7.2}$$

where ρ represents the mass density.

The displacement components u and v behind the crack tip given from Equations (2.78) can be put in the form

$$u = \frac{\sigma}{E}\sqrt{ar}\,f_1(\theta), \qquad v = \frac{\sigma}{E}\sqrt{ar}\,f_2(\theta). \tag{7.3}$$

For a fixed element the distance r from the moving crack tip is proportional to the crack length and the displacements u and v can be put in the form

$$u = \frac{c_1\sigma a}{E}, \qquad v = \frac{c_2\sigma a}{E}. \tag{7.4}$$

Thus, Equation (7.2) becomes

$$K = \tfrac{1}{2}\rho\dot{a}^2\frac{\sigma^2}{E}\iint_R (c_1^2 + c_2^2)\,dx\,dy. \tag{7.5}$$

For the case of the infinite plate the crack length $2a$ is the only characteristic length dimension. Thus, the area integral in Equation (7.5) must be proportional to a and K becomes

$$K = \tfrac{1}{2}k^2\rho\dot{a}^2 a^2 \frac{\sigma^2}{E^2} \tag{7.6}$$

where k is a constant.

Dynamic fracture

Integrating the energy balance equation (Equation (4.1)) with respect to time and omitting the strain energy dissipated to plastic deformation we obtain

$$W - U^e - K - \Gamma = c \tag{7.7}$$

where c is a constant.

For a crack propagating under either fixed-grips or constant stress conditions and introducing the value of $W - U^e$ from Equation (4.23) we obtain, for conditions of generalized plane stress,

$$\frac{\pi \sigma^2 a^2}{E} - \tfrac{1}{2} k^2 \rho \dot{a}^2 a^2 \frac{\sigma^2}{E^2} - 4\gamma a = c. \tag{7.8}$$

Differentiating Equation (7.8) with respect to a and making use of the second assumption above ($\partial \dot{a}/\partial a = 0$) we find, for the crack speed, that

$$\dot{a} = \sqrt{\frac{2\pi}{k^2}} \sqrt{\frac{E}{\rho}} \sqrt{1 - \frac{a_0}{a}} \tag{7.9}$$

where from Griffith's equation (Equation (4.25))

$$\gamma = \frac{\sigma^2 \pi a_0}{2E} \tag{7.10}$$

with a_0 being the half crack length at $t = 0$.

For $a \gg a_0$ Equation (7.9) predicts a limiting crack velocity

$$\dot{a} = \sqrt{\frac{2\pi}{k}} v_s, \quad v_s = \sqrt{\frac{E}{\rho}} \tag{7.11}$$

where v_s is the velocity of longitudinal waves in the material which is equal to the velocity of sound.

Roberts and Wells [7.14] found that $\sqrt{2\pi/k} = 0.38$ for a Poisson's ratio $\nu = 0.25$. Thus Equation (7.9) becomes

$$\dot{a} = 0.38 v_s \sqrt{1 - \frac{a_0}{a}}. \tag{7.12}$$

Berry [7.15] and Dulaney and Brace [7.16] re-examined Mott's theory by relieving the assumption $\partial \dot{a}/\partial a = 0$. Assuming that the applied stress σ is greater than the critical stress σ_0 for the initiation of crack extension, Equation (7.8) becomes

$$\frac{\pi \sigma^2 a^2}{E} - \frac{k^2 \rho \dot{a}^2 a^2 \sigma^2}{2E^2} - \frac{2\pi a_0 a \sigma_0^2}{E} = c. \tag{7.13}$$

Putting $\dot{a} = 0$ and $a = a_0$ at $t = 0$, Equation (7.13) takes the form

$$\frac{\pi \sigma^2 a^2}{E} - \frac{2\pi a_0^2 \sigma_0^2}{E} = c. \tag{7.14}$$

From Equations (7.13) and (7.14) by eliminating the constant c we obtain

$$\dot{a}^2 = \frac{2\pi E}{k^2 \rho} \left(1 - \frac{a_0}{a}\right) \left[1 - (2n^2 - 1)\frac{a_0}{a}\right] \tag{7.15}$$

where

$$\sigma_0 - n\sigma, \quad n \leq 1. \tag{7.16}$$

For $n = 1$ Equation (7.15) gives

$$\dot{a} = 0.38 v_s \left(1 - \frac{a_0}{a}\right) \tag{7.17}$$

using the result obtained by Roberts and Wells to determine k.

Experimentally obtained crack velocities are below the theoretical values given by Equation (7.17). Roberts and Wells [7.14] reported values of $\dot{a}/v_s$ in the range 0.20–0.37 for various materials. Kanninen [7.17], based on the Dugdale model, predicted a crack velocity $\dot{a} = 0.1 v_s$ in steel plate. Duffy et al. [7.18] measured cleavage fracture velocities in steel pipes of the order of 700 m/s and shear fracture of the order of 200 m/s.

7.3. Stress field around a rapidly propagating crack

(a) Introductory remarks

The first mathematical analysis of the stress field in the neighborhood of a moving crack was made by Yoffe [7.8]. She considered a crack of constant length propagating with constant speed in an infinite elastic medium under a uniform applied stress perpendicular to the plane of the crack. Using a Fourier method approach she obtained a singular crack-tip stress field which reduces to the Inglis solution as the speed of the crack approached zero. The interesting result that the circumferential stress in the vicinity of the crack tip presents a maximum at an angle to the crack plane for large enough crack velocities contrary to the static case was obtained. The critical crack velocity for this behavior was of about $0.6C$, where C is the shear wave velocity which is equal to $\sqrt{\mu/\rho}$. Thus, an explanation was offered for the crack-curving phenomenon that was experimentally observed at high crack speeds. The analysis, however, does not provide an estimate of the forces required to maintain the crack motion.

Craggs [7.19] considered the problem of a semi-infinite crack and obtained a solution based on the much more elementary complex variable method which agrees with Yoffe's solution near the end of the crack. He also obtained the force required to extend the crack. Broberg [7.20] solved the dynamic problem of a crack expanding from zero length at a uniform rate. Baker [7.21] included a finite initial crack in Broberg's solution. After these pioneering works a number of solutions for handling constant velocity crack problems have appeared [7.22–7.29]. When the crack speed varies with time the problem becomes complicated and it is difficult to obtain realistic solutions. Kostrov [7.30] first solved the problem of an accelerating shear crack. Eshelby [7.31] considered the antiplane motion of a crack with one end stationary and the other end moving at a nonuniform rate. Freund [7.32] used a semi-inverse approach to obtain the stress intensity factor for a semi-infinite crack accelerating under in-plane static loads. The problem of a finite length crack propagating with acceleration under plane extensional loading was treated by Kostrov [7.33] and Tsai [7.34].

In the following a formulation of the problem of a rapidly propagating crack is given and the near crack-tip stress and displacement fields are presented.

(b) Formulation of the problem

An infinite elastic body in a state of plane strain with a semi-infinite crack is considered. The body is referred to the Cartesian system $x_1 x_2$, where the x_1-axis is oriented along the crack direction and it is assumed that the crack propagates in a self-similar manner. The Navier equations of motion take the form

$$(\lambda + \mu) u_{j,ij} + \mu u_{i,jj} = \rho \ddot{u}_i \tag{7.18}$$

where

$$\lambda = \frac{2\mu\nu}{1 - 2\nu}. \tag{7.19}$$

In Equations (7.18) and (7.19) μ denotes the shear modulus, ν Poisson's ratio and the dot symbol the partial differentiation with respect to time.

The in-plane displacements u_1 and u_2 for plane strain conditions may be expressed in terms of the two potentials Φ and Ψ by

$$u_1 = \frac{\partial \Phi}{\partial x_1} + \frac{\partial \Psi}{\partial x_2}, \qquad u_2 = \frac{\partial \Phi}{\partial x_2} - \frac{\partial \Psi}{\partial x_1}. \tag{7.20}$$

Introducing the values of u_1 and u_2 from Equation (7.20) into Equation (7.19) gives

$$c_1^2 \nabla^2 \Phi = \ddot{\Phi}, \qquad c_2^2 \nabla^2 \Psi = \ddot{\Psi} \tag{7.21}$$

where c_1 and c_2, given by

$$c_1^2 = \frac{\kappa + 1}{\kappa - 1} \frac{\mu}{\rho}, \qquad c_2^2 = \frac{\mu}{\rho} \tag{7.22}$$

denote the dilatational wave speed and the shear wave speed, respectively.

The stresses σ_{11}, σ_{22}, σ_z and σ_{12} in terms of the potentials Φ and Ψ are given by

$$\sigma_{11} = \lambda \nabla^2 \Phi + 2\mu \left(\frac{\partial^2 \Phi}{\partial x_1^2} + \frac{\partial^2 \Psi}{\partial x_1 \partial x_2} \right) \tag{7.23a}$$

$$\sigma_{22} = \lambda \nabla^2 \Phi + 2\mu \left(\frac{\partial^2 \Phi}{\partial x_2^2} - \frac{\partial^2 \Psi}{\partial x_1 \partial x_2} \right) \tag{7.23b}$$

$$\sigma_z = \lambda \nabla^2 \Phi \tag{7.23c}$$

$$\sigma_{12} = \mu \left(2 \frac{\partial^2 \Phi}{\partial x_1 \partial x_2} - \frac{\partial^2 \Psi}{\partial x_1^2} + \frac{\partial^2 \Psi}{\partial x_2^2} \right). \tag{7.23d}$$

When the crack moves at a uniform velocity it is convenient to introduce a new coordinate system x, y attached to the crack tip. We have

$$\begin{aligned} x &= x_1 - a(t) \\ y &= x_2. \end{aligned} \tag{7.24}$$

(c) Singular stress field

When Equations (7.21) are referred to the moving coordinate system and attention is paid to the singular stress field in the vicinity of the crack tip the following equations are obtained

$$\begin{aligned} \beta_1^2 \frac{\partial^2 \Phi}{\partial x^2} + \frac{\partial^2 \Phi}{\partial y^2} &= 0 \\ \beta_2^2 \frac{\partial^2 \Psi}{\partial x^2} + \frac{\partial^2 \Psi}{\partial y^2} &= 0 \end{aligned} \tag{7.25}$$

where

$$\beta_1^2 = 1 - \frac{v^2}{c_1^2}, \qquad \beta_2^2 = 1 - \frac{v^2}{c_2^2} \tag{7.26}$$

with $v = \dot{a}$ denoting the crack speed.

When the new variables y_1 and y_2 are introduced such that

$$y_1 = \beta_1 y, \qquad y_2 = \beta_2 y \tag{7.27}$$

Equations (7.25) become

$$\begin{aligned} \frac{\partial^2 \Phi}{\partial x^2} + \frac{\partial^2 \Phi}{\partial y_1^2} &= 0 \\ \frac{\partial^2 \Psi}{\partial x^2} + \frac{\partial^2 \Psi}{\partial y_2^2} &= 0. \end{aligned} \tag{7.28}$$

Introduce now the complex variables

$$\begin{aligned} z_1 &= x + i y_1 = x + i \beta_1 y = r_1 \, e^{i\theta_1} \\ z_2 &= x + i y_2 = x + i \beta_2 y = r_2 \, e^{i\theta_2}. \end{aligned} \tag{7.29}$$

Equations (7.28) suggest that the function Φ or Ψ is the real or the imaginary part of a complex function of the variable z_1 or z_2. When attention is paid to the singular stress field it may be put

$$\Phi = A_1 \operatorname{Re} z_1^\lambda, \qquad \Psi = A_2 \operatorname{Re} z_2^\lambda, \tag{7.30}$$

where A_1, A_2 and λ are real constants.

Following the eigenvalue expansion method for a semi-infinite static crack, described in Section 2.3, the asymptotic stress and displacement fields are obtained. From the boundary condition along the crack faces, expressed by $\sigma_{22} = \sigma_{12} = 0$, two linear homogeneous equations for the constants A_1 and A_2 are obtained.

It follows that $A_2 = -2A_1\beta_2/(1+\beta_1^2)$ and $\lambda = 3/2$. When now the dynamic opening-mode stress intensity factor $K(t)$ is introduced by

$$K(t) = \lim_{r \to 0}[\sqrt{2\pi r}\,\sigma_{22}(r,0,t)] \qquad (7.31)$$

as in the static case, the following equations for the singular stress field are obtained:

$$\sigma_{11} = \frac{K(t)B}{\sqrt{2\pi r}}\left[(1+2\beta_1^2-\beta_2^2)\sqrt{\frac{r}{r_1}}\cos\frac{\theta_1}{2} - \frac{4\beta_1\beta_2}{1+\beta_2^2}\sqrt{\frac{r}{r_2}}\cos\frac{\theta_2}{2}\right] \qquad (7.32a)$$

$$\sigma_{22} = \frac{2K(t)B}{\sqrt{2\pi r}}\left[-(1+\beta_2^2)\sqrt{\frac{r}{r_2}}\cos\frac{\theta_1}{2} + \frac{4\beta_1\beta_2}{1+\beta_2^2}\sqrt{\frac{r}{r_2}}\cos\frac{\theta_2}{2}\right] \qquad (7.32b)$$

$$\sigma_{12} = \frac{2K(t)B\beta_1}{\sqrt{2\pi r}}\left[\sqrt{\frac{r}{r_1}}\sin\frac{\theta_1}{2} - \sqrt{\frac{r}{r_2}}\sin\frac{\theta_2}{2}\right] \qquad (7.32c)$$

where

$$B = \frac{1+\beta_2^2}{4\beta_1\beta_2 - (1+\beta_2^2)^2}. \qquad (7.33)$$

Equations (7.32) for $v \to 0$ coincide with Equations (2.76) of the stress field around a static crack.

The particle velocity field in the vicinity of the crack tip is given by

$$\dot{u}_1 = -\frac{K(t)Bv}{\sqrt{2\pi r}\,\mu}\left[\sqrt{\frac{r}{r_1}}\cos\frac{\theta_1}{2} - \frac{2\beta_1\beta_2}{1+\beta_2^2}\sqrt{\frac{r}{r_2}}\cos\frac{\theta_2}{2}\right] \qquad (7.34a)$$

$$\dot{u}_2 = -\frac{K(t)Bv}{\sqrt{2\pi r}\,\mu}\left[\sqrt{\frac{r}{r_1}}\sin\frac{\theta_1}{2} - \frac{2}{1+\beta_2^2}\sqrt{\frac{r}{r_2}}\sin\frac{\theta_2}{2}\right]. \qquad (7.34b)$$

Relations for the dynamic stress intensity factor have been developed by various investigators. Broberg [7.20] and Baker [7.21] gave the following equation

$$K(t) = k(v)K(0) \qquad (7.35)$$

where $K(0)$ is the static stress intensity factor for a static crack of length equal to the length of the moving crack and $k(v)$ is a geometry independent function of crack speed. The quantity $k(v)$ decreases monotonically with crack speed and can be approximated by

$$k(v) = 1 - \frac{v}{c_R} \qquad (7.36)$$

where c_R denotes the Rayleigh wave velocity. Observe that the dynamic stress intensity factor becomes equal to zero when the crack speed v becomes equal to c_R.

Rose [7.35] gave the following approximation for $k(v)$:

$$k(v) = \left(1 - \frac{v}{c_R}\right)(1-hv)^{-1/2} \qquad (7.37)$$

where

$$h = \frac{2}{c_1}\left(\frac{c_2}{c_R}\right)^2\left[1 - \left(\frac{c_2}{c_1}\right)^2\right] \qquad (7.38)$$

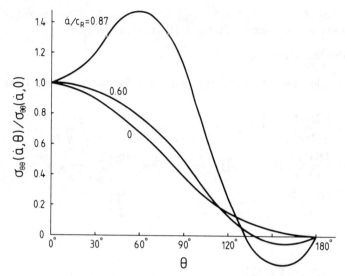

Figure 7.1. Normalized circumferential stress $\sigma_{\theta\theta}(\dot{a},\theta)/\sigma_{\theta\theta}(\dot{a},0)$ versus polar angle θ for different values of $\dot{a}/c_R$.

where c_1 and c_2 are given by Equation (7.22).

(d) *Observations on the singular stress field*

Two fundamental properties of the singular elastodynamic stress field resulting from Equations (7.32) will be referred to. The first is related to the angular variation of the circumferential stress σ_θ and the second to the stress triaxiality ahead of the crack tip.

Figure 7.1 presents the angular variation of the stress σ_θ normalized to its value for $\theta = 0$ for different values of crack speed to the Rayleigh wave speed ratio $\dot{a}/c_R$. Note that c_R is somewhat smaller than the shear wave speed and observe that σ_θ presents a maximum for an angle θ different from zero when the crack speed becomes large. This result, which was first indicated by Yoffe [7.8], explains the experimentally observed phenomenon of crack branching at large crack speeds.

The ratio of the principal stresses σ_{22} and σ_{11} ahead of the crack ($\theta = 0$) is expressed by

$$\frac{\sigma_{22}}{\sigma_{11}} = \frac{4\beta_1\beta_2 - (1+\beta_2^2)^2}{(1+\beta_2^2)(1+2\beta_1^2 - \beta_2^2) - 4\beta_1\beta_2}. \tag{7.39}$$

The variation of σ_{22}/σ_{11} with $\dot{a}/c_R$ is presented in Figure 7.2. Observe that σ_{22}/σ_{11} decreases continuously from 1 at zero crack speed to 0 at the Rayleigh speed. Thus, the stress triaxiality ahead of a rapidly moving crack decreases as the crack speed increases. This results in higher plastic deformation and explains the observed phenomenon of increasing crack growth resistance at high crack speeds.

Dynamic fracture

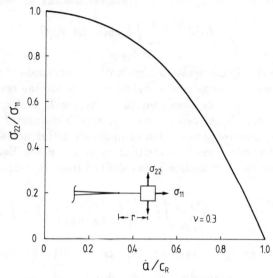

Figure 7.2. Variation of σ_{22}/σ_{11} versus $\dot{a}/c_R$.

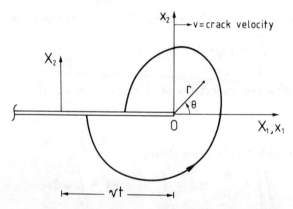

Figure 7.3. A fixed and a moving coordinate system attached to the crack tip.

7.4. Strain energy release rate

Consider a crack in a two-dimensional elastic body moving with constant velocity v in the X_1 direction (Figure 7.3). The body is referred to the fixed coordinate system $X_1 X_2$ while a set of moving rectangular coordinates x_1, x_2 are attached to the crack tip. We have

$$x_1 = X_1 - vt, \qquad x_2 = X_2. \tag{7.40}$$

Let us now consider an arbitrary contour C which encompasses the crack tip and travels at the same speed as the crack. The balance of energy flow across the moving region bounded by C may be mathematically written as [7.36]:

$$\int_C T_i u_i \, ds = \frac{d}{dt} \iint_R \rho E \, dA + \frac{1}{2}\frac{d}{dt} \iint_R \rho \dot{u}_i \dot{u}_i \, dA + vG. \tag{7.41}$$

The left-hand side of this equation represents the work rate of tractions across C, the first and second terms on the right-hand side are the rate of increase of internal and kinetic energies stored inside the region R enclosed by the curve C and the third term is the rate at which energy is dissipated by the moving crack. $T_i = \sigma_{ij} n_j$ denotes the traction components acting across C and n_j the components of the unit normal vector $\mathbf{n}$ of curve C. When Equation (7.41) is referred to the moving coordinates x_1, x_2 defined from Equation (7.40) it takes the form

$$G = \iint_R \frac{\partial}{\partial x_1}(\rho E) \, dA + \frac{1}{2} v^2 \iint_R \frac{\partial}{\partial x_1}\left(\rho \frac{\partial u_i}{\partial x_1} \frac{\partial u_i}{\partial x_1}\right) dA - \int_C T_i \frac{\partial u_i}{\partial x_1} \, ds. \tag{7.42}$$

Using Green's convergence theorem, Equation (7.42) takes the form

$$G = \int_C \left(\rho E + \frac{1}{2}\rho v^2 \frac{\partial u_i}{\partial x_1} \frac{\partial u_i}{\partial x_1}\right) dx_2 - T_i \frac{\partial \dot{u}_i}{\partial x_1} \, ds. \tag{7.43}$$

Conservation of the mechanical energy of the system results in

$$\iint_R \rho \dot{E} \, dA = \iint_R \sigma_{ij} \dot{\epsilon}_{ij} \, dA. \tag{7.44}$$

Introducing the strain energy density function $w = w(\epsilon_{ij})$ from

$$\sigma_{ij} = \frac{\partial w}{\partial \epsilon_{ij}} \tag{7.45}$$

and applying Green's theorem, we obtain

$$\int_C \rho E \, dx_2 = \iint_R \sigma_{ij} \frac{\partial \epsilon_{ij}}{\partial x_1} \, dA = \iint_R \frac{\partial w}{\partial x_1} \, dA = \int_C w \, dx_2. \tag{7.46}$$

Using Equation (7.46), Equation (7.43) becomes

$$G = \int_C \left(w + \frac{1}{2}\rho v^2 \frac{\partial u_i}{\partial x_1} \frac{\partial u_i}{\partial x_1}\right) dx_2 - T_i \frac{\partial u_i}{\partial x_1} \, ds. \tag{7.47}$$

For a static crack ($v = 0$), Equation (7.47) reduces to the J-integral (Equation (5.16)) which is equal to the strain energy release rate G for elastic behavior (Equation (5.39)). Using analogous derivations as in the case of the J-integral (Section 5.2) it can easily be proven that G is path independent – that is, it retains its value for an arbitrary choice of the integration path C surrounding the crack tip.

As in the case of the J-integral (Section 5.3), G can easily be calculated by choosing C as a circle of radius R centered at the crack tip and using the asymptotic expressions of stresses and displacements given from Equation (7.32) and (7.34):

$$G = R \int_{-\pi}^{\pi} \left[w \cos\theta + \tfrac{1}{2}\rho c^2 \left[\left(\frac{\partial u_1}{\partial x_1}\right)^2 + \left(\frac{\partial u_2}{\partial x_1}\right)^2 \right] \cos\theta - \right.$$
$$\left. - (\sigma_{11}\cos\theta + \sigma_{12}\sin\theta)\frac{\partial u_1}{\partial x_1} + (\sigma_{12}\cos\theta + \sigma_{22}\sin\theta)\frac{\partial u_2}{\partial x_1} \right] dt \quad (7.48)$$

or

$$G = \left[\frac{\beta_1(1-\beta_2^2)}{4\beta_1\beta_2 - (1+\beta_2^2)^2} \frac{1}{2\mu} \right] K^2(t). \tag{7.49}$$

Equation (7.49) establishes a relation between the strain energy release rate or crack driving force G and the dynamic stress intensity factor $K(t)$. For the static crack problem ($v=0$) Equation (7.49) reduces to Equation (4.47).

A fracture criterion for dynamic crack propagation based on Equation (7.49) can be established in a manner analogous to the static case. It is assumed that dynamic crack propagation occurs when the strain energy release rate G becomes equal to a critical value that is equivalent to a critical stress intensity factor fracture criterion. For small-scale yielding the concept of K-dominance for stationary cracks can be extended to dynamic cracks. In these circumstances the fracture criterion for a propagating crack takes the form

$$K(t) = K_{ID}(v,t) \tag{7.50}$$

where $K_{ID}(v,t)$ represents the resistance of the material to dynamic crack propagation and is assumed to be a material property. The dynamic stress intensity factor $K(t)$ is determined from the solution of the corresponding elastodynamic problem and is a function of loading, crack length and geometrical configuration of the cracked body. The material parameter K_{ID} can be determined experimentally and depends on crack speed and environmental conditions. A brief description of the available experimental methods used for the determination of K and K_{ID} will be presented in Section 7.9. For further information on the results of this section the reader is referred to [7.36].

7.5. Transient response of cracks to impact loads

This section briefly discusses the effect of loads that are applied suddenly on cracked bodies. This type of loading is known as *shock* or *impact* and common examples include explosive propulsion or impact of projectiles. As a rule, when the time required to increase the load from zero to its maximum value is less than half the fundamental natural period of vibration of the structure, the dynamic effect becomes significant. In such cases, the interaction of stress waves with structure geometry determines the location and magnitude of stresses and displacements that are greater than those associated with the corresponding static loading.

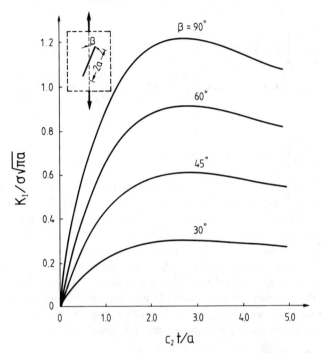

Figure 7.4. Opening-mode stress intensity factor versus time for different values of the crack inclination angle. $\nu = 0.29$ [7.10].

The case of a single crack in a mixed-mode stress field which is governed by the values of the opening-mode K_I and sliding-mode K_{II} stress intensity factors is considered to show the influence of impact loading on the stress field and fracture characteristics of the body. Chen and Sih [7.10, pp. 1–58] showed that the asymptotic stress field in the vicinity of the crack tip is expressed by the same equations as for the stationary crack (Equations (2.76) for opening mode and Equations (2.91) for sliding mode) where the dynamic effect enters only through the stress intensity factor.

Values of $K_I/\sqrt{\pi a}$ and $K_{II}/\sqrt{\pi a}$ for a crack of length $2a$ in an infinite plate subjected to a uniform uniaxial stress σ subtending an angle β with the crack axis are shown in Figures 7.4 and 7.5. Observe that the dynamic stress intensity factor K_I rises sharply at first, reaching a maximum at $c_2 t/a \simeq 3.0$ and then decays in amplitude, oscillating about its static value. An analogous behavior is observed for the stress intensity factor K_{II} which, however, reaches a maximum sooner, at $c_2 t/a \simeq 2$.

For the determination of the fracture angle θ_0 and the critical stress σ_m for crack growth, the strain energy density theory presented in Chapter 6 is used. θ_0 is plotted in Figure 7.6 versus β for three different values of $c_2 t/a = 1.0$, 2.0 and 3.0 representing the time elapsed after impact. The variation of the critical stress σ_m normalized to $\sigma_S = \sqrt{4\mu S_c (1 - 2\nu)a}$ against time $c_2 t/a$ for

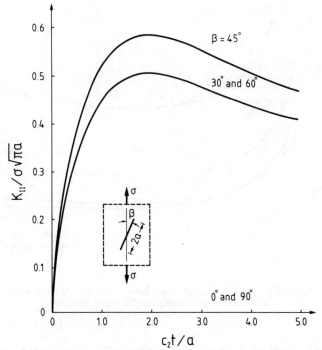

Figure 7.5. Sliding-mode stress intensity factor versus time for different values of the crack inclination angle. $\nu = 0.29$ [7.10].

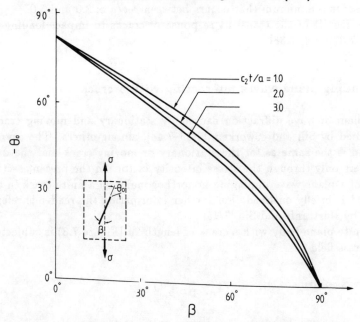

Figure 7.6. Crack extension angle $-\theta_0$ versus crack inclination angle β for different values of $c_2 t/a$. $\nu = 0.29$ [7.10].

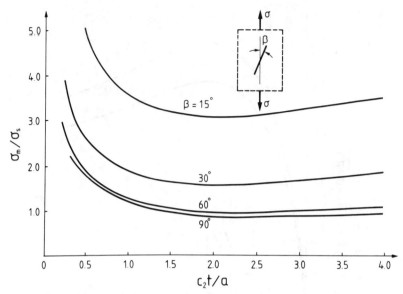

Figure 7.7. Critical impact stress versus time for different values of the crack inclination angle. $\nu = 0.29$ [7.10].

various crack orientations is shown in Figure 7.7. The curves represent the critical stresses required for crack growth at various times. Observe that each curve presents a minimum that occurs between $c_2 t/a \simeq 2.0$ and 3.0.

Further results of the transient response of cracks to impact loading can be found in [7.37] and [7.38].

7.6. Standing plane waves interacting with a crack

The problem of wave diffraction caused by stationary and moving cracks has been studied by Sih and coworkers [7.39–7.44], among others. The asymptotic stress field is the same as for the stationary or moving crack and the dynamic effect enters only through the stress intensity factor. In the present section the problem of a plane wave impinging nonorthogonally on a finite crack in a plane strain field is briefly outlined. For further information the reader is referred to the work by Hartranft and Sih [7.45].

An infinite plane body with a crack of length $2a$ (Figure 7.8) is subjected to a biaxial stress field

$$\sigma_y = \sigma \cos \alpha y \cos \omega t, \qquad \sigma_x = \frac{\nu}{1-\nu}\sigma_y \qquad (7.51)$$

where the nonuniform stress σ_x prevents the body from displacing in the horizontal direction. At locations remote from the crack the vertical displacement is

Dynamic fracture

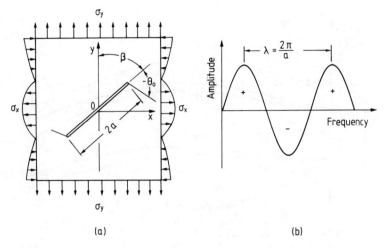

Figure 7.8. An inclined crack in a plate subjected to cyclic load [7.10].

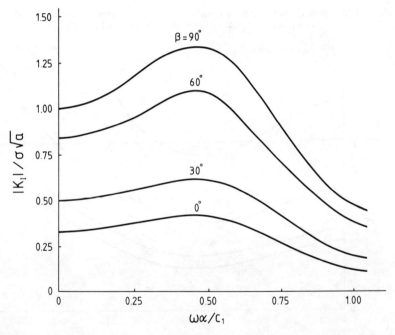

Figure 7.9. Opening-mode stress intensity factor versus input frequency for different values of the crack inclination angle. $\nu = 0.25$ [7.10].

given by

$$u_y = \frac{\alpha}{\rho\omega^2}\sigma \sin \alpha y \cos \omega t \tag{7.52}$$

while the horizontal displacement is zero. The wave number α and the circular frequency ω are related by $\omega = \alpha c_1$.

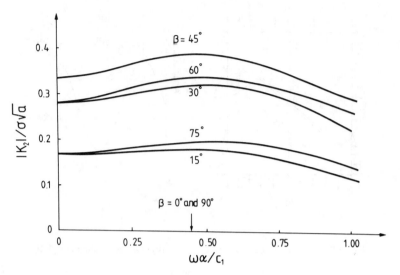

Figure 7.10. Sliding-mode stress intensity factor versus input frequency for different values of the crack inclination angle. $\nu = 0.25$ [7.10].

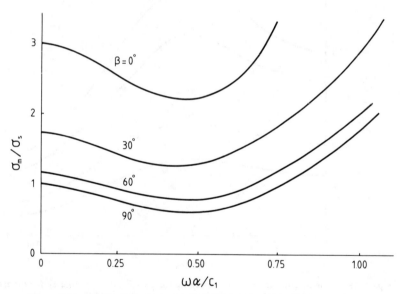

Figure 7.11. Critical stress versus input frequency for different values of the crack inclination angle. $\nu = 0.25$ [7.10].

The stress intensity factors K_I and K_{II} which are time dependent can be put into the form

$$K_I = |K_I| \cos \omega(t - \delta_1)$$
$$K_{II} = |K_{II}| \cos \omega(t - \delta_2)$$

(7.53)

where the phase delays δ_j ($j = 1, 2$) between the peak applied stress and the peak stress intensity factor are introduced. The variation of the normalized K_I and K_{II} with the input frequency ω for various crack angles β is shown in Figures 7.9 and 7.10. Observe that K_I and K_{II} initially increase and then decrease in magnitude after reaching their maximum values. The peak values of K_{II} do not coincide with those of K_I.

As in the previous case, the strain energy density fracture criterion can be used to determine the fracture angle and the critical applied stress σ_m for crack growth. The variation of σ_m/σ_s ($\sigma_s = \sqrt{4\mu S_c/(1-2\nu)a}$) with $\omega a/c_1$ is presented in Figure 7.11. Observe that the fracture stress decreases at first with the frequency passing through a minimum and then increases. The critical stress also increases as the crack is moved to the direction $\beta = 0$.

7.7. Crack branching

When a crack propagates at a high velocity it may be divided into two branches which in many cases are further divided until a multiple crack branching pattern is obtained. The phenomenon of crack branching has attracted the interest of early investigators in dynamic fracture mechanics [7.8, 7.46]. Cotterell [7.47], among others, suggested that when a crack reaches a sufficiently high velocity further increase of the crack driving force is responsible for crack branching with no increase in crack velocity. Clark and Irwin [7.48] suggested that crack branching takes place when a critical stress intensity factor has been attained rather than a limiting crack velocity. This result has been verified from experiments in glass plates containing initial notches of various root curvatures. Crack branching is greater in specimens with blunter notches which imply higher stress intensity factors to initiate fracture. When the crack driving force is higher than a limiting value the excess energy in the vicinity of the original crack tip initiates a new crack.

Attempts have been made to explain the crack branching pattern. Yoffe [7.8] assumed that the crack branches at a direction which coincides with the maximum of the local circumferential stress ahead of the moving crack. She predicted a half branch angle of 26° for $\nu = 0.25$, which deviated significantly from the experimentally observed value of approximately 15° in glass plates. At this point it should be mentioned that the maximum stress criterion is incapable of explaining the phenomenon of growth of a rapidly moving crack. Indeed, the ratio of the principal stress σ_{22} and σ_{11} along the crack direction given by Equation (7.39) is always smaller than unity. This means that the crack propagates in a direction parallel to the maximum stress rather than normal to it, which contradicts the original assumption. Andersson [7.49], from a static calculation of the stress intensity factors at branched cracks and based on the assumption that crack bifurcation occurs at the angle which maximizes K_I, found half crack branch angles of about 30°. Using a static analysis and the maximum stress criterion Kalthoff [7.50] gave a half branch angle of $\theta = -\tan^{-1} 2(K_{II}/K_I)$ which predicted branch angles close to experimental results.

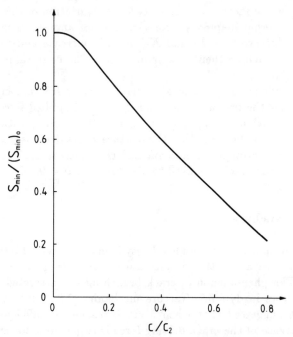

Figure 7.12. Strain energy density factor versus crack speed [7.10].

The model of a finite crack spreading at both ends at constant velocity in conjunction with the strain energy density criterion has been used by Sih [7.10, pp. XXXIX–XLIV] to predict the crack bifurcation angle. For this case the strain energy density factor is given by

$$S = \frac{K_I^2}{8\pi\mu} F^2(\beta_1, \beta_2) [2(1+\beta_2^2)^2 [2(1-\nu)(1+\beta_1^2)^2 -$$
$$- 2(1-2\nu)(2\beta_1^2 + 1 - \beta_2^2)(1+\beta_2^2)] g^2(\beta_1) +$$
$$+ 32\beta_1^2 \beta_2^2 g^2(\beta_2) - 16\beta_1 \beta_2 (1+\beta_1^2)(1+\beta_2^2) g(\beta_1) g(\beta_2) +$$
$$+ 8\beta_1^2 (1+\beta_2^2) [h(\beta_1) - h(\beta_2)]^2] \qquad (7.54)$$

where

$$g^2(\beta_j) + h^2(\beta_j) = \sec\theta (1 + \beta_j^2 \tan^2\theta)^{-1/2}$$
$$g^2(\beta_j) - h^2(\beta_j) = \sec\theta (1 + \beta_j^2 \tan^2\theta)^{-1} \qquad (7.55)$$

and

$$F(\beta_1, \beta_2) = \beta_1 [[(1+\beta_2^2)^2 - 4\beta_1^2\beta_2^2] K(\beta_1) - 4\beta_1^2 (1-\beta_2^2) K(\beta_2) -$$
$$- [4\beta_1^2 + (1+\beta_2^2)^2] E(\beta_1) + 8\beta_1^2 E(\beta_2)]^{-1} \qquad (7.56)$$

in which K and E are complete elliptic integrals of the first and second kind, respectively.

Table 7.1. Angles of crack bifurcation

ν	ν/β	θ_0	$S_{\min}/(S_{\min})_0$
0.21	0.45	0°	0.56209
	0.46	±18.84°	0.55199
0.22	0.46	0°	0.55196
	0.47	±17.35°	0.54204
0.23	0.47	0°	0.54185
	0.48	±16.27°	0.53210
0.24	0.48	0°	0.53177
	0.49	±15.52°	0.52217

According to the strain energy density failure criterion the crack extends in the direction which makes S a minimum. The critical value $S_{\min}$ is then computed from Equation (7.54) as a function of crack velocity. The variation of $S_{\min}/(S_{\min})_0$ with v/β_2, where $(S_{\min})_0 = (1-2\nu)\sigma\sqrt{a}/4\mu$ and v represents the crack velocity, is shown in Figure 7.12. Observe that $S_{\min}$ decreases smoothly from its largest value at $v = 0$ as the crack velocity increases. Values of the half branch angle θ_0 with the corresponding values of $S_{\min}/(S_{\min})_0$ for different values of crack velocity and Poisson's ratio are shown in Table 7.1. Note that as ν is varied from 0.21 to 0.24 θ_0 changes from ±18.84° to ±15.52° which is very close to experimental observations.

7.8. Crack arrest

The problem of arrest of a rapidly propagating crack is of major theoretical and practical importance. The load transmission characteristics of the system play a significant role in the arrest of a crack. When energy is constantly supplied to the crack-tip region, continuing crack motion generally occurs. This is the situation of a crack in a uniform stress tensile field. On the other hand, crack growth under constant displacement conditions eventually leads to crack arrest since the energy supplied to the crack-tip region progressively decreases with time. The crack arrest capability of a system increases when the distance between the energy source and crack tip increases with time, as occurs, for example, in the splitting of a long cantilever beam specimen.

A crack arrest criterion based on the stress intensity factor can be put in the form

$$K(t) = K_{IA} = \min\left[K_{ID}(\dot{a})\right] \tag{7.57}$$

where $K(t)$ is the dynamic stress intensity factor and K_{ID} the material fracture toughness for dynamic crack propagation. Experimental studies indicated that K_{ID} depends on crack speed $\dot{a}$ [7.51–7.53]. A typical form of the curve $K_{ID} = K_{ID}(\dot{a})$ for many metals and polymers is shown in Figure 7.13. Observe that K_{ID} is nearly speed independent at low crack speeds and increases as the crack speed increases.

Crack arrest in practice is promoted when part of the load is taken up and

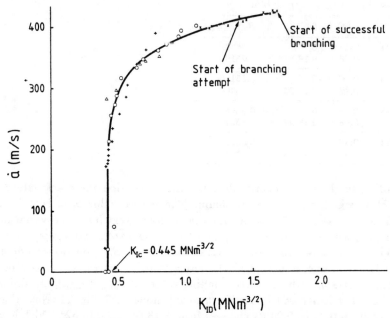

Figure 7.13. Dynamic fracture toughness versus crack velocity for Homalite 100 [7.51].

transmitted to other structural elements. Usually arrest strips are used, a method that finds application in aircraft structures. For further information on crack arrest procedures used in structural design refer to [7.54].

7.9. Experimental determination of crack velocity and dynamic stress intensity factor

Experimental studies played a key role in improving our understanding of dynamic fracture behavior of materials and structures and in the measurement of the relevant dynamic fracture material properties. In this section the most widely used experimental methods for measuring the crack velocity and the dynamic stress intensity factor are briefly presented.

(a) Crack velocity

Initial measurements of crack velocity were conducted using a series of conducting wires placed at certain intervals along the crack path and perpendicular to the direction of the crack propagation. The wires form one leg of a bridge which is connected to an oscilloscope. Due to the propagating crack the wires break and the corresponding times are obtained from the trace on the oscilloscope. This technique allows measurement of the average velocity over the gage length between the wires.

High-speed photography is perhaps the most widely used method for recording rapid crack propagation. The multiple-spark Cranz–Schardin camera, which is capable of operating at rates of up to 10^6 frames per second, is widely employed. Although best results are obtained for transparent materials the method can also be used for nontransparent materials by polishing the surface of the specimens.

(b) *Dynamic stress intensity factor*

The method of dynamic photoelasticity was first used by Wells and Post [7.55] to determine the state of stress and the velocity of a rapidly propagating crack. From the analysis of the isochromatic pattern around the crack tip the stress intensity factor was obtained. Further studies of this problem have been performed by Kobayashi and coworkers [7.56, 7.57]. These investigations were based on the static solution of the stress field near the crack tip. Kobayashi and Mall [7.58] estimated that the error introduced when the static stress field was used is small for crack propagating velocities less than $0.15c_1$. Extensive studies on the dynamic photoelastic investigation of crack problems have been performed at the Photomechanics Laboratory of the University of Maryland [7.51, 7.52, 7.59, 7.60] using the dynamic stress field around a moving crack. A K versus $\dot{a}$ relationship was established and it was found for Homalite 100 to be independent of the specimen geometry for crack velocities below 300 m/s.

The optical method of caustics has also been used extensively for the experimental study of crack initiation, rapid crack growth, crack arrest and crack branching [7.61–7.69]. A dynamic correction for the determination of the stress intensity factor from the obtained optical pattern was used. This method was proved to be very efficient and powerful for the study of dynamic crack problems.

For the determination of dynamic fracture toughness, K_{ID}, several types of specimens – including the double cantilever beam specimen, the single edge-notched specimen and the wedge-loaded specimen – have been proposed. The last type of specimen presents a number of advantages over the others and is mainly used in dynamic fracture testing. Duplex specimens with crack initiation taking place in a hardened starter section welded into the test material are sometimes preferred in situations where large monolithic specimens are needed. The dynamic fracture toughness K_{ID} as a function of crack velocity is determined by measuring the critical stress intensity factor for crack initiation and the crack length at arrest, and using appropriate dynamic analysis for the specific type of specimen used. The critical value K_{IA} of the stress intensity factor at crack arrest is determined as the minimum value of K_{ID} taken from a number of measurements at various crack speeds.

Besides the dynamic fracture toughness K_{ID}, which depends on crack velocity, the critical value K_{Id} of the stress intensity factor for crack initiation under a rapidly applied load is of interest in practical applications. K_{Id} depends on the loading rate and temperature and is considered to be a material parameter. For the experimental determination of K_{Id} the three-point bend specimen used in static analysis is mainly employed. The specimen is loaded by a falling weight and static analysis is employed for the determination of K_{Id}. The trend is that

K_{Id} diminishes with increasing loading rate below the transition temperature, while the reverse happens above the transition temperature.

References

7.1. Hopkinson, J., On the rupture of iron wire by a blow, *Original Papers by the Late John Hopkinson II* (ed. B. Hopkinson), Cambridge Univ. Press, Cambridge pp. 316–320 (1901).
7.2. Hopkinson, B., The effects of momentary stresses in metals, *Proceedings of the Royal Society of London*, Ser. **A74**, 498–506 (1905).
7.3. Hopkinson, B., The effects of the detonation of guncotton, *Scientific Papers*, Cambridge Univ. Press, Cambridge (1921).
7.4. Mott, N. F., Fracture of metals: Theoretical considerations, *Engineering* **165**, 16–18 (1948).
7.5. Schardin, H., Elle, D. and Struth, W., Über den zeitlichen Ablauf des Bruchvorganges in Glas und Kunstglas, *Z. Tech. Physic* **21**, 393–400 (1940).
7.6. Kerkhof, F., Analyse des sproden Zugbruchs von Glasern mittels Ultraschall, *Naturwiss.* **40**, 478 (1953).
7.7. Kerkhof, F., Über den Bruchvorgang beim Manteldruckversuch, *Glastechnische Berichte* **33**, S456–459 (1960).
7.8. Yoffe, E. H., The moving Griffith crack, *Philosophical Magazine* **42**, 739–750 (1951).
7.9. Wells, A. A. and Post, D., The dynamic stress distribution surrounding a running crack – A photoelastic analysis, *Proceedings of the Society for Experimental Stress Analysis* **16**, 69–92 (1958).
7.10. Sih, C. C. (ed.), *Mechanics of Fracture*, Vol. 4, *Elastodynamic Crack Problems*, Noordhoff Int. Publ., The Netherlands (1977).
7.11. Erdogan, F., Crack-propagation theories, in *Fracture – An Advanced Treatise*, Vol. II (ed. H. Liebowitz), Academic Press, pp. 497–590 (1968).
7.12. Achenbach, J. D., Dynamic effects in brittle fracture, in *Mechanics Today*, Vol. 1 (ed. S. Nemat-Nasser), Pergamon Press, pp. 1–57 (1972).
7.13. Freund, L. B., The analysis of elastodynamic crack tip stress fields, in *Mechanics Today*, Vol. 3 (ed. S. Nemat-Nasser), Pergamon Press, pp. 55–91 (1976).
7.14. Roberts, D. K. and Wells, A. A., The velocity of brittle fracture, *Engineering* **178**, 820–821 (1954).
7.15. Berry, J. P., Some kinetic considerations of the Griffith criterion for fracture, *Journal of the Mechanics and Physics of Solids* **8**, 194–216 (1960).
7.16. Dulaney, E. N. and Brace, W. F., Velocity behavior of a growing crack, *Journal of Applied Physics* **31**, 2233–2236 (1960).
7.17. Kanninen, M. F., An estimate of the limiting speed of a propagating ductile crack, *Journal of the Mechanics and Physics of Solids* **16**, 215–228 (1968).
7.18. Duffy, A. R., McClure, G. M., Eiber, R. J. and Maxey, W. A., Fracture design practices for pressure piping, in *Fracture – An Advanced Treatise*, Vol. V (ed. H. Liebowitz), Academic Press, pp. 159–232 (1969).
7.19. Craggs, J. W., On the propagation of a crack in an elastic-brittle material, *Journal of the Mechanics and Physics of Solids* **8**, 66–75 (1960).
7.20. Broberg, K. B., The propagation of a brittle crack, *Arkiv for Fysik* **18**, 159–192 (1960).
7.21. Baker, B. R., Dynamic stresses created by a moving crack, *Journal of Applied Mechanics, Trans. ASME* **29**, 449–458 (1962).
7.22. Sih, G. C., Some elastodynamic problems of cracks, *International Journal of Fracture* **4**, 51–58 (1968).
7.23. Sih, G. C. and Chen, E. P., Moving crack in a finite strip under tearing action, *Journal of the Franklin Institute* **290**, 25–35 (1970).
7.24. Sih, G. C. and Chen, E. P., Crack propagating in a strip of material under plane extension, *International Journal of Engineering Sciences* **10**, 537–551 (1972).
7.25. Nillson, F., Dynamic stress-intensity factor for finite strip problems, *International Journal of Fracture* **8**, 403–411 (1972).
7.26. Craggs, J. W., The growth of a disk-shaped crack, *International Journal of Engineering*

Sciences **4**, 113–124 (1966).
7.27. Webb, D. and Atkinson, C., A note on a penny-shaped crack expanding under a non-uniform internal pressure, *International Journal of Engineering Sciences* **7**, 525–530 (1969).
7.28. Kostrov, B. V., The axisymmetric problem of propagation of a tensile crack, *Applied Mathematics and Mechanics* **28**, 793–803 (1964).
7.29. Cherepanov, G. P. and Afanesev, E. F., Some dynamic problems of the theory of elasticity – a review, *International Journal of Engineering Sciences* **12**, 665–690 (1975).
7.30. Kostrov, B. V., Unsteady crack propagation for longitudinal shear cracks, *Applied Mathematics and Mechanics* **30**, 1042–1049 (1966).
7.31. Eshelby, J. D., The elastic field of a crack extending non-uniformly under general antiplane loading, *Journal of the Mechanics and Physics of Solids* **17**, 177–199 (1969).
7.32. Freund, L. B., Crack propagation in an elastic solid subjected to general loading – II non-uniform rate of extension, *Journal of the Mechanics and Physics of Solids* **20**, 141–152 (1972).
7.33. Kostrov, B. V., On the crack propagation with variable velocity, *International Journal of Fracture* **11**, 47–56 (1975).
7.34. Tsai, Y. M., Propagation of a brittle crack at constant and accelerating speeds, *International Journal of Solids and Structures* **9**, 625–642 (1973).
7.35. Rose, L. R. F., An approximate (Wiener–Hopf) kernel for dynamic crack problems in linear elasticity and viscoelasticity, *Proceedings of the Royal Society of London* **A349**, 497–521 (1976).
7.36. Sih, G. C., Dynamic aspects of crack propagation, in *Inelastic Behavior of Solids* (eds M. F. Kanninen, W. F. Adler, A. R. Rosenfield and R. I. Jaffee), McGraw-Hill, pp. 607–639 (1970).
7.37. Sih, G. C., Ravera, R. S. and Embley, G. T., Impact response of a finite crack in plane extension, *International Journal of Solids and Structures* **8**, 977–993 (1972).
7.38. Embley, G. T. and Sih, G. C., Response of a penny-shaped crack to impact waves, *Proceedings of the 12th Midwestern Mechanics Conference* **6**, 473–487 (1971).
7.39. Sih, G. C. and Loeber, J. F., Wave propagation in an elastic solid with a line of discontinuity or finite crack, *Quarterly of Applied Mathematics* **27**, 193–213 (1969).
7.40. Sih, G. C. and Loeber, J. F., Normal compression and radial shear waves scattering at a penny-shaped crack in an elastic solid, *Journal of the Acoustical Society of America* **46**, 711–721 (1969).
7.41. Sih, G. C. and Loeber, J. F., A class of diffraction problems involving geometrically induced singularities, *Journal of Mathematics and Mechanics* **19**, 327–350 (1969).
7.42. Sih, G. C. and Loeber, J. F., Interaction of horizontal shear waves with a running crack, *Journal of Applied Mechanics* **37**, 324–330 (1970).
7.43. Chen, E. P. and Sih, G. C., Running crack in an incident wave field, *International Journal of Solids and Structures* **9**, 897–919 (1973).
7.44. Chen, E. P. and Sih, G. C., Scattering of plane waves by a propagating crack, *Journal of Applied Mechanics* **42**, 705–711 (1975).
7.45. Hartranft, R. J. and Sih, G. C., Application of the strain energy density fracture criterion to dynamic crack problems, in *Prospects of Fracture Mechanics* (eds G. C. Sih, H. C. van Elst and D. Broek), Noordhoff Int. Publ., The Netherlands, pp. 281–297 (1974).
7.46. Schardin, H., Velocity effects in fracture, in *Fracture*, M.I.T. Press, Cambridge, pp. 297–330 (1959).
7.47. Cotterell, B., On brittle fracture paths, *International Journal of Fracture Mechanics* **1**, 96–103 (1965).
7.48. Clark, A. B. J. and Irwin, G. R., Crack propagation behaviors, *Experimental Mechanics* **6**, 321–330 (1966).
7.49. Andersson, H., Stress intensity factors at the tips of a star-shaped contour in an infinite tensile sheet, *Journal of the Mechanics and Physics of Solids* **17**, 405–417 (1969), Errata, *op. cit.* **18**, 437 (1970).
7.50. Kalthoff, J. F., On the characteristic angle for crack branching in ductile materials, *International Journal of Fracture Mechanics* **7**, 478–480 (1971).
7.51. Irwin, G. R., Dally, J. W., Kobayashi, T., Fourney, W. F., Etheridge, M. J. and Rossmanith, H. P., On the determination of the $\dot{a}-K$ relationship for birefringent polymers, *Experimental Mechanics* **19**, 121–128 (1979).
7.52. Dally, J. W., Dynamic photoelastic studies of fracture, *Experimental Mechanics* **19**, 349–361 (1979).
7.53. Rosakis, A. J. and Zehnder, A. T., On the dynamic fracture of structural metals, *Inter-*

national Journal of Fracture **27**, 169–186 (1985).
7.54. Bluhm, J. I., Fracture arrest, in *Fracture – An Advanced Treatise* (ed. H. Liebowitz), Academic Press, Vol. 5, pp. 1–63 (1969).
7.55. Wells, A. and Post, D., The dynamic stress distribution surrounding a running crack – A photoelastic analysis, *Proceedings of the Society for Experimental Stress Analysis* **16**, 69–92 (1958).
7.56. Bradley, W. B. and Kobayashi, A. S., An investigation of propagating cracks by dynamic photoelasticity, *Experimental Mechanics* **10**, 106–113 (1970).
7.57. Kobayashi, A. S. and Chan, C. F., A dynamic photoelastic analysis of dynamic-tear-test specimen, *Experimental Mechanics* **16**, 176–181 (1976).
7.58. Kobayashi, A. S. and Mall, S., Dynamic fracture toughness of Homalite 100, *Experimental Mechanics* **18**, 11–18 (1978).
7.59. Rossmanith, H. P. and Irwin, G. R., Analysis of dynamic isochromatic crack-tip stress patterns, *University of Maryland Report* (1979).
7.60. Dally, J. W. and Kobayashi, T., Crack arrest in duplex specimens, *International Journal of Solids and Structures* **14**, 121–129 (1979).
7.61. Katsamanis, F., Raftopoulos, D. and Theocaris, P. S., Static and dynamic stress intensity factors by the method of transmitted caustics, *Journal of Engineering Materials and Technology* **99**, 105–109 (1977).
7.62. Theocaris, P. S. and Katsamanis, F., Response of cracks to impact by caustics, *Engineering Fracture Mechanics* **10**, 197–210 (1978).
7.63. Theocaris, P. S., Dynamic propagation and arrest measurements by the method of caustics on overlapping skew parallel cracks, *International Journal of Solids and Structures* **14**, 639–653 (1978).
7.64. Theocaris, P. S. and Milios, J., Dynamic crack propagation in composites, *International Journal of Fracture* **16**, 31–51 (1980).
7.65. Theocaris, P. S. and Papadopoulos, G., Elastodynamic forms of caustics for running cracks under constant velocity, *Engineering Fracture Mechanics* **13**, 683–698 (1980).
7.66. Theocaris, P. S. and Pazis, D., Crack deceleration and arrest phenomena at an oblique bimaterial interface, *International Journal of Solids and Structures* **19**, 611–623 (1983).
7.67. Rosakis, A. J., Experimental determination of the fracture initiation and dynamic crack propagation resistance of structural steels by the optical method of caustics, Ph.D. Thesis, Brown University (1982).
7.68. Rosakis, A. J. and Freund, L. B., The effect of crack-tip plasticity on the determination of dynamic stress-intensity factors by the optical method of caustics, *Journal of Applied Mechanics, Trans. ASME* **48**, 302–308 (1981).
7.69. Beinert, J. and Kalthoff, J. F., Experimental determination of dynamic stress intensity factors by shadow patterns, in *Mechanics of Fracture*, Vol. 7 (ed. G. C. Sih), Martinus Nijhoff Publ., pp. 281–330 (1981).

8

Fatigue and environment-assisted fracture

8.1. Introduction

It was first realized in the middle of the nineteenth century that engineering components and structures often fail when subjected to repeated fluctuating loads whose magnitude is well below the critical fracture load under monotonic loading. Early investigations were primarily concerned with axle and bridge failures which occurred at cyclic load levels less than half their corresponding monotonic load magnitudes. Failure due to repeated loading was called 'fatigue failure' and it is traditionally considered to be a separate and distinct failure mechanism. This mainly arises from the inability of the current continuum mechanics theories to explain adequately the fatigue behavior of solids. What changes, after all, is only the loading; instead of a monotonically increasing load, the load is repeated many times with a lower amplitude. Thus, fatigue should not be treated as a separate failure mechanism but rather as the process of cumulative damage at different scale levels caused by repeated fluctuating loads.

Early studies in fatigue have not accounted for the details of the failure mode nor for the existence and growth of initial imperfections in the material, but have tried to determine the fatigue life in terms of global measurable quantities, like stress, strain, mean stress, etc. The results of tests performed on small laboratory specimens subjected to repeated sinusoidally fluctuating loads were interpreted in diagrams expressing the stress amplitude versus the number of cycles to failure, known as $S-N$ curves. The fatigue life was found to increase with decreasing stress level, and below a certain stress level, known as the fatigue limit, failure will not occur for an arbitrary number of loading cycles. The mean stress level, defined as the average of the maximum and minimum stress on the cyclic loading, plays an important role on fatigue life. The trend of a decreasing cyclic life with increasing mean stress for a given maximum applied stress level was found. A number of empirical relationships for fatigue life, derived from the curve fitting of test data, have been proposed in the literature. The $S-N$ curve method and other available procedures based on gross specimen quantities lead to an inaccurate prediction of the fatigue life of engineering components due to the large scatter of experimental results as influenced by specimen size and geometry, material and the nature of the fluctuating load. Furthermore, the physical phenomena and mechanisms governing the fatigue process are completely ignored.

A better understanding of the fatigue phenomenon can be obtained by modeling the fatigue crack initiation and propagation processes. Crack initiation is analyzed within the microscopic scale level, while for crack propagation the continuum mechanics approach is used. The necessity for addressing these two processes separately arises from the inability of the current theory to bridge the gap between material damage that occurs at microscopic and macroscopic scale levels. It is generally accepted that, when a structure is subjected to repeated external load, energy is accumulated in the neighborhood of voids and microscopic defects which grow and coalesce forming microscopic cracks. Eventually larger macroscopic cracks are formed. A macrocrack may be defined as one that is large enough to permit the application of the principles of homogeneous continuum mechanics. A macrocrack is usually referred to as a fatigue crack. The number of cycles required to initiate a fatigue crack is the fatigue crack initiation life N_i.

Following the initiation of a fatigue crack, slow stable crack propagation begins until the crack reaches a critical size corresponding with the onset of global instability leading to catastrophic failure. Thus, the fatigue life of an engineering component may be considered to be composed of three stages, namely the initiation or stage I, the propagation or stage II and the fracture or stage III, in which the crack growth rate increases rapidly as global instability is approached. The number of cycles required to propagate a fatigue crack until it reaches its critical size is the fatigue crack propagation life N_p. Depending on the material, the amplitude of the fluctuating load and environmental conditions, the fatigue crack initiation life may be a small or a substantial part of the total fatigue life.

The phenomenon of fatigue crack initiation, which is studied within the context of microscopic scale level, is a very complicated problem and a few quantitative theories have been proposed for its study. Fatigue cracks generally initiate at sites of high stress concentrations where the localized stress exceeds the yield stress of the material and zones of plastic deformation are developed. Within the microscopic scale level, cracks initiate at the points of intersection of a slip band with a free surface, at inclusion and grain boundaries and near initial flaws. The basic mechanicsm of crack nucleation and propagation, according to the most widely accepted microstructural fatigue theories, is the cyclic slip and the resulting extrusions and intrusions.

It has long been recognized that the strength of solids depends greatly on the environment in which they are located. Under the influence of the environment, a body may behave in a more brittle or ductile manner and its strength may increase or decrease. The failure of engineering components subjected to an aggressive environment may occur under applied stresses well below the strength of the material. Environmental conditions greatly influence the processes of local failure at the tip of a crack and cause subcritical crack growth and gradual failure of structural components. Failure under such conditions involves an interaction of complex chemical, mechanical and metallurgical processes. The basic subcritical crack growth mechanisms include stress corrosion cracking, hydrogen embrittlement and liquid embrittlement. An aggressive environment has a deleterious effect on the fatigue life of an engineering component. The corrosion-fatigue behavior of a structural system subjected to a fluctuating load in the presence of an environment is extremely complicated.

Because of its practical importance the problems of fatigue and environment-assisted crack growth have been given much attention in the literature. A vast number of data are available and a number of different theories have been proposed, based mainly on experimental data correlations. It is the purpose of the present chapter to briefly present the phenomena of fatigue and environment-assisted crack growth within the framework of the macroscopic scale level. The basic fatigue crack growth laws based on the stress intensity factor are described. The strain energy density theory is also applied to study mixed-mode fatigue crack growth. A phenomenological analysis of the problem of stress corrosion cracking is also presented.

8.2. Fatigue crack propagation laws

(a) General considerations

Fatigue crack propagation, referred to as stage II, represents a large portion of the fatigue life of many materials and engineering structures. Accurate prediction of the fatigue crack propagation stage is of utmost importance for determining the fatigue life. The main question of fatigue crack propagation may be stated in this form: *Determine the number of cycles N_c required for a crack to grow from a certain initial crack size a_0 to the maximum permissible crack size a_c, and the form of this increase $a = a(N)$, where the crack length a corresponds to N loading cycles.* Figure 8.1 presents a plot of a versus N which is required for predicting, say, the life of a particular engineering component. a_i represents the crack length that is big enough for fracture mechanics to apply but too small for detection, while a_1 is the non-destructive inspection detection limit. The crack first grows slowly until the useful life of the component is reached. The crack then begins to propagate very rapidly reaching a length a_f at which catastrophic failure begins.

Fatigue crack propagation data are obtained from precracked specimens subjected to fluctuating loads and the change in crack length is recorded as a function of loading cycles. The crack length is plotted against the number of loading cycles for different load amplitudes. The stress intensity factor is used as a correlation parameter in analyzing the fatigue crack propagation results. The experimental results are usually plotted in a $\log(\Delta K)$ versus $\log(da/dN)$ diagram, where ΔK is the amplitude of the stress intensity factor and da/dN is the crack propagation rate. The load is usually sinusoidal with constant amplitude and frequency (Figure 8.2). Two of the four parameters K_{max}, K_{min}, $\Delta K = K_{max} - K_{min}$ or $R = K_{max}/K_{min}$ are needed to define the stress intensity factor variation during a loading cycle.

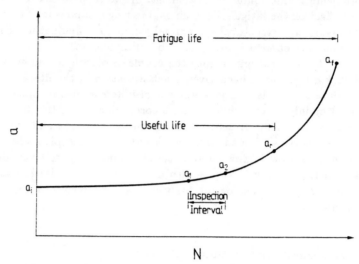

Figure 8.1. Typical form of crack size versus number of cycles curve for constant amplitude loading.

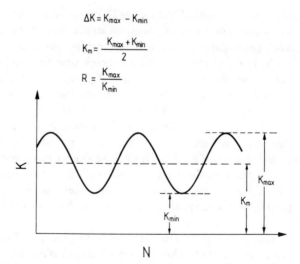

Figure 8.2. A sinusoidal load with constant amplitude and frequency.

A typical plot of the characteristic sigmoidal shape of a $\log(\Delta K)$–$\log(da/dN)$ fatigue crack growth rate curve is shown in Figure 8.3. Three regions according to the curve shape can be distinguished. In region I, da/dN diminishes rapidly to a vanishingly small level and for some materials there is a threshold value of the stress intensity factor amplitude ΔK_{th} meaning that for $\Delta K < \Delta K_{th}$ no crack propagation takes place. In region II there is a linear $\log(\Delta K)$–$\log(da/dN)$ relation. Finally, in region III the crack growth rate curve rises and the maximum

Fatigue and environment-assisted fracture

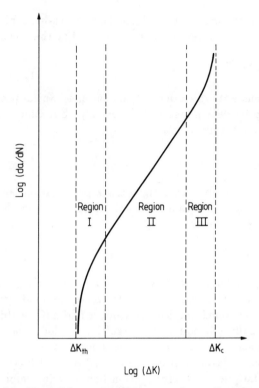

Figure 8.3. Typical form of the fatigue crack growth rate curve.

stress intensity factor $K_{\max}$ in the fatigue load cycle becomes equal to the critical stress intensity factor K_c, leading to catastrophic failure. Experimental results indicate that the fatigue crack growth rate curve depends on the ratio R and is shifted toward higher da/dN values as R increases.

It should be mentioned that all previous discussion concerns situations where the applied stress level is so small that the fatigue crack growth process can be characterized by the stress intensity factor. This assumption will be discussed in detail later in this chapter.

(b) Crack propagation laws

A number of different quantitative continuum mechanics models of fatigue crack propagation have been proposed in the literature. All these models lead to relations mainly based on experimental data correlations which relate da/dN to such variables as the external load, the crack length, the geometry and the material properties. Representative examples of such relations will be analyzed in this section.

One of the earlier mathematical models of fatigue crack propagation was proposed by Head [8.1]. He considered an infinite plate with a central crack of length $2a$ subjected to a sinusoidally applied stress $\pm\sigma$. Modeling the material elements ahead of the crack tip as rigid-plastic work-hardening tensile bars and the remaining elements as elastic bars he arrived at the relation

$$\frac{\mathrm{d}a}{\mathrm{d}N} = C_1 \sigma^3 a^{3/2} \tag{8.1}$$

where C_1 is a constant which depends on the mechanical properties of the material and has to be determined experimentally. Equation (8.1) can be written in terms of stress intensity factor as

$$\frac{\mathrm{d}a}{\mathrm{d}N} = C K_\mathrm{I}^3. \tag{8.2}$$

One of the most widely used fatigue crack propagation laws is that proposed by Paris and Erdogan [8.2] and is usually referred to in the literature as the 'Paris law'. It has the form

$$\frac{\mathrm{d}a}{\mathrm{d}N} = C(\Delta K)^m \tag{8.3}$$

where $\Delta K = K_\mathrm{max} - K_\mathrm{min}$, with K_max and K_min referring to the maximum and minimum values of the stress intensity factor in the load cycle. The constants C and m are determined empirically from a $\log(\Delta K)$–$\log(\mathrm{d}a/\mathrm{d}N)$ plot. The value of m is usually put equal to 4, resulting in the so-called '4th power law' while the coefficient C is assumed to be a material constant. Equation (8.3) represents a linear relationship between $\log(\Delta K)$ and $\log(\mathrm{d}a/\mathrm{d}N)$ and is used to describe the fatigue crack propagation behavior in region II of the diagram of Figure 8.3. Fatigue crack propagation data are well predicted from Equation (8.3) for specific geometrical configurations and loading conditions. The effect of mean stress, loading and specimen geometry is included in the constant C. Despite these drawbacks, Equation (8.3) has been widely used to predict the fatigue crack propagation life of engineering components.

Equation (8.3) does not, however, account for the crack growth characteristics at both low and high levels of ΔK. At high ΔK values, as K_max approaches the critical level K_c, an increase in crack growth rate is observed. For this case (region III of Figure 8.3) Forman et al. [8.3] proposed the relation

$$\frac{\mathrm{d}a}{\mathrm{d}N} = \frac{C(\Delta K)^n}{(1-R)K_c - \Delta K} \tag{8.4}$$

where $R = K_\mathrm{min}/K_\mathrm{max}$ and C and n are material constants. Equation (8.4) resulted from the modification of Equation (8.3) by the term $(1-R)K_c - \Delta K$ which decreases with increasing load ratio R and decreasing fracture toughness K_c, both of which give rise to increasing crack growth rates at a given ΔK level. Note that for $K_\mathrm{max} = K_c$, corresponding to instability, Equation (8.4) predicts an unbounded value of $\mathrm{d}a/\mathrm{d}N$.

For low values of ΔK (region I of Figure 8.3) Donahue et al. [8.4] have suggested the relation

$$\frac{\mathrm{d}a}{\mathrm{d}N} = K(\Delta K - \Delta K_\mathrm{th})^m \tag{8.5}$$

where ΔK_{th} denotes the threshold value of ΔK. According to Klesnil and Lucas [8.5], ΔK_{th} is given by

$$\Delta K_{th} = (1 - R)^\gamma \Delta K_{th(0)} \tag{8.6}$$

where $\Delta K_{th(0)}$ is the threshold value at $R = 0$ and γ is a material parameter.

A generalized fatigue crack propagation law which can describe the sigmoidal response exhibited by the data of Figure 8.3 has been suggested by Erdogan and Ratwani [8.6], and has the form

$$\frac{da}{dN} = \frac{C(1+\beta)^m (\Delta K - \Delta K_{th})^n}{K_c - (1+\beta)\Delta K} \tag{8.7}$$

where C, m, n are empirical material constants, and

$$\beta = \frac{K_{max} + K_{min}}{K_{max} - K_{min}}. \tag{8.8}$$

The factor $(1+\beta)^m$ has been introduced to account for the effect of the mean stress level on fatigue crack propagation, while the factor $[K_c - (1+\beta)\Delta K]$ takes care of the experimental data at high stress levels. Finally, the factor $(\Delta K - \Delta K_{th})^n$ accounts for the experimental data at low stress levels and the existence of a threshold value ΔK_{th} of ΔK at which no crack propagation occurs. By proper choice of constants, Equation (8.8) can be made to fit the experimental data over a range from 10^{-8} to 10^{-2} in/cycle.

Attempts have been made to apply the J-integral concept to elastic–plastic fatigue crack propagation. A relation of the form

$$\frac{da}{dN} = C(\Delta J)^m \tag{8.9}$$

in complete analogy to the Paris law has been suggested [8.7, 8.8]. However, the J-integral concept does not apply to elastic–plastic problems where unloading occurs and, therefore, use of Equation (8.9) cannot be justified on theoretical grounds.

8.3. Fatigue life calculations

When a structural component is subjected to fatigue loading, a dominant crack reaches a critical size under the peak load during the last cycle leading to catastrophic failure. The basic objective of the fatigue crack propagation analysis is the determination of the crack size, a, as a function of the number of cycles, N (Figure 8.1). Thus, the fatigue crack propagation life N_p is obtained. When the type of applied load and the expression of the stress intensity factor are known, application of one of the foregoing fatigue laws enables a realistic calculation of the fatigue crack propagation life of the components.

As an example, consider a plane fatigue crack of length $2a_0$ in a plate subjected to a uniform uniaxial stress σ perpendicular to the plane of the crack. The stress intensity factor K is given by

$$K = f(a)\sigma\sqrt{\pi a} \tag{8.10}$$

where $f(a)$ is a geometry dependent function.

Integrating the fatigue crack propagation law expressed by Equation (8.3), gives

$$N - N_0 = \int_{a_0}^{a} \frac{da}{C(\Delta K)^m} \tag{8.11}$$

where N_0 is the number of load cycles corresponding to the half crack length a_0. Introducing the stress intensity factor range ΔK, where K is given from Equation (8.10), into Equation (8.11) we obtain

$$N - N_0 = \int_{a_0}^{a} \frac{da}{C\left[f(a)\Delta\sigma\sqrt{\pi a}\right]^m}. \tag{8.12}$$

Assuming that the function $f(a)$ is equal to its initial value $f(a_0)$ so that

$$\Delta K = \Delta K_0 \sqrt{\frac{a}{a_0}}, \qquad \Delta K_0 = f(a_0)\Delta\sigma\sqrt{\pi a} \tag{8.13}$$

Equation (8.12) gives

$$N - N_0 = \frac{2a_0}{(m-2)C(\Delta K_0)^m}\left[1 - \left(\frac{a_0}{a}\right)^{m/2-1}\right] \quad \text{for } m \neq 2. \tag{8.14}$$

Unstable crack propagation occurs when

$$K_{\max} = f(a)\sigma_{\max}\sqrt{\pi a} \tag{8.15}$$

from which the critical crack length a_c is obtained. Then Equation (8.14) for $a = a_c$ gives the fatigue crack propagation life $N_p = N_c - N_0$.

Usually, however, $f(a)$ varies with the crack length a and the integration of Equation (8.12) cannot be performed directly but only through the use of numerical methods.

8.4. Variable amplitude loading

The fatigue crack propagation results discussed so far have been concerned with constant amplitude load fluctuation. Although this type of loading occurs frequently in practice the majority of engineering structures are subjected to complex fluctuating loading. Unlike the case of constant cyclic load where ΔK increases gradually with increasing crack length, abrupt changes take place in ΔK due to changes in applied load. Thus, there occur load interaction effects which greatly influence the fatigue crack propagation behavior.

It was first recognized empirically in the early 1960s that the application of a tensile overload in a constant amplitude cyclic load results in crack retardation following the overload; that is, the crack growth rate is smaller than it would have been under constant amplitude loading. This effect is shown schematically

Fatigue and environment-assisted fracture 263

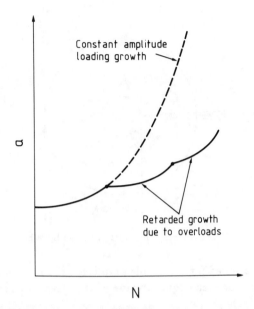

Figure 8.4. Typical form of crack length versus number of cycles curve for constant amplitude loading and constant amplitude plus overloading.

in Figure 8.4. The amount of crack retardation is dramatically decreased when a tensile-compressive overload follows a constant amplitude cyclic load.

An explanation of the crack retardation phenomenon may be obtained by examining the behavior of the plastic zone ahead of the crack tip. The overload has left a large plastic zone behind. The elastic material surrounding this plastic zone after unloading acts like a clamp on this zone causing compressive residual stresses. As the crack propagates into the plastic zone the residual compressive stresses tend to close the crack at some distance. Hence the crack will propagate at a decreasing rate into the zone of residual stresses. When these stresses are overcome and the crack is opened again, subsequent fluctuating loading causes crack growth.

Due to the crack retardation phenomenon, the determination of the fatigue life under a variable amplitude loading by simply summing the fatigue lives of the various constant amplitude loads in the loading history results in conservative predictions. Of the various methods proposed for this reason, we will briefly present the root-mean-square model and the models based on crack retardation and crack closure.

The root-mean-square model proposed by Barsom [8.9] applies to variable amplitude narrow-band random loading spectra. It is assumed that the average fatigue crack growth rate under a variable amplitude random loading fluctuation is approximately equal to the rate of fatigue crack growth under constant amplitude cyclic load which is equal to the root-mean-square of the variable amplitude loading. Thus, the fatigue crack propagation laws presented in Section

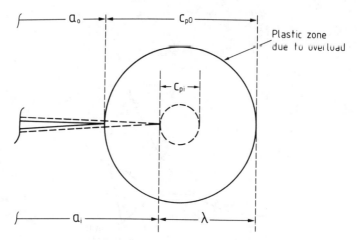

Figure 8.5. Crack retardation model proposed by Wheeler [8.10].

8.2 can be equally applied for a variable amplitude random loading when ΔK is substituted by the root-mean-square value of the stress intensity factor ΔK_{rms} given by

$$\Delta K_{rms} = \sqrt{\frac{\sum (\Delta K_i)^2 n_i}{\sum n_i}} \qquad (8.16)$$

where n_i is the number of loading amplitudes with a stress intensity factor range of ΔK_i.

The model based on crack retardation proposed by Wheeler [8.10] assumes that after a peak load there is a load interaction effect when the crack-tip plastic zones for the subsequent loads are smaller than the plastic zone due to the peak load. Consider that at a crack length a_0 an overload stress σ_0 creates a crack-tip plastic zone of length c_{po} which according to Equation (3.59) or (3.60) is given by

$$c_{po} = \frac{1}{A} \frac{\sigma_0^2 a_0}{\sigma_Y^2} \qquad (8.17)$$

where $A = 1$ or 3 for plane strain or generalized plane stress conditions, respectively (Figure 8.5). When the crack has propagated to a length a_i a stress σ_i will produce a plastic zone of length c_{pi} given by

$$c_{pi} = \frac{1}{A} \frac{\sigma_i^2 a_i}{\sigma_Y^2} \qquad (8.18)$$

The plastic zone due to the stress σ_i is included inside the plastic zone due to the overload. Then a retardation factor Φ is introduced given by

$$\Phi = \left(\frac{c_{pi}}{\lambda}\right)^m \qquad (8.19)$$

where $\lambda = a_0 + c_{po} - a_i$ and m is an empirical parameter. Then, the crack growth increment during load cycle i is

$$\Delta a_i = \Phi\left(\frac{da}{dN}\right) \qquad (8.20)$$

where da/dN is the constant amplitude crack growth rate corresponding to the stress intensity factor range ΔK_i of load cycle i.

When $a_i + c_{pi} > a_0 + c_{po}$ the crack has propagated through the overload plastic zone and the retardation factor is $\Phi = 1$.

The model based on crack closure has been introduced by Elber [8.11, 8.12]. It is based on the observation that the faces of fatigue cracks subjected to zero-tension loading close during unloading, and compressive residual stresses act on the crack faces at zero load. Thus, the crack closes at a tensile rather than zero or compressive load. An effective stress intensity factor range is defined by

$$(\Delta K)_{\text{eff}} = K_{\max} - K_{\text{op}} \qquad (8.21)$$

where K_{op} corresponds to the point at which the crack is fully open. Then, one of the crack propagation laws of Section 8.2 can be employed. Using, for example, the Paris law (Equation (8.3)) this can be stated in the form

$$\frac{da}{dN} = C(U \Delta K)^m \qquad (8.22)$$

where

$$U = \frac{K_{\max} - K_{\text{op}}}{K_{\max} - K_{\min}}, \qquad \Delta K = K_{\max} - K_{\min}. \qquad (8.23)$$

For the determination of U a number of empirical relations have been proposed. Elber suggested that U can be given in the form

$$U = 0.5 + 0.4R,$$

where

$$R = \frac{K_{\min}}{K_{\max}} \quad \text{for } -0.1 \le R \le 0.7. \qquad (8.24)$$

Schijve [8.13] proposed the relation

$$U = 0.55 + 0.33R + 0.12R^2 \qquad (8.25)$$

which extends the previous equation to negative R ratios in the range $-1.0 < R < 0.54$. Relations (8.24) and (8.25) were obtained for a 2024-T3 aluminum. A method of determining K_{op} has been advanced by De Koning [8.14].

A common characteristic of the above fatigue crack propagation models is that they mainly rely on experimental data correlations for specific loading and geometry conditions and have therefore limited predictive capability under different loading histories, geometrical configuration and material types.

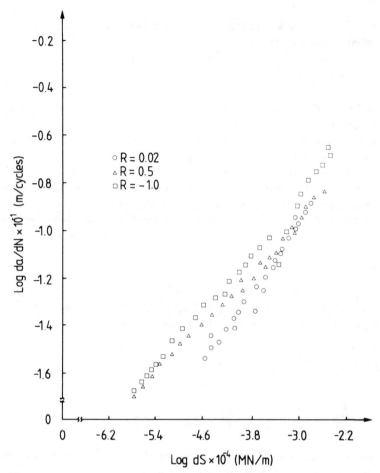

Figure 8.6. Crack growth rate versus change of strain energy density factor for 7075 aluminum obtained from a center cracked specimen [8.15].

8.5. Mixed-mode fatigue crack propagation

One of the basic shortcomings of the previous analysis of fatigue crack propagation based on the stress intensity factor is the inability to include complex crack geometries and loading fluctuations where the direction of crack growth is not known in advance. The conventional stress intensity factor approach is restricted to symmetrical configurations between the applied stress and crack plane and self-similar crack growth. Such an idealization is seldom encountered in service and must be regarded as the exception rather than the rule. The majority of failures are of the mixed-mode type, where the crack does not propagate in a direction normal to the applied load because of the lack of geometric symmetry.

The concept of the strain energy density factor developed in Chapter 6 has been extended to predict the growth of fatigue cracks under mixed-mode loading

Fatigue and environment-assisted fracture

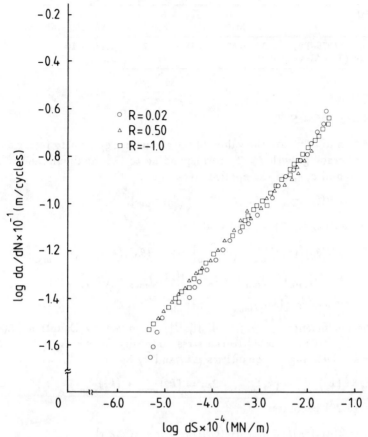

Figure 8.7. Crack growth rate versus change of strain energy density factor for Ti–6Al–4V titanium obtained from a center cracked specimen [8.15].

conditions [8.15, 8.16]. An empirical relation of the form

$$\frac{da}{dN} = C(\Delta S)^n \qquad (8.26)$$

has been proposed by experimental data correlations. The change of the strain energy density factor ΔS is calculated from the linear theory of elasticity. In a typical $\log(da/dN)$ versus $\log(\Delta S)$ plot the parameter C represents the y-intercept and n the slope of the line drawn through the data points. Results for 7075–T6 aluminum and Ti–6Al–4V titanium are shown in Figures 8.6 and 8.7. The material and fatigue parameters for these metal alloys are shown in Table 8.1.

For cracks that do not grow in a self-similar manner the hypotheses of the strain energy density theory referred to in Chapter 6 for monotonic loading can be extended to the case of cyclic loading by assuming that the crack will grow in the direction of the minimum strain energy density factor. In Equation (8.26)

Table 8.1. Material and fatigue parameters for two metal alloys.

Material	ν	$E(10^5)$ MPa	σ_{ys} MPa	K_{ic} MPa$\sqrt{m}$	C (m/cycle) (MN/m)$^{-n}$	n
Aluminum (7075–T6)	0.33	0.65	538	32.97	1.148×10	2.188
Titanium (Ti–6Al–4V)	0.31	1.28	917	71.53	7.391×10^{-1}	1.901

ΔS is given by

$$\Delta S = S^{\max} - S^{\min} \tag{8.27}$$

where $S^{\max}$ and $S^{\min}$ are the values of the strain energy density factor along the direction of crack growth $\theta = \theta_0$ corresponding to the maximum and minimum values $\sigma_{\max}$ and $\sigma_{\min}$ of the applied stress, i.e.

$$S^{\max} = S(\theta_0, \sigma_{\max}), \quad S^{\min} = S(\theta_0, \sigma_{\min}). \tag{8.28}$$

Using Equation (6.31) we find that

$$\Delta S = a_{11}(\theta_0)[(k_\mathrm{I})^2_{\max} - (k_\mathrm{I})^2_{\min}] + 2a_{12}(\theta_0)[(k_\mathrm{I})_{\max}(k_\mathrm{II})_{\max} - $$
$$- (k_\mathrm{I})_{\min}(k_\mathrm{II})_{\min}] + a_{22}(\theta_0)[(k_\mathrm{II})^2_{\max} - (k_\mathrm{II})^2_{\min}] + $$
$$+ a_{33}(\theta_0)[(k_\mathrm{III})^2_{\max} - (k_\mathrm{III})^2_{\min}] \tag{8.29}$$

where the coefficients a_{ij} $(i, j = \mathrm{I, II, III})$ are given by Equation (6.32) and k_j $(j = \mathrm{I, II, III})$ are related to the stress intensity factors K_j $(j = 1, 2, 3)$ by $K_j = \sqrt{\pi} k_j$. Defining the quantities Δk_j and $\overline{k}_j$ by

$$\Delta k_j = (k_j)_{\max} - (k_j)_{\min}, \quad 2\overline{k}_j = (k_j)_{\max} + (k_j)_{\min} \tag{8.30}$$

Equation (8.29) becomes

$$\Delta S = 2[a_{11}(\theta_0)\overline{k}_\mathrm{I}\Delta k_\mathrm{I} + a_{12}(\theta_0)(\overline{k}_\mathrm{II}\Delta k_\mathrm{I} + \overline{k}_\mathrm{I}\Delta k_\mathrm{II}) + $$
$$+ a_{22}(\theta_0)\overline{k}_\mathrm{II}\Delta k_\mathrm{II} + a_{33}(\theta_0)\overline{k}_\mathrm{III}\Delta k_\mathrm{III}]. \tag{8.31}$$

Note that Equation (8.31) includes not only the stress range but also the mean stress which is consistent with experimental data that da/dN depends not only on the stress amplitude but also on the stress ratio R.

The problem of a crack of length $2a_0$ in an infinite plate subjected to a remote uniaxial cyclic stress $\pm \sigma$ which subtends an angle β to the crack axis is considered as an example of mixed-mode fatigue crack growth. The stress intensity factors k_I and k_II are given by

$$k_\mathrm{I} = \sigma\sqrt{a_0}\sin^2\beta, \quad k_\mathrm{II} = \sigma\sqrt{a_0}\sin\beta\cos\beta. \tag{8.32}$$

The angle of crack initiation for the first increment of growth is determined from Equation (6.15a) where the values of k_I and k_II are given from Equation (8.32). Consider a small increment of crack from both tips A and B so that the newly developed bend crack is $B'BOAA'$ (Figure 8.8(a)). Determination of stress intensity factors k_I and k_II at the tips A' and B' of the crack $B'BOAA'$ necessitates complicated and lengthy calculations. To overcome this difficulty the bend crack $B'BOAA'$ is approximated by the straight crack $A'B'$. The new

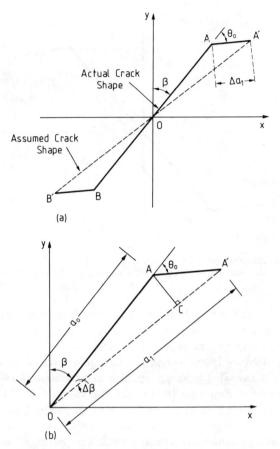

Figure 8.8. Incremental growth of an inclined crack. (a) First increment of growth and (b) straight line approximation.

crack angle is $\beta_1 = \beta + \Delta\beta$ and the new half-crack length is $a_1 = a_0 + CA'$. From geometrical considerations it follows that

$$\beta_1 = \beta + \frac{\Delta a_1 \sin\theta_0}{a_0 + \Delta a_1 \cos\theta_0} \tag{8.33a}$$

and

$$a_1 = a_0 + \frac{\Delta a_1 + a_0 \cos\theta_0}{a_0 + \Delta a_1 \cos\theta_0}\Delta a_1. \tag{8.33b}$$

Thus the stress intensity factors k_I, k_II at the tips A', B' of the crack $B'BOAA'$ are calculated from Equations (8.32) by replacing a_0 and β by a_1 and β_1 respectively. The same procedure is used to calculate k_I, k_II at the tip of the crack after the second increment of growth by replacing the actual bend crack by a straight crack, and so on. This allows the determination of the stress intensity factors during the process of fatigue crack growth. Equation (8.31) with $k_\mathrm{III} = 0$ is then

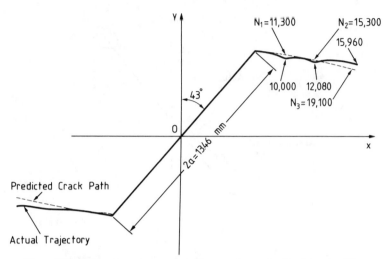

Figure 8.9. Fatigue crack growth trajectory for an inclined crack with $\beta = 43°$ and $\sigma_{\max} = 172.38$ MPa and $\sigma_{\min} = 17.24$ MPa [8.15].

used to determine ΔS for each increment of crack growth Δa_1, Δa_2, etc. The number of cycles N_1, N_2, etc., corresponding to Δa_1, Δa_2, etc., is calculated from (8.26), the fatigue crack growth equation.

The above procedure is applied to the fatigue growth of an inclined crack in a Ti–6Al–4V mill annealed titanium studied in references [8.15] and [8.16]. The fatigue parameters of Equation (8.26) take the values $C = 0.412$ and $n = 1.87$ and the crack has an initial length of $2a_0 = 13.46$ mm and an inclination angle $\beta = 43°$. Analytical results for the fatigue crack trajectory and the number of cycles for each increment of crack growth are presented in Figure 8.9 and compared with the existing experimental data. Observe that the experimental results are close to the theoretical predictions.

8.6. Nonlinear fatigue analysis based on the strain energy density theory

The fatigue crack propagation models discussed in Sections 8.2 or 8.5 based, respectively, on the stress intensity factor range or the strain energy density factor range present a major disadvantage: they both rely on linear elasticity to characterize the fatigue fracture which is mainly a process of damage accumulation. The basic assumption of linear elastic fracture mechanics that there is a K-dominance region around the crack tip is violated in fatigue. The growing crack leaves a wake of residual plastic zone behind it which makes questionable the existence of a K-dominance region. As the crack grows material elements ahead of it undergo damage at each load cycle. This process is repeated until a threshold point is reached at which the elements break and the crack advances. The strain energy density failure criterion, as developed in Chapter 6, will be

Fatigue and environment-assisted fracture

applied to study the fatigue crack propagation process [8.17–8.19]. The nonlinear material response and the accumulated damage in the crack-tip region are taken into account.

In the strain energy density theory attention is focused on a material element ahead of the crack tip and on the direction of crack growth. During the fatigue load energy is accumulated in the element due to material damage. It is assumed that crack growth occurs when the accumulated strain energy density reaches a critical value that is characteristic of the material. A crack growth increment Δr will occur after ΔN loading cycles in accordance with the relation

$$\left(\frac{dW}{dV}\right)_c = \left(\frac{dW}{dV}\right)_0 + \sum_{j=1}^{\Delta N} \Delta\left(\frac{dW}{dV}\right)_j^{r=\Delta r} \tag{8.34}$$

where $(dW/dV)_c$ is the critical amount of strain energy density, which is a material constant, $(dW/dV)_0$ represents the static response due to the applied load and $\Delta(dW/dV)_j$ the amount of energy accumulated during the load cycle j. The symbol Δ stands for the change in a quantity from the beginning to the end of a cycle. From Equation (6.9) it follows that

$$\Delta\left(\frac{dW}{dV}\right)_j^{r=\Delta r} = \frac{(\Delta S)_j^{r=\Delta r}}{\Delta r} \tag{8.35}$$

where $(\Delta S)_j^{r=\Delta r}$ is defined as

$$(\Delta S)_j^{r=\Delta r} = S(r = \Delta r;\ t = t_f^j) - S(r = \Delta r;\ t = t_0^j). \tag{8.36}$$

In Equation (8.36) t_f^j is the time at the completion of the jth cycle and t_0^j is the time at the start of the jth cycle. Using Equations (8.35) and (8.36), Equation (8.34) takes the form

$$\left(\frac{dW}{dV}\right)_c - \left(\frac{dW}{dV}\right)_0^{r=\Delta r} = \frac{1}{\Delta r}\sum_{j=1}^{\Delta N}(\Delta S)_j^{r=\Delta r} \tag{8.37}$$

or

$$\frac{\Delta r}{\Delta N} = \frac{dr}{dN} = \frac{\sum_{j=1}^{\Delta N}(\Delta S)_j^{r=\Delta r}}{\Delta N[(dW/dV)_c - (dW/dV)_0^{r=\Delta r}]}. \tag{8.38}$$

Equation (8.38) expresses the rate of crack growth, $\Delta r/\Delta N$, in terms of the quantities ΔS_j and $(dW/dV)_0$ which are determined from the local stress and strain fields in the vicinity of the crack tip.

To simplify evaluation of Equation (8.38) the assumption is made that the damage accumulated during each cycle is constant and equal to the amount accumulated during the first cycle of the current crack growth increment. Denoting ΔS as the change in S for the first cycle of energy accumulation, Equation (8.38) becomes

$$\frac{\mathrm{d}r}{\mathrm{d}N} = \frac{(\Delta S)^{r=\Delta r}}{(\mathrm{d}W/\mathrm{d}V)_c - (\mathrm{d}W/\mathrm{d}V)_0^{r=\Delta r}}. \tag{8.39}$$

The static response of the material, $(\mathrm{d}W/\mathrm{d}V)_0^{r=\Delta r}$, may be defined as the average response during a load cycle. For a constant amplitude loading $(\mathrm{d}W/\mathrm{d}V)_0^{r=\Delta r}$ can be calculated from

$$\left(\frac{\mathrm{d}W}{\mathrm{d}V}\right)_0^{r=\Delta r} = \tfrac{1}{2}\left[\tfrac{1}{2}\left[\left(\frac{\mathrm{d}W}{\mathrm{d}V}\right)_{t_0}^{r=\Delta r} + \left(\frac{\mathrm{d}W}{\mathrm{d}V}\right)_{t_f}^{r=\Delta r}\right] + \left(\frac{\mathrm{d}W}{\mathrm{d}V}\right)_{\bar{t}}^{r=\Delta r}\right] \tag{8.40}$$

where t_0 is the time at the initiation of the load cycle; t_f at the completion of the load cycle; and $\bar{t}$ at mid-cycle.

Based on the above method, the fatigue crack propagation behavior of a center and edge-cracked panel made of a material which presents large inelastic deformation prior to fracture was analyzed in references [8.18–8.20]. The stress analysis of the panel was performed by an elastic–plastic finite element code using the J_2 flow theory of plasticity and a singular crack-tip element. Finite increments of crack growth Δr were chosen to determine the corresponding number of loading cycles ΔN, or ΔN was assumed to obtain Δr. The results of the nonlinear analysis were compared with fatigue crack propagation models described by Equations (8.3) and (8.26) which were based on linear elasticity. It was found that the parameters entered in the linear models are extremely path dependent, that is, they are sensitive to the history of the crack growth steps and they vary by several orders of magnitude. This proves the inadequacy of the linear models to describe the fatigue crack growth process which is mainly due to their unrealistic basic assumption in which the mechanism of damage accumulation was excluded. The methodology based on the strain energy density theory provides a consistent way for studying the path-dependent nature of fatigue crack growth by taking into consideration the irreversible material damage in the vicinity of the crack-tip region. Fatigue data have been obtained to show the size effect of cylindrical specimens [8.21] in addition to prediction of failure initiation size by fatigue in unnotched specimens [8.22].

Much remains to be done for the complete understanding of the nonlinear fatigue crack growth process. Improvements are needed on constitutive equations that can better describe the irreversible character of material damage. In this respect the strain energy density damage theory developed by Sih [8.23] provides a basic tool for the understanding of the irreversible material behavior without presuming the constitutive relations of the material. This theory has successfully been applied to analyze the case of a slowly moving crack in a two-dimensional specimen and the cooling and heating phenomenon according to which a solid, when loaded, undergoes a state of cooling prior to heating. It is believed that use of this theory will open the door for a complete understanding of the nonlinear path dependent nature of fatigue crack growth process.

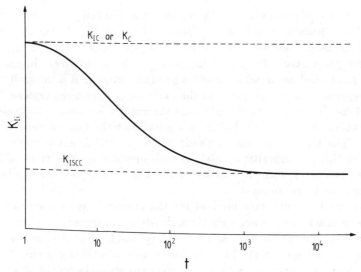

Figure 8.10. Initial stress intensity factor, K_{I_i}, versus time to failure, t, under environment assisted fracture.

8.7. Environment-assisted fracture

It has long been recognized that failure of engineering components subjected to an aggressive environment may occur under applied stresses well below the strength of the material. Failure under such conditions involves an interaction of complex chemical, mechanical and metallurgical processes. The basic subcritical crack growth mechanisms include stress corrosion cracking, hydrogen embrittlement and liquid metal embrittlement. Although a vast number of experiments have been performed and a number of theories have been proposed, a general mechanism for environment-assisted cracking is still lacking. It is not the intent of this section to cover the material in depth, but rather to present a brief account of the phenomenological aspect of the problem of stress corrosion cracking. For further information on the subject the interested reader is referred to the books of Cherepanov [8.24], Williams [8.25], Barsom and Rolfe [8.26] and Hertzberg [8.27]. The related subject of time-dependent fracture with an emphasis on crack growth is adequately treated in the book by Kanninen and Popelar [8.28].

The experimental methods for the evaluation of the stress corrosion susceptibility of a material under given environmental conditions fall into two categories: the time-to-failure tests and the crack growth rate tests. Both kinds of tests are performed on fatigue precracked specimens. The most widely used specimens are the cantilever beam specimen subjected to constant load and the wedge-loaded specimen subjected to constant displacement. The stress intensity factors for the specimens are calculated by appropriate calibration formulas.

In the time-to-failure tests the specimens are loaded to various initial stress intensity factor levels K_{I_i} and the time required to failure is recorded. The test results are represented in a K_{I_i} versus time diagram, a representative form of which is shown in Figure 8.10. Observe that as K_{I_i} decreases the time to failure increases. The maximum value of K_{I_i} is equal to K_{I_c} or K_I, the valid or invalid plane strain fracture toughness. A threshold stress intensity factor K_{ISCC} is obtained below which there is no crack growth. It is generally accepted that K_{ISCC} is a unique property of the material-environment system. The time required for failure can be divided into the incubation time (the time interval during which the initial crack does not grow) and the time of subcritical crack growth. The incubation time depends on the material, environment and K_{I_i}, while the time of subcritical crack growth depends on the type of load, the specimen geometry and the kinetics of crack growth caused by the interaction of material and environment.

In the crack growth rate method for the study of stress corrosion cracking the rate of crack growth per unit time, da/dt, is measured as a function of the instantaneous stress intensity factor, K_I. A typical form of the curve $\log(da/dt)$–K_I is shown in Figure 8.11. This can be divided into three regions. In regions I and III the rate of crack growth, da/dt, depends strongly on the stress intensity factor, K_I, while in region II da/dt is almost independent of K_I. This behavior in region II indicates that crack growth is not of a mechanical nature, but it is caused by chemical, metallurgical and other processes occurring at the crack tip. Note that in region I the threshold stress intensity factor corresponds to K_{ISCC}.

Environment-assisted cracking is a complicated phenomenon and depends on many factors, including the material, the environment, temperature and pressure. A stress-assisted diffusion theory has been recently proposed by Aifantis [8.29] and was applied to a variety of problems ranging from the bending of beams and 'anelastic' effect to stress corrosion cracking and hydrogen embrittlement [8.30–8.32]. The theory is based on a differential statement expressing balance of momentum and appropriate constitutive equations for the stress that the solute exerts on itself and the drag exerted upon it by the lattice.

For the case of slow diffusion motions and neglecting the effect of nonhydrostatic stress components, the following differential equation was derived

$$\frac{\partial \rho}{\partial t} = (D + N\sigma^0)\nabla^2 \rho - (M - N)\nabla \sigma^0 \nabla \rho \qquad (8.41)$$

where ρ denotes solute concentration, σ^0 is the trace of the mechanical stress tensor and D, N, M are phenomenological coefficients. This equation generalizes Cottrell's stress-assisted diffusion theory in replacing his constant diffusivity by an effective one $(D + N\sigma^0)$ dependent linearly on the hydrostatic stress. Experimental evidence and microscopic arguments support this conclusion. The equilibrium solution of Equation (8.41) gives the following expression for the distribution of the concentration ρ around the crack tip:

$$\rho = \rho_0(1 + \beta\sigma^0)^\alpha \qquad (8.42)$$

where ρ_0 represents the concentration on the boundary and $\alpha = N/M$, $\beta = N/D$.

Equation (8.42) was used to derive fracture criteria under coupled environmental and mode I loading conditions. A crack growth condition was developed

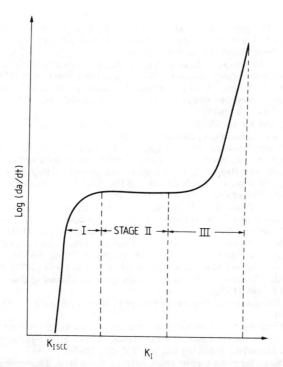

Figure 8.11. Logarithm of subcritical crack growth rate, $\log(da/dt)$, versus stress intensity factor, K_I, under environment assisted cracking.

according to which cracking is governed by the level of concentration at a specific point in front of the crack. Crack growth begins when the concentration at the critical point reaches a critical value. A criterion of the form

$$P = P_0 K_I^{-2\alpha} \qquad (8.43)$$

was established, where P is the pressure. In an analogous manner the subcritical crack velocity V was expressed by

$$V = V_0 K_I^{2\gamma}. \qquad (8.44)$$

The above results were extended to the case of crack growth under mixed-mode conditions by Gdoutos and Aifantis [8.33]. Equations analogous to Equations (8.43) and (8.44) were obtained where K_I was replaced by $(K_I^2 + K_{II}^2)^{1/2}$. A step-by-step procedure was adopted for the determination of the crack path during subcritical crack growth. Furthermore, Gdoutos and Aifantis [8.34] extended the optical method of caustics to environmental cracking under mixed-mode conditions. It was found that the axis of symmetry of the caustic represents the direction of the crack growth, while the condition of the crack growth is expressed in terms of the diameter of the caustic. When this diameter reaches a critical value, crack growth takes place.

References

8.1. Head, A. K., The growth of fatigue crack, *Philosophical Magazine* **44**, 925–938 (1953).
8.2. Paris, P. and Erdogan, F., A critical analysis of crack propagation laws, *Journal of Basic Engineering, Trans. ASME* **85**, 528–534 (1963).
8.3. Forman, R. G., Kearney, V. E. and Engle, R. M., Numerical analysis of crack propagation in cyclic-loaded structures, *Journal of Basic Engineering, Trans. ASME* **89**, 459–464 (1967).
8.4. Donahue, R. J., Clark, H. M., Atanmo, P., Kumble, R. and McEvily, A. J., Crack opening displacement and the rate of fatigue crack growth, *International Journal of Fracture Mechanics* **8**, 209–219 (1972).
8.5. Klesnil, M. and Lucas, P., Effect of stress cycle asymmetry on fatigue crack growth, *Material Science Engineering* **9**, 231–240 (1972).
8.6. Erdogan, F. and Ratwani, M., Fatigue and fracture of cylindrical shells containing a circumferential crack, *International Journal of Fracture Mechanics* **6**, 379–392 (1970).
8.7. Dowling, N. E. and Begley, J. A., Fatigue crack growth during gross plasticity and the J-integral, in *Mechanics of Crack Growth*, ASTM STP 590, American Society for Testing and Materials, Philadelphia, pp. 82–103 (1976).
8.8. Dowling, N. E., Crack growth during low-cycle fatigue of smooth axial specimens, in *Cyclic Stress-Strain and Plastic Deformation Aspects of Fatigue Crack Growth*, ASTM STP 637, American Society for Testing and Materials, Philadelphia, pp. 97–121 (1977).
8.9. Barsom, J. M., Fatigue-crack growth under variable-amplitude loading in ASTM A514-B steel, in *Progress in Flaw Growth and Fracture Toughness Testing*, ASTM STP 536, American Society for Testing and Materials, Philadelphia, pp. 147–167 (1973).
8.10. Wheeler, O.E., Spectrum loading and crack growth, *Journal of Basic Engineering, Trans. ASME* **94**, 181–186 (1972).
8.11. Elber, W., Fatigue crack closure under cyclic tension, *Engineering Fracture Mechanics* **2**, 37–45 (1970).
8.12. Elber, W., Equivalent constant-amplitude concept for crack growth under spectrum loading, *Fatigue Crack Growth Under Spectrum Loads*, ASTM STP 595, American Society for Testing and Materials, Philadelphia, pp. 236–250 (1976).
8.13. Schijve, J., Some formulas for the crack opening stress level, *Engineering Fracture Mechanics* **14**, 461–465 (1981).
8.14. De Koning, A. U., A simple crack closure model for prediction of fatigue crack growth rates under variable-amplitude loading, in *Fracture Mechanics – Thirteenth Conference*, ASTM STP 743, American Society for Testing and Materials, Philadelphia, pp. 63–85 (1981).
8.15. Sih, G. C. and Barthelemy, B. M., Mixed mode fatigue crack growth predictions, *Engineering Fracture Mechanics* **13**, 439–451 (1980).
8.16. Sih, G. C., Fracture mechanics of engineering structural components, in *Fracture Mechanics Methodology* (eds G. C. Sih and L. Faria), Martinus Nijhoff Publ., The Netherlands, pp. 35–101 (1984).
8.17. Sih, G. C., Fatigue crack growth and macroscopic damage accumulation, in *Defects and Fracture* (eds G. C. Sih and H. Zorski), Martinus Nijhoff Publ., The Netherlands pp. 53–62 (1982).
8.18. Moyer, E. T. Jr, Nonlinear aspects of fatigue crack propagation – a fracture mechanics approach, Ph.D. Thesis, Lehigh University (1982).
8.19. Sih, G. C. and Moyer, E. T. Jr, Path dependent nature of fatigue crack growth, *Engineering Fracture Mechanics* **17**, 261–280 (1983).
8.20. Sih G. C. and Chao C. K., Size effect of cylindrical specimens with fatigue cracks, *Theoretical and Applied Fracture Mechanics* **1**, 239–247 (1984).
8.21. Sih G. C. and Chao C. K., Influence of load amplitude and uniaxial tensile properties on fatigue crack growth *Theoretical and Applied Fracture Mechanics* **2**, 247–257 (1984).
8.22. Sih G. C. and Chao C. K., Failure initiation in unnotched specimens subjected to monotonic and cyclic loading, *Theoretical and Applied Fracture Mechanics*, **2**, 67–73 (1984).
8.23. Sih G. C., Thermomechanics of solids: nonequilibrium and irreversibility, *Theoretical and Applied Fracture Mechanics* **9**, pp. 175–198 (1988).
8.24. Cherepanov, G. P., *Mechanics of Brittle Fracture*, McGraw-Hill, pp. 398–511 (1979).
8.25. Williams, J. G., *Fracture Mechanics of Polymers*, Ellis Horwood Limited Publ., pp. 189–206 (1984).
8.26. Barsom, J. M. and Rolfe, S. T., *Fracture and Fatigue Control in Structures* (2nd edn),

Prentice-Hall, pp. 345–424 (1987).
8.27. Hertzberg, R. W., *Deformation and Fracture Mechanics of Engineering Materials* (2nd ed), Wiley, pp. 425–456 (1983).
8.28. Kanninen, M. F. and Popelar, C. H., *Advanced Fracture Mechanics*, Oxford Univ. Press, pp. 437–497 (1985).
8.29. Aifantis, E. C., On the problem of diffusion in solids, *Acta Mechanica* **37**, 265–296 (1980).
8.30. Wilson, R. K. and Aifantis, E. C., On the theory of stress-assisted diffusion I, *Acta Mechanica* **45**, 273–296 (1982).
8.31. Unger, D. J. and Aifantis, E. C., On the theory of stress-assisted diffusion II, *Acta Mechanica* **47**, 117–151 (1983).,
8.32. Aifantis, E. C., Elementary physicochemical degradation processes, in *Mechanics of Structured Media* (ed. A. P. S. Selvadurai), Elsevier, pp. 79–91 (1981).
8.33. Gdoutos, E. E. and Aifantis, E. C., Environmental cracking under mixed-mode conditions, *Engineering Fracture Mechanics* **23**, 431–439 (1986).
8.34. Gdoutos, E. E. and Aifantis, E. C., The method of caustics in environmental cracking, *Engineering Fracture Mechanics* **23**, 423–430 (1986).

9

Engineering applications

9.1. Introduction

The objective of engineering design is the determination of the geometry and dimensions of the machine or structural elements and the selection of the material in such a way that the elements perform their operating function in an efficient, safe and economic manner. To achieve these objectives, use is made of an appropriate failure criterion which consists of comparing a critical quantity depending on the loading and environmental conditions with a material characteristic parameter. For the selection of the appropriate failure criterion the designer should know the most probable mode of failure. Possible failure modes consist of: (i) yielding or excessive plastic deformations; (ii) general instability (e.g. buckling); and (iii) fracture. In engineering design, however, all possible failure modes should be taken into consideration. For example, structures designed according to the first two failure modes failed in a sudden, catastrophic manner due to unstable crack propagation. This, as was explained in Chapter 1, gave impetus to the development of the discipline of fracture mechanics.

This chapter is devoted to the application of fracture mechanics to engineering design. Fracture mechanics technology has received much attention in recent years and has been successfully applied to determine the useful life of structural components. Contrary to the conventional approach, this discipline is based on the principle that all materials contain initial defects that can affect the load-carrying capacity of engineering structures. Defects are initiated in the material by manufacturing procedures or can be created during the service life, like fatigue, environment-assisted or creep cracks. The quantitative assessment of these defects requires a complete knowledge of the local stress and strain fields coupled with a suitable failure criterion. Although considerable progress has been made in evaluating the stresses and strains in solids weakened by cracks, the search for an appropriate failure criterion to explain the process of material damage is one of the most controversial topics of fracture mechanics. This is mainly because too much emphasis has been placed merely on the fitting of experimental data with analytical prediction from one special situation to another. A coupled stress and failure analysis has recently been proposed by Sih [9.1, 9.2] based on the interdependence of volume and surface energy through the rate of change of volume with surface area. Application of this theory to crack problems showed the inadequacies of the theory of plasticity, which is based on the neglect of the dilatational energy component.

The present chapter starts with a brief outline of the fracture mechanics design methodology and its differences with the conventional approach based on the strength of materials. The basic fracture criteria presented in previous chapters are briefly discussed and example problems follow. The applicability of fracture mechanics which has initially been developed for isotropic materials to composites and concrete is discussed. The chapter concludes with a brief description of the more widely used nondestructive testing methods for defect detection. For an in-depth study of the use of fracture mechanics to machine and structural design the interested reader is referred to references [9.3] and [9.4].

9.2. Fracture mechanics design philosophy

Conventional design analysis of engineering components assumes a defect-free structural geometry and determines a relationship between the applied loading and the maximum stress that is developed in the component. For this reason a stress analysis is performed, based on the theory of elasticity and strength of materials. Safe design is achieved by making sure that the maximum stress is less than the ultimate stress of the material (divided by a factor of safety).

Fracture mechanics design methodology is based on the realistic assumption that all materials contain crack-like flaws. An analysis is then performed of the component with a crack placed in the most probable or dangerous site and a characteristic quantity defining the propensity of the crack to extend is determined. Usually, one dominant crack is assumed. The characteristic quantity depends on the particular failure criterion used, which may be the strain energy release rate, the stress intensity factor, the J-integral, the crack opening displacement or the strain energy density factor. For the determination of this quantity an analysis of the cracked component is performed, based on the methods developed in Chapters 2 and 3. The characteristic quantity is compared with its critical value which represents the material resistance to crack growth. This is a material parameter and can be determined experimentally by the methods used in Chapters 4–6. By this procedure the maximum allowable applied loads for a specified crack size, or the maximum permissible crack size for specified applied loads, are determined.

Most of the failure criteria proposed for linear and nonlinear fracture mechanics analyses – namely, the strain energy release rate (or the stress intensity factor), the J-integral, the crack opening displacement and the strain energy density criterion – suffer serious limitations. The usefulness of a failure criterion should be tested by its consistency and ability to explain the physical phenomena in general. At this point it seems appropriate to distinguish linear elastic and ductile fracture mechanics.

Linear elastic fracture mechanics assumes that the onset of crack growth and final instability leading to catastrophic fracture coincide. Hence, it suffices to have a single parameter to describe the fracture process. The critical value of this parameter represents the fracture toughness and corresponds to the onset of the unstable fracture process. Thus, the strain energy release rate, the stress intensity factor, the J-integral, the crack opening displacement and the strain energy density criteria based on the quantities G, K, J, δ and S, whose critical values are G_c, K_c, J_c, δ_c and S_c, are equivalent. The characterizing quantities G, K, J, δ and S, as was shown in Chapters 4-6, are linearly related. It should be pointed out that the criteria based on G, K, J and δ are restricted to symmetry between the applied load and crack plane and self-similar crack growth. In other words, the direction and shape of the crack growth must be known *a priori*. Such an idealization is seldom encountered in service and must be regarded as the exception rather than the rule. The ability of a criterion to treat mixed mode fracture cannot be overemphasized. Mixed mode fracture can occur in the plane of the crack specimen when load and crack are not symmetrically aligned or in the thickness direction when ductile fracture modes are present, such as cup-and-cone failure or the development of shear lips near the specimen surfaces. The strain energy density criterion proved to be very instrumental in solving problems of mixed-mode crack growth. This criterion has been used by Gdoutos [9.5] to determine the allowable load corresponding to crack initiation and the crack path for a host of engineering problems of practical interest.

Nonlinear fracture mechanics involves the study of crack initiation, slow growth and final instability in the presence of both material and structural nonlinearity. The crack first grows slowly prior to the onset of unstable propagation leading to catastrophic failure. The extent of the process of slow crack growth depends on many factors, including the material of the body, the loading rate and history, the size of the structure and environmental conditions. Of the five more widely used failure criteria mentioned previously, the strain energy release rate and stress intensity factor criteria failed to recognize the importance of stable crack growth prior to the onset of catastrophic failure. The J-integral and crack opening displacement criteria suffer serious limitations in predicting the slow crack growth behavior prior to the onset of rapid fracture. They basically invoke separate conditions for initiation, slow growth and termination. For the J-integral, for instance, crack initiation is related to J_{ri}, unstable crack growth to J_c and stable crack growth involves an assumption that $dJ/da = \text{const}$ (usually put in the form $dJ/da = T\sigma_Y^2/E$, where T is the tearing modulus, σ_Y the yield stress and E the modulus of elasticity) which implies a constant slope of the J-a curve. Such an assumption, however, is contradicted by experimental results which show that J varies nonlinearly with a and dJ/da is not constant. Analogous arguments are valid for δ. Furthermore, these criteria attempt to circumvent the real fracture problem and try to extend the concepts of linear elastic fracture mechanics in situations of extensive crack-tip plastic deformation and substantial crack growth. Use, therefore, of these criteria in nonlinear fracture mechanics should be made with great caution.

On the other hand, the strain energy density criterion can consistently address the whole history of crack propagation from initiation to final instability through the intermediate stage of stable crack growth. The process of crack initiation and stable growth is described by the strain energy density function dW/dV whose critical value $(dW/dV)_c$ can be obtained from uniaxial specimen tests. Global instability corresponding to the sudden release of energy is described by the strain energy density factor S, whose critical value S_c is related to the critical stress intensity factor K_{Ic}. The onset of unstable crack growth occurs when the crack growth increment r during stable growth takes a critical value r_c. This represents the maximum increment of growth along the crack border corresponding to incipient fracture. The three material parameters $(dW/dV)_c$, S_c and r_c are uniquely related as $r_c(dW/dV)_c = S_c$ and the processes of crack initiation, stable and unstable crack growth are described in a consistent and unified manner. Note that the quantities $(dW/dV)_c$ and S_c are geometry- and load- independent material parameters. The validity of the strain energy density approach has been established by applying it to a host of nontrivial problems of fundamental importance. This includes nonhomogeneous and composite materials, plates and shells with cracks, three-dimensional crack problems, dynamic crack propagation, plates with straight and curved cracks, three-dimensional crack problems, fatigue crack growth, problems of mixed-mode crack propagation, ductile fracture involving the prediction of flat and slanted fracture surfaces, etc. Furthermore, this criterion can be used to evaluate material damage with or without the presence of initial defects.

The previously mentioned failure criteria based on fracture mechanics design methodology have extensively been used in engineering applications. A few examples are given below to demonstrate the use of these criteria to engineering design.

9.3. Design example problems

The use of fracture mechanics in the design of machine and structural elements follows the following steps: (i) selection of an appropriate failure criterion; (ii) a stress analysis (mathematical, numerical, experimental) for the determination of the characteristic quantity (stress intensity factor, J-integral, crack opening displacement, strain energy density factor) relevant to the failure criterion; and (iii) an experimental measurement of the critical value of the characteristic quantity. Once this information is available it is then possible to compute the maximum allowable load that a machine or structural element can tolerate for a given defect size, or the maximum permissible defect size for a given applied load.

Three simple example problems of mode-I and mixed-mode brittle fracture are considered to demonstrate this procedure. The critical stress intensity factor and the strain energy density failure criteria are used.

(a) Rotating disk [9.4]

Consider a disk of radius c with a hole of radius b rotating at an angular velocity $\omega = 2\pi N/60$, where N is the number of revolutions per minute. The maximum stress at the rim of the hole is given by

$$\sigma_{\max} = \frac{3+\nu}{4}\rho\omega^2 c^2 \left[1 + \frac{1}{3+\nu}\frac{\nu}{c}\left(\frac{b}{c}\right)^2\right] \tag{9.1}$$

where ν is Poisson's ratio and ρ is the mass density of the material.

The value of the critical speed N_c at failure, based on the conventional approach, is determined from

$$\sigma_{\max} = \sigma_u \tag{9.2}$$

where σ_u is the ultimate strength. For a rotating disk with $c = 2b$ made of steel with $\nu = 0.3$, $\rho = 0.104$ lb/s^2/in^2 and $\sigma_u = 280$ kip/in^2, Equations (9.1) and (9.2) give

$$N_c = 8405 \text{ rev/min}. \tag{9.3}$$

Calculation of N_c based on fracture mechanics assumes that the disk contains a crack which is located where the maximum stress occurs. Based on the critical stress intensity factor failure criterion, the critical condition is expressed by

$$K_{\mathrm{I}} = K_{\mathrm{I}c} \tag{9.4}$$

where K_{I} is the stress intensity factor at the crack tip and $K_{\mathrm{I}c}$ is its critical value.

For small crack lengths a $(a \ll b)$, K_{I} can be approximated

$$K_{\mathrm{I}} = 1.12\sigma_{\max}\sqrt{\pi a}. \tag{9.5}$$

Using Equations (9.1), (9.4) and (9.5), the critical speed N_c is given by

$$N_c = 13.56 \frac{\sqrt{K_{\mathrm{I}c}}}{a^{1/4}\sqrt{\rho[(3+\nu)c^2 + (1-\nu)b^2]}}. \tag{9.6}$$

For $c = 2b$, $b/a = 10$, and $K_{\mathrm{I}c} = 40$ ksi$\sqrt{\text{in}}$ the critical speed for an initial crack of $a = 0.1$ in. is found to be

$$N_c = 4020 \text{ rev/min}. \tag{9.7}$$

Comparing the values of N_c from Equations (9.3) and (9.7) it is observed that the prediction based on the maximum stress criterion differs by more than 100 percent from that obtained by fracture mechanics, which assumes that the disk contains a crack.

The variation of N_c versus crack length a for $b/a = 10$ and various values of c/b is shown in Figure 9.1. Observe that N_c reduces substantially as the size of the initial crack increases. This reduction of N_c is more pronounced at small crack lengths.

(b) *Slanted crack in the thickness direction* [9.4]

Consider a center crack of length $2a$ in a plane specimen tilted at an angle ω with the specimen surface and subjected to a uniaxial stress σ (Figure 9.2). The

Engineering applications

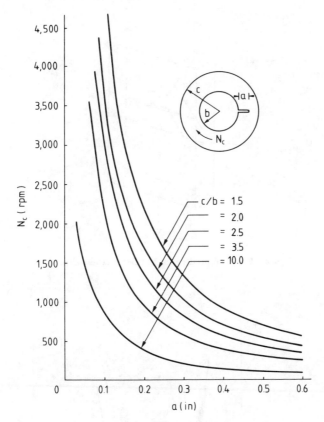

Figure 9.1. Critical speed versus crack length for a rotating disk with different values of outer-to-inner diameter ratio.

stress field in the neighborhood of the crack tip consists of a combination of opening-mode (mode-I) and tearing-mode (mode-III) loading. By ignoring the influence of the plate surfaces the stress intensity factors K_I and K_{III} are given by

$$K_I = \sigma\sqrt{\pi a}\,\sin^2\omega, \qquad K_{III} = \sigma\sqrt{\pi a}\,\sin\omega\cos\omega. \tag{9.8}$$

This is a problem of mixed-mode crack growth and for the determination of the crack path and critical stress at instability use is made of the strain energy density failure criterion. From the first hypothesis of the strain energy density theory (p. 202) it is found that the crack propagates along its own plane tilted at an angle ω with the plane of loading. Equation (6.12) gives for the critical stress

$$\sigma_c = \frac{2\sqrt{\mu S_c}}{\sqrt{a}\,\sin\omega\sqrt{1 - 2\nu\sin^2\omega}} \tag{9.9}$$

where μ is the shear modulus and S_c is the critical value of the strain energy density factor, which is a material constant.

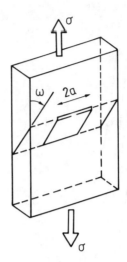

Figure 9.2. A slanted crack in the thickness direction.

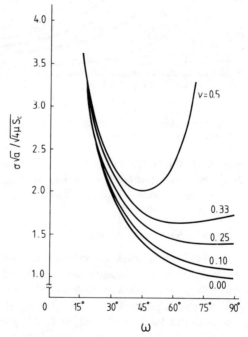

Figure 9.3. Critical stress versus crack angle for the slanted crack of Figure 9.2 for various values of Poisson's ratio.

The variation of $\sigma_c \sqrt{a}/\sqrt{4\mu S_c}$ versus the angle ω is plotted in Figure 9.3 for various values of ν. Observe that, for low values of ν, σ_i decreases with increasing angle and tends to a minimum at $\omega = 90°$. However, for $\nu > 0.25$ the minimum

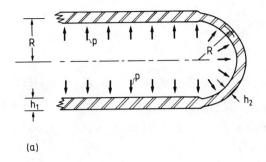

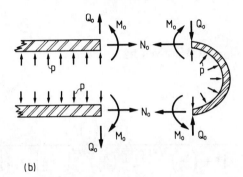

Figure 9.4. (a) Geometry and (b) free-body diagram of a half of a cylindrical shell with hemispherical heads.

stress σ_i for unstable crack propagation occurs at an angle ω_0 given by

$$\omega_0 = \sin^{-1}\left(\frac{1}{\sqrt{4\nu}}\right). \tag{9.10}$$

This example shows that the critical stress for crack growth does not always becomes minimum under mode-I loading and demonstrates the importance of mixed-mode crack propagation.

(c) Cylindrical shell with hemispherical heads [9.5, 9.6]

The problem of a solid rocket motor featuring a long cylindrical shell with hemispherical heads both of radius R (Figure 9.4(a)) is considered. Let h_1 and h_2 be the thicknesses of the cylindrical and hemispherical parts respectively. The free-body diagram showing the forces and moments at the shell juncture is shown in Figure 9.4(b), where p is the value of the applied internal stress and Q_0 and M_0 the shear force and bending moment at the shell juncture. The meridional stress σ_ξ and the hoop stress σ_θ at a distance ξ from the juncture will be given by the relations

$$\sigma_\xi = \frac{pR}{2h_i} \pm \frac{6M_\xi}{h_i^2} \tag{9.11a}$$

$$\sigma_\theta = \frac{pR}{k_i h_i} + \frac{N_\theta}{h_i} + \frac{6M_\theta}{h_i^2} \tag{9.11b}$$

where

$$M_\xi = \frac{Q_0}{B_i} e^{-B_i \xi} \sin B_i \xi + M_0 e^{-B_i \xi}(\sin B_i \xi + \cos B_i \xi) \tag{9.12a}$$

$$M_\theta = \nu M_\xi \tag{9.12b}$$

$$N_\theta = 2Q_0 B_i R\, e^{-B_i \xi} \cos B_i \xi + 2M_0 B_i^2 R\, e^{-B_i \xi}(\cos B_i \xi - \sin B_i \xi) \tag{9.12c}$$

$$B_i^4 = \frac{3(1-\nu^2)}{R^2 h_i^2} \tag{9.12d}$$

$$D_i = \frac{E h_i^3}{12(1-\nu^2)} \tag{9.12e}$$

with the quantities Q_0 and M_0 given by

$$\frac{Q_0}{\frac{pR^2 B_1^3 D_1}{Eh_2}\left[(2-\nu)\left(\frac{h_2}{h_1} - 1\right) + 1\right]} = \frac{1 + \left(\frac{h_1}{h_2}\right)^{5/2}}{\left[1 + \left(\frac{h_1}{h_2}\right)^{5/2}\right]\left[1 + \left(\frac{h_1}{h_2}\right)^{3/2}\right] - \frac{1}{2}\left[1 - \left(\frac{h_1}{h_2}\right)^2\right]^2} \tag{9.13a}$$

$$\frac{M_0}{\frac{pR^2 B_1^2 D_1}{2Eh_2}\left[(2-\nu)\left(\frac{h_1}{h_2} - 1\right) + 1\right]} = \frac{1 - \left(\frac{h_1}{h_2}\right)^2}{\left[1 + \left(\frac{h_1}{h_2}\right)^{5/2}\right]\left[1 + \left(\frac{h_1}{h_2}\right)^{3/2}\right] - \frac{1}{2}\left[1 - \left(\frac{h_1}{h_2}\right)^2\right]^2}. \tag{9.13b}$$

In the above equations k_i is equal to 1 for the cylindrical shell and 2 for the hemispherical shell; E is the modulus of elasticity; and ν is Poisson's ratio. The upper signs in Equations (9.11) refer to the inner surfaces of the shells, and the lower signs to the outer surfaces.

The previous developments show that an element of the shell in either the cylindrical or the hemispherical regions is generally subjected to membrane and bending stresses. Figure 9.5 illustrates a crack of length $2a$ subjected to the normal extensional stresses σ_1, σ_2 and bending moments M_1, M_2. The crack is inclined at an angle β to the direction of the stress σ_1 (or the plane of the moment M_1).

Considering the biaxial extensional stresses and bending moments, we obtain the stress intensity factors k_I, k_II and k_III as

$$k_\mathrm{I} = \left[\frac{12z}{h^3}\Phi(1)(M_1 \sin^2\beta + M_2 \cos^2\beta) + \sigma_1 \cos^2\beta + \sigma_2 \sin^2\beta\right]\sqrt{a} \tag{9.14a}$$

$$k_\mathrm{II} = \left[\frac{12z}{h^3}\Psi(1)(M_1 - M_2)\sin\beta\cos\beta + (\sigma_1 - \sigma_2)\sin\beta\cos\beta\right]\sqrt{a} \tag{9.14b}$$

Engineering applications

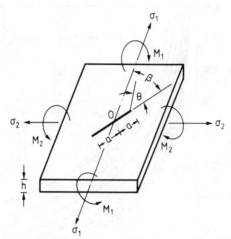

Figure 9.5. An inclined crack subjected to combined biaxial in-plane stresses and bending moments.

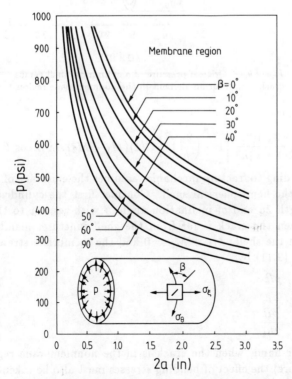

Figure 9.6. Critical pressure of a cylindrical shell versus crack length for an element in the membrane region.

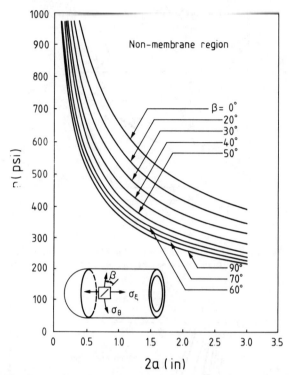

Figure 9.7. Critical pressure of a cylindrical shell versus crack length for an element in the nonmembrane region.

$$k_{III} = -\frac{3\sqrt{10}}{2(1+\nu)h^2}\left[1-\left(\frac{2z}{h}\right)^2\right]\Omega(1)(M_1-M_2)\sin\beta\cos\beta\sqrt{a}. \quad (9.14c)$$

Design according to fracture mechanics assumes the existence of a crack in the cylindrical or the hemispherical shell. Consider first the cylindrical shell with a crack of length $2a$ and of inclination angle β with respect to the meridional direction. When the crack is far from the shell juncture, membrane stresses predominate in the shell. $M_1 = M_2 = 0$ and the membrane stresses σ_1 and σ_2 are given from (9.11) by:

$$\sigma_1 = \sigma_\theta = \frac{pR}{h_1} \quad (9.15a)$$

$$\sigma_2 = \sigma_\xi = \frac{pR}{2h_1} \quad (9.15b)$$

On the other hand, when the crack is in the nonmembrane region (close to the shell juncture) the effect of bending stresses must also be taken into account. Numerical results are obtained by considering an element at a distance $\xi = 4$ in. from the shell juncture, when $h_1 = h_2 = \frac{3}{8}$ in. and $R = 60$ in. These values

Engineering applications

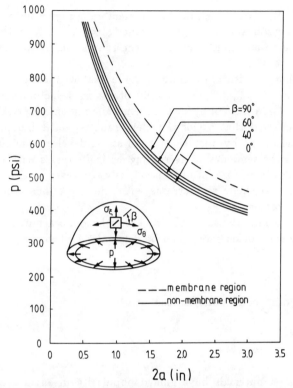

Figure 9.8. Critical pressure versus crack length for an element of the hemispherical heads of the shell of Figure 9.4 located in the membrane (dashed curve) and in the nonmembrane (solid curves) region.

allow computation of the extensional stresses $\sigma_{1,2}$, the bending moments $M_{1,2}$ and, consequently, the stress intensity factors k_I, k_II and k_III (Equations (9.14)).

Using the strain energy density theory the critical pressure p_c for unstable crack propagation can be estimated. The variation of p_c versus the crack length $2a$ for various values of crack inclination angle β is given by Figure 9.6 for the membrane and by Figure 9.7 for the non-membrane region. These figures illustrate the trend of decreasing critical pressure with increasing crack length and decreasing crack inclination angle. Furthermore, we see that for the same crack length and crack angle the critical pressure is higher for the membrane region. It is also observed that the critical pressure is more sensitive to the crack-angle orientation in the membrane region than in the nonmembrane region.

Consider now a crack of length $2a$ and inclination angle β in the hemispherical shell. When the crack exists in the membrane region we obtain from Equations (9.11) that

$$\sigma_\xi = \sigma_\theta = \sigma_1 = \sigma_2 = \frac{pR}{2h}. \tag{9.16}$$

This relation indicates that a material element in the membrane region is in a state of hydrostatic tension and therefore the crack propagates in a self-similar manner ($\theta_0 = 0$). For a crack in the nonmembrane region numerical results are obtained for a material element at a distance $\xi = 4.19$ in. from the shell juncture (corresponding to an angle ϕ made by the radius aiming at the point and the equator equal to 4°).

The variation in critical pressure p_c for unstable crack growth versus crack length for values $\beta = 0, 40°, 60°$ and $90°$ of the crack inclination angle is shown in Figure 9.8. The values of p_c for the membrane region are presented by dashed lines. It is observed that p_c for the membrane region is independent of crack inclination due to the hydrostatic state of stress developed in the shell. Figure 9.8 establishes the tendency of decreasing critical stresses with increasing crack lengths seen also in all cases studied previously. However, unlike results obtained for the cylindrical shell, p_c increases with the crack inclination angle in the nonmembrane region.

Using these results, the tolerable crack lengths and crack inclination angles for a structurally safe rocket motor can be determined when the allowable pressure p_c is known.

9.4. Fiber-reinforced composites

(a) Introductory remarks

Fiber composite materials have gained popularity in engineering applications during the past few years due to their flexibility in obtaining the desired mechanical and physical properties in combination with lightweight components. For this reason they are now being widely used for aerospace and other applications where high strengths and high stiffness-to-weight ratios are required. These materials are usually made of glass or boron, graphite or other nonglass (for the so-called advanced composites) fibers embedded in a matrix. The way in which they are fabricated is very important for their efficient and reliable use in design. Quality control during the fabrication process is crucial and must be enforced.

Modeling the mechanical behavior of fiber composites is not a simple task. The materials are heterogeneous and are characterized by the presence of several types of inherent flaws. The basic failure modes are manifold, including broken fibers, flaws in the matrix, debonded interfaces or their combination. For laminated composites another failure mechanism is delamination between the layers. The load transmission characteristics along the interface between the fibers and the matrix is crucial and it is not always possible to be accurately modeled.

Due to their importance in engineering applications a great amount of effort has been spent in the study of the mechanical and physical properties of fiber composites. Early studies considered fiber composites as homogeneous anisotropic media and used the principles of continuum mechanics. Although much progress has been made using this approach, the failure process in a composite is extremely complex and cannot be adequately described without taking into consideration the inherent flaws of the composite.

Attempts have lately been made to apply the principles of fracture mechanics to composites. This discipline has had great success in analyzing the fracture behavior of polycrystals by considering them as homogeneous and isotropic materials in the macroscopic scale level and by introducing a dominant macrocrack. Application of the same procedure to characterize the toughness of composites by ignoring the heterogeneity of the material and the existing flaws has had limited success. Unless the failure mechanism is properly analyzed, reliable predictions of the allowable stresses in service from small-scale laboratory tests cannot meet with success.

A great number of journals, conferences and books are now devoted to composites. The increasing stream of publications reflects the importance of composites in engineering applications. This section is not intended to overwhelm the reader with the vast number of investigations into the fracture mechanics of composites, but rather the purpose is to discuss briefly the applicability of the principles of fracture mechanics to composites. The anisotropic fracture mechanics approach and some micromechanical failure models are discussed. For more information on the use of the concepts of fracture mechanics to anlyze the failure behavior of composites, the interested reader is referred to references [9.7–9.15].

(b) *Homogeneous anisotropic model*

Attempts have been made to model fiber-reinforced composites as homogeneous anisotropic elastic materials. Consider a homogeneous rectilinearly anisotropic material whose principal axes of material symmetry coincide with the x and y directions. For a state of generalized plane stress the stress–strain relationship may be written as [9.10]

$$\begin{bmatrix} \epsilon_x \\ \epsilon_y \\ \gamma_{xy} \end{bmatrix} = \begin{bmatrix} a_{11} & a_{12} & a_{16} \\ a_{12} & a_{22} & a_{26} \\ a_{16} & a_{26} & a_{66} \end{bmatrix} \begin{bmatrix} \sigma_x \\ \sigma_y \\ \tau_{xy} \end{bmatrix} \qquad (9.17)$$

where σ_x, σ_y, τ_{xy} and ϵ_x, ϵ_y, γ_{xy} are the inplane stresses and strains and a_{ij} ($i, j = 1, 2, 6$) are referred to as compliance coefficients.

For plane strain conditions the stress–strain relationship is expressed by

$$\begin{bmatrix} \epsilon_x \\ \epsilon_y \\ \gamma_{xy} \end{bmatrix} = \begin{bmatrix} b_{11} & b_{12} & b_{16} \\ b_{12} & b_{22} & b_{26} \\ b_{16} & b_{26} & b_{66} \end{bmatrix} \begin{bmatrix} \sigma_x \\ \sigma_y \\ \tau_{xy} \end{bmatrix} \qquad (9.18a)$$

with

$$\sigma_z = -(a_{33})^{-1}(a_{13}\sigma_x + a_{23}\sigma_y + a_{26}\tau_{xy}) \qquad (9.18b)$$

where the constants b_{ij} are given by

$$b_{ij} = a_{ij} - \frac{a_{i3}a_{j3}}{a_{33}} \quad (i, j = 1, 2, 6). \qquad (9.18c)$$

The general solution of the stress and displacement fields in the vicinity of the crack tip in rectilinearly anisotropic bodies was first derived by Sih *et al.*

[9.16]. The most general situation was reduced to a sum of plane and antiplane problems, as in the case of isotropic bodies. The plane problem was further reduced to two loading modes characterized by their corresponding stress intensity factors. The basic modes are defined from the orientation of the applied loads with respect to the crack plane. The first mode corresponds to symmetric loads, the second mode to skew-symmetric plane loads and the third mode to antiplane shear loads relative to the crack plane. The stresses for the three modes are given by:

Mode I

$$\sigma_x = \frac{K_\mathrm{I}}{\sqrt{2\pi r}} \operatorname{Re}\left[\frac{s_1 s_2}{s_1 - s_2}\left(\frac{s_2}{(\cos\theta + s_2 \sin\theta)^{1/2}} - \frac{s_1}{(\cos\theta + s_1 \sin\theta)^{1/2}}\right)\right] \quad (9.19\mathrm{a})$$

$$\sigma_y = \frac{K_\mathrm{I}}{\sqrt{2\pi r}} \operatorname{Re}\left[\frac{1}{s_1 - s_2}\left(\frac{s_1}{(\cos\theta + s_2 \sin\theta)^{1/2}} - \frac{s_2}{(\cos\theta + s_1 \sin\theta)^{1/2}}\right)\right] \quad (9.19\mathrm{b})$$

$$\tau_{xy} = \frac{K_\mathrm{I}}{\sqrt{2\pi r}} \operatorname{Re}\left[\frac{s_1 s_2}{s_1 - s_2}\left(\frac{1}{(\cos\theta + s_1 \sin\theta)^{1/2}} - \frac{1}{(\cos\theta + s_2 \sin\theta)^{1/2}}\right)\right]; \quad (9.19\mathrm{c})$$

Mode II

$$\sigma_x = \frac{K_\mathrm{II}}{\sqrt{2\pi r}} \operatorname{Re}\left[\frac{1}{s_1 - s_2}\left(\frac{s_2^2}{(\cos\theta + s_2 \sin\theta)^{1/2}} - \frac{s_1^2}{(\cos\theta + s_1 \sin\theta)^{1/2}}\right)\right] \quad (9.20\mathrm{a})$$

$$\sigma_y = \frac{K_\mathrm{II}}{\sqrt{2\pi r}} \operatorname{Re}\left[\frac{1}{s_1 - s_2}\left(\frac{1}{(\cos\theta + s_2 \sin\theta)^{1/2}} - \frac{1}{(\cos\theta + s_1 \sin\theta)^{1/2}}\right)\right] \quad (9.20\mathrm{b})$$

$$\tau_{xy} = \frac{K_\mathrm{II}}{\sqrt{2\pi r}} \operatorname{Re}\left[\frac{1}{s_1 - s_2}\left(\frac{s_1}{(\cos\theta + s_1 \sin\theta)^{1/2}} - \frac{s_2}{(\cos\theta + s_2 \sin\theta)^{1/2}}\right)\right]; \quad (9.20\mathrm{c})$$

Mode III

$$\tau_{xz} = -\frac{K_\mathrm{III}}{\sqrt{2\pi r}} \operatorname{Re}\left[\frac{s_3}{(\cos\theta + s_3 \sin\theta)^{1/2}}\right] \quad (9.21\mathrm{a})$$

$$\tau_{yz} = \frac{K_\mathrm{III}}{\sqrt{2\pi r}} \operatorname{Re}\left[\frac{1}{(\cos\theta + s_3 \sin\theta)^{1/2}}\right]. \quad (9.21\mathrm{b})$$

In these equations s_1 and s_2 are the roots of the characteristic equation

$$a_{11}\mu^4 - 2a_{16}\mu^3 + (2a_{12} + a_{66})\mu^2 - 2a_{26}\mu + a_{22} = 0. \quad (9.22)$$

It was shown by Lekhnitskii [9.17] that s_1 and s_2 are either complex or purely imaginary and cannot be real. Putting

$$s_1 = \mu_1 = \alpha_1 + i\beta_1, \quad s_2 = \mu_2 = \alpha_2 + i\beta_2, \quad \mu_3 = \overline{\mu}_1, \quad \mu_4 = \overline{\mu}_2 \quad (9.23)$$

where α_j, β_j $(j = 1, 2)$ are real constants, it is always possible, without loss of generality, to have $\beta_1 > 0$, $\beta_2 > 0$, $\beta_1 \neq \beta_2$.

Engineering applications

The stress intensity factors associated with each mode have the same values as in the corresponding isotropic problem when the applied loads on the crack surfaces are self-equilibrating. However, when the resultant force on the crack surfaces does not vanish the stress intensity factors depend on the anisotropic material properties.

When it is assumed that the crack propagates in a self-similar manner, knowledge of the stress and displacement fields in the vicinity of the crack tip allows determination of the energy release rates, as in the case of isotropic materials. For mode I we obtain

$$G_I = \tfrac{1}{2} \int_{-\Delta a}^{\Delta a} \sigma_y(x,0)(u_y^+ - u_y^-) \, dx. \tag{9.24}$$

Equation (9.24) gives

$$G_I = \frac{K_I^2}{2} a_{22} \, \text{Re}\left[i\left(\frac{s_1 + s_2}{s_1 s_2}\right) \right]. \tag{9.25}$$

For orthotropic materials, and when the crack is placed along the direction of minimum resistance to crack propagation ($a_{16} = a_{26} = 0$), G_I can be simplified to

$$G_I = K_I^2 \left(\frac{a_{11} a_{22}}{2}\right)^{1/2} \left[\left(\frac{a_{22}}{a_{11}}\right)^{1/2} + \frac{2a_{12} + a_{66}}{2a_{12}}\right]^{1/2}. \tag{9.26}$$

For mode II we obtain in a similar manner

$$G_{II} = K_{II}^2 \frac{a_{11}}{\sqrt{2}} \left[\left(\frac{a_{22}}{a_{11}}\right)^{1/2} + \frac{2a_{12} + a_{66}}{2a_{12}}\right]^{1/2}. \tag{9.27}$$

Self-similar crack growth in anisotropic materials is the exception rather than the rule. Realistic description of material failure by crack propagation necessitates a fracture criterion that can cope with mixed-mode crack growth. As such, the strain energy density theory has been introduced by Sih [9.10]. The strain energy density factor for crack growth under mode I and mode II takes the form

$$S = A_{11} k_I^2 + 2 A_{12} k_I k_{II} + A_{22} k_{II}^2 \tag{9.28}$$

where

$$A_{11} = \tfrac{1}{4}\left[a_{11} A^2 + a_{22} C^2 + a_{66} E^2 + 2a_{12} AC + 2a_{16} AE + 2a_{26} CE\right] \tag{9.29a}$$

$$A_{12} = \tfrac{1}{4}\left[a_{11} AB + a_{22} CD + a_{66} EF + a_{12}(AD + BC) + a_{16}(AF + BE) + a_{26}(CF + DE)\right] \tag{9.29b}$$

$$A_{22} = \tfrac{1}{4}\left[a_{11} B^2 + a_{22} D^2 + a_{66} F^2 + 2a_{12} BD + 2a_{16} BF + 2a_{26} DF\right] \tag{9.29c}$$

with

$$A = \text{Re}\left[\frac{s_1 s_2}{s_1 - s_2}\left(\frac{s_2}{z_2} - \frac{s_1}{z_1}\right)\right], \quad B = \text{Re}\left[\frac{1}{s_1 - s_2}\left(\frac{s_2^2}{z_2} - \frac{s_1^2}{z_1}\right)\right] \tag{9.30a}$$

$$C = \text{Re}\left[\frac{1}{s_1 - s_2}\left(\frac{s_1}{z_2} - \frac{s_2}{z_1}\right)\right], \quad D = \text{Re}\left[\frac{1}{s_1 - s_2}\left(\frac{1}{z_2} - \frac{1}{z_1}\right)\right] \tag{9.30b}$$

$$E = \text{Re}\left[\frac{s_1 s_2}{s_1 - s_2}\left(\frac{1}{z_1} - \frac{1}{z_2}\right)\right], \qquad F = \text{Re}\left[\frac{1}{s_1 - s_2}\left(\frac{s_1}{z_1} - \frac{s_2}{z_2}\right)\right] \qquad (0.30c)$$

and

$$K_1 = \frac{K_1}{\sqrt{\pi}}, \qquad k_2 = \frac{K_2}{\sqrt{\pi}}. \qquad (9.31)$$

In deriving these equations it is assumed that the crack line is parallel to the x-principal orthotropic direction which coincides with the fiber orientation. Equations (9.28)–(9.30), in conjunction with the strain energy density theory, were used by Sih [9.10] to obtain the crack growth angle and the critical fracture load for fiber-reinforced composites. Results can be found in [9.10, 9.18, 9.19].

The validity of the homogeneous anisotropic elasticity theory for modeling the failure of fiber composites depends on the degree to which the discrete nature of the composite affects the failure modes. There is no general answer. Each particular situation must be analyzed separately and the results obtained should be compared with predictions by other models or experiments.

Besides the homogeneous anisotropic model a number of continuum models based on linear elastic fracture mechanics have been proposed for the study of fiber composites. Among them, the theory proposed by Waddoups et al. [9.20] is based on the generalized concept of the process zone. The actual crack length is extended by the length of the process zone which is taken equal to an intense energy zone at the crack tip. For a crack of length $2a$ the critical stress σ_c, according to the stress intensity factor criterion, is expressed by

$$\sigma_c = \frac{K_{Ic}}{\sqrt{\pi(a+\ell)}} \qquad (9.32)$$

where ℓ is the length of the intense energy zone at each crack tip. ℓ could be determined by experiment.

Other continuum models can be found in references [9.20–9.25]. These models are mainly empirical and they are based on the determination of some parameters by fitting experimental data. It should therefore be expected that they have limited predictive capability.

(c) Discrete models

Realistic characterization of the failure of fiber composites necessitates consideration of the heterogeneous nature of the materials. A number of local failure modes (Figure 9.9) such as fiber breaking, matrix cracking and interface debonding precede catastrophic fracture. They absorb a large amount of the energy supplied to the system and delay the formation of a large crack leading to instability. This is a major advantage of fiber composites as compared to metals.

Modeling of local failures is based on the solution of a number of problems involving the interaction of cracks with boundaries or interfaces in isotropic or anisotropic, single or multiphase media. For a study of the solution of such problems the reader is referred to [9.10] and [9.26].

Engineering applications

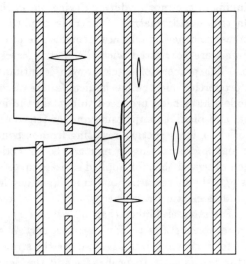

Figure 9.9. Local failure modes of a fiber composite.

A discrete model for the characterization of failure of fiber composites has been proposed by Sih and coworkers [9.10, 9.18, 9.19]. This assumes that failure takes place in a layer of matrix material sandwiched between the edges of two anisotropic solids which have the bulk mechanical properties of the fiber and the matrix. A crack in the matrix is placed parallel to fiber and the composite is subjected to an off-axis load. The strain energy density theory is used to determine the crack propagation angle and the critical failure load. The results of this model were compared with those obtained from the homogeneous anisotropic model presented previously.

Other models which try to incorporate the various micromechanical failure processes are referenced in [9.27–9.30].

9.5. Concrete

(a) Introductory remarks

Portland cement concrete is a heterogeneous multiphase system which consists of a mixture of cement paste, sand particles and aggregates and sets and hardens upon addition of water. It can be considered as a two-phase composite in which the aggregates are embedded into the mortar consisting of cement gel and sand particles. Mortar is regarded as a homogeneous and isotropic continuum functioning as the matrix binding the aggregates.

Defects play a vital role on the mechanical macroscopic behavior of concrete. Microcracks are usually present even before loading at regions of high material porosity near the interface between the coarse aggregate and the mortar. They are caused by shrinking of the mortar during drying out of the concrete. Cracks are also present in the mortar matrix. Under an applied load both types of cracks start to increase and new cracks are formed. The interface cracks extend inside the mortar and are connected with the matrix cracks. Aggregates act as crack arrestors. The process of crack growth is intimately related to the phenomenological properties and the mechanical failure of concrete.

A number of models have been proposed to explain the inelastic response of concrete to uniaxial and biaxial compression and/or tension. They are based on the deterioration of the microstructure resulting from debonding between mortar and aggregate and microcracking in mortar. For uniaxial compression, for example, it has been observed that for applied stresses up to about 30 percent of the ultimate stress (f_c) the increase in bond cracking is negligible and the deformation work is stored as elastic strain energy. Up to this point the stress–strain response is linear. For stresses higher than 30 percent, f_c nonlinearity in the global stress–strain response starts to appear and a portion of the elastic strain energy is consumed for increasing bond failure. As the stress is increased above 70 percent f_c mortar cracks start to propagate and the deviation from nonlinearity of the stress–strain diagram becomes more pronounced. For a description of the damage mechanism of concrete refer to [9.31].

The model proposed by Testa and Stubbs [9.32] considers the effect of bond failure on the stress–strain diagram of concrete. It consists of a circular rigid inclusion modeling the aggregate with two symmetrical interface cracks in an infinite plate modeling the mortar. Propagation of the interface cracks is considered around the inclusion up to the point of branching into the mortar. The global stress–strain diagram is obtained by calculating the total strain energy density of the system using the principles of fracture mechanics and applying Castigliano's theorem. The nonlinear stress–strain response results from bond failure and depends on various parameters including the strength of bond and mortar and the size of the inclusion. Along the same lines, Gdoutos et al. [9.33] modeled the aggregates as square rigid inclusions partially bonded to an elastic matrix and obtained the nonlinear stress–strain response of concrete under biaxial compression and/or tension. For further details on the mechanical modeling of concrete refer to [9.34] and [9.35].

The following section discusses briefly the applicability of the principles of fracture mechanics to concrete along with two models, the cohesive and damage model for simulating the state of affairs near the crack tip.

(*b*) *Fracture mechanics*

Although the cracking of concrete is the basic failure mechanism and has been studied extensively for many years, it was not until 1961 that Kaplan [9.36] applied the concepts of fracture mechanics to concrete. From three- and four-point bending experiments Kaplan measured the elastic strain energy release rate G_c and found that it varied greatly with the specimen size. Glucklish [9.37] obtained that G_c is much higher than twice the surface energy of concrete (2γ). This is attributed to microcracking at the crack tip, which creates an energy

Engineering applications

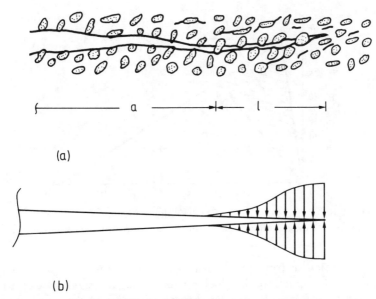

Figure 9.10. Cohesive model in concrete.

absorption mechanism in a manner analogous to crack-tip plastic deformation in metals. The dependence of G_c (or K_c) on various concrete parameters including the volume, size and roughness of aggregates, the water and air content of mortar as well as the age, loading rate, etc., has been studied by several investigators. For a thorough study of the work done for characterizing the fracture toughness of concrete by linear elastic fracture mechanics concepts, the reader is referred to [9.38].

A great amount of effort has been spent in determining the effect of specimen geometry and size on G_c or K_c. Unfortunately, the obtained results are contradictory. This created much controversy as to whether linear elastic fracture mechanics can be applied to concrete. The obtained discrepancy is due to the ignorance of energy dissipation during slow crack growth and local material damage prior to instability. A number of investigators applied nonlinear fracture criteria including the J-integral, R-curve and crack-tip opening displacement to characterize the fracture toughness of concrete. The obtained values of the critical parameters in these criteria were again geometry and size dependent.

For an in-depth study of the application of the principles of fracture mechanics to concrete the reader is referred to [9.39-9.41]. In the remainder of this section the cohesive and damage models for characterizing the fracture behavior of concrete are briefly presented.

(c) Cohesive model

The nonlinear and dissipative phenomena occurring in the neighborhood of the tip of a mode-I crack in concrete are modeled in a manner analogous to the

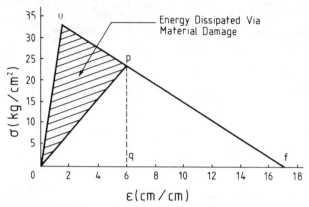

Figure 9.11. Uniaxial bilinear elastic-softening stress–strain curve of concrete.

plastic deformation in front of a crack in metals by the Dugdale model. The region ahead of the crack tip, in which microcracking and aggregate interlocking and bridging takes place, is considered as an extension of the physical crack in which a closing pressure applies (Figure 9.10). The actual crack of length a is replaced by a fictitious crack of length $(a+\ell)$, where ℓ is the length of the process zone. For the determination of the length and the pressure distribution in the process zone a constitutive model for the material in the zone is needed. This model, relating the stress to the crack opening displacement, can be obtained from a displacement-controlled tension test. The determination of the stress distribution for the fictitious crack takes place using the concepts of linear elastic fracture mechanics. The characteristics of the process zone are determined by canceling the stress singularity created by the external loading and the stress distribution in the process zone at the tip of the fictitious crack, as in the Dugdale model.

The cohesive model for concrete was first proposed by Hillerborg et al. [9.42]. For more details on this model the reader should consult references [9.43–9.45].

(d) Damage model

The problem of incremental damage of concrete by stable crack growth prior to material separation was addressed by Sih [9.46, 9.47]. He used a uniaxial bilinear elastic-softening stress–strain curve (Figure 9.11) in conjunction with the strain energy density theory. When loading is removed at a point p of the stress–strain curve the unloading path is assumed to follow the line pO. The new bilinear stress–strain curve of the material is the line Opf. The material then experiences a reduction in the modulus of elasticity as the slope of the line Op is smaller than that of Ou. This model does not allow permanent deformation in unloading. For an undamaged material element with an equivalent stress along the line Ou the critical strain energy density $(dW/dV)_c$ is equal to the area Ouf. For a damaged material with representative point p along the line uf the

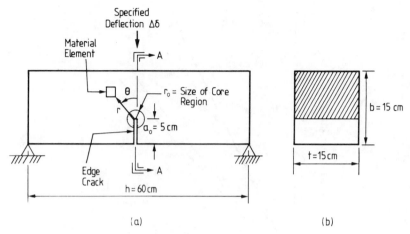

Figure 9.12. Three-point bend specimen.

Table 9.1. Three different softening rates for concrete.

Material type	Softening strain ϵ_f (cm/cm × 10^{-3})	Critical strain energy density $(dW/dV)_c$ (kg/cm² × 10^{-3})
A	16.0	24.90
B	8.0	14.14
C	4.0	7.7

dissipated strain energy density $(dW/dV)_p$ for material damage is represented by the area Oup, while the energy available for release $(dW/dV)^*$ is equal to the area Opf.

The three-point bending specimen with an edge crack (Figure 9.12) was analyzed. Three different types of softening behavior, referred to as materials A, B and C, were considered. The values of ϵ_f and $(dW/dV)_c$ for these materials are shown in Table 9.1. Applying the strain energy density theory the resistance curves for materials A, B and C were obtained by plotting the strain energy density factor S as a function of the crack length for stable slow crack growth. Figure 9.13 presents results for a constant deflection increment $\Delta\delta = 4 \times 10^{-3}$ cm for materials A, B and C, while Figures 9.14 and 9.15 explain the effect of loading step and specimen size. Observe from Figures 9.14 and 9.15 that the resistance curves are straight lines that rotate about a common point in a counterclockwise sense as the load step is increased, while they are moving upward with increasing specimen size. Thus, the combined interaction of material properties, load step and specimen geometry and size can be easily analyzed. For example, there is an upper limit in specimen size beyond which no stable crack growth occurs and failure takes place in a catastrophic manner. These results are of major importance to the designer who can extrapolate and use these curves for situations different from those for which they have been extracted.

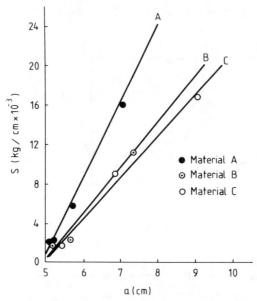

Figure 9.13. Strain energy density factor versus crack length for materials A, B and C with constant deflection increment $\Delta\delta = 4 \times 10^{-3}$ cm [9.46].

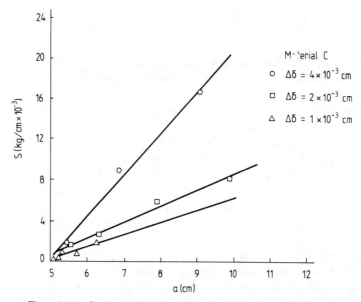

Figure 9.14. Strain energy density factor versus crack length for material C with three different deflection increments [9.46].

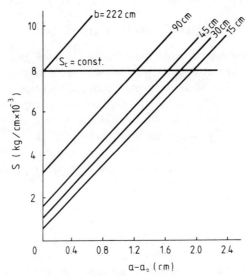

Figure 9.15. Strain energy density factor versus crack length increment for material C with different size scales and $\Delta\delta/b = 2.6 \times 10^{-4}$ [9.46].

For further information on the damage model refer to [9.46–9.48].

9.6. Crack detection methods

(a) Introductory remarks

The fracture process of machine or structural parts involves crack initiation, subcritical growth and final termination. Fracture mechanics methodology is based on the realistic assumption that all materials contain initial defects which constitute the nuclei of fracture initiation. Design for the prevention of crack initiation is physically unrealistic. Initial defects appear in a material due to its composition or they can be introduced in a structure during fabrication or service life. The detection of defects in structures plays an essential role in design using the discipline of fracture mechanics. A number of nondestructive testing (NDT) methods for the detection, positioning and sizing of defects have been developed. Our ability to use fracture mechanics in design is largely due to the reliability of the NDT methods. At the production or service inspection stage, parts containing flaws larger than those determined according to fracture mechanics design must be rejected or replaced.

Six NDT methods that are widely used for defect detection will be briefly described below. These are dye penetrant, magnetic particles, eddy currents, radiography, ultrasonics and acoustic emission. Each of these methods posesses advantages and disadvantages depending on the application. For further details on these methods the reader should consult references [9.49–9.54].

(b) *Dye penetrant*

This technique is commonly used for detecting surface flaws. It involves application of a colored or fluorescent dye onto a cleaned surface of the component. After allowing sufficient time for penetration, the excess penetrant is washed off and the surface is dusted with a post-penetrant material (developer) such as chalk. The developer acts as a blotter and the defects are detected as colored lines. The reliability of the method mainly depends on the surface preparation of the component. The method is widely used and can detect small cracks. It has the advantage of fast inspection at low cost. It applies, however, only to surface flaws.

(c) *Magnetic particles*

This method is based on the principle that flaws in a magnetic material produce a distortion to an induced magnetic field. Measuring this distortion provides information on the existing defects. The magnetic field is induced by passing a current through the component or using permanent or electromagnets. For detecting the distortion of the magnetic field the surface under inspection is coated with a fluorescent liquid that contains magnetic particles in suspension. The method can easily be applied and is speedy and economical. As in the dye penetrant method, it can be used only for detecting cracks on or near the surface.

(d) *Eddy currents*

When a coil carrying alternating current is placed near a conducting surface, eddy currents are induced in the surface. The eddy currents create a magnetic field that links to the coil, whose impedance changes when a defect is present. By measuring this change information about the defect can be provided. The induced eddy currents concentrate near the surface of the conductor, according to the so-called 'skin effect'. The penetration depth is influenced by the frequency of the current, the magnetic permeability and electrical conductivity of the conductor and the coil and conductor geometry. In a ferritic conductor the penetration depth is smaller than 1 mm at most frequencies, while in nonmagnetic conductors it may be several millimetres. The sensitivity of the method is higher for defects near the surface and decreases with increasing depth. Problems in the method arise from the difficulty of relating the defect size to the change in impedance and the influence on impedance of a number of factors, including the relative position of the coil and the conductor and the presence of structure variations or material inhomogeneities. Measurement of defect size is made by comparing its response to that observed from a standard defect.

(e) *Radiography*

Radiography is one of the oldest NDT methods for detecting subsurface defects. A source of X- or γ-rays is transmitted through the specimen. If the specimen has variations in thickness or density – due, for example, to the presence of defects – the emerging radiation will not be of uniform intensity. Since defects absorb less X-rays than surrounding material their presence can be revealed by using a sensitive photographic film where they appear as dark lines. The method is particularly suitable for finding volumetric defects. The method may be used to detect cracks, but in order to obtain substantial differential absorption between rays passing through the cracks and those passing through the surrounding material they should be oriented parallel to the plane of the incident radiation. Thus, the method is insensitive to cracks unless their orientations are known beforehand. This may involve a number of exposures at different positions of the X-rays.

(f) *Ultrasonics*

This method is based on the transmission of ultrasonic waves into the material by a transducer containing a piezoelectric crystal. Metallurgical defects and/or surface boundaries reflect the incident pulse which is monitored on an oscilloscope. The distance between the first pulse and the reflection gives the position of the crack. The size of the crack can also be estimated. The method is characterized by high sensitivity for detection of cracks, at all positions, ability to measure crack position and size, fast response for rapid inspection, economy, inspection of thick material sections and portability of equipment for *in-situ* inspection. The application of the method is, however, limited by unfavorable specimen geometry and the difficulty in distinction of cracks and other types of defects, such as inclusions. The method is also characterized by the subjective way of interpreting the echoes by the operator.

(g) *Acoustic emission*

This method involves the use of a sensing transducer and sophisticated electronic equipment to detect sounds and stress waves emitted inside the material during the process of cracking. The detected emissions are then amplified, filtered and interpreted. The method is capable of locating flaws without resorting to a point-by-point search over the entire surface of interest. It can be used to detect crack initiation and growth. A disadvantage of the method lies in the difficulty of interpreting the obtained signals.

References

9.1. Sih, G. C., Mechanics and physics of energy density theory, *Theoretical and Applied Fracture Mechanics* **4**, 157–173 (1985).
9.2. Sih, G. C., Thermomechanics of solids: Nonequilibrium and irreversibility, *Theoretical and Applied Fracture Mechanics* **9**, 175–198 (1988).
9.3. Liebowitz, H. (ed.), *Fracture – An Advanced Treatise*, Vol. V, *Fracture Design of Structures*, Academic Press (1969).
9.4. Sih, G. C. and Faria, L. (eds), *Fracture Mechanics Methodology: Evaluation of Structural*

Components Integrity, Martinus Nijhoff Publ. (1984).
9.5. Gdoutos, E. E., *Problems of Mixed Mode Crack Propagation*, Martinus Nijhoff Publ. (1984).
9.6. Au, N. N. and Lin, S. R., Mixed mode fracture in space launch pressure vessels, *Proceedings of the U.S.-Greece Conference on Mixed Mode Crack Propagation* (eds G. C. Sih and P. S. Theocaris), Sijthoff and Noordhoff, pp. 55–76 (1981).
9.7. Dharan, C. K. H., Fracture mechanics of composite materials, *Journal of Engineering Materials and Technology, Trans. ASME* 100, 233–246 (1978).
9.8. Sih, G. C. and Tamuzs, V. P. (eds), *Fracture of Composite Materials*, Sijthoff and Noordhoff, The Netherlands (1979).
9.9. Liebowitz, H. (ed.), *Fracture – An Advanced Treatise*, Vol. VII, *Fracture of Nonmetals and Composites*, Academic Press (1972).
9.10. Sih, G. C. and Chen, E. P., *Mechanics of Fracture*, Vol. 6, *Cracks in Composite Materials*, Martinus Nijhoff Publ. Co., The Netherlands (1981).
9.11. Sih, G. C. and Skudra, A. M. (eds), *Failure Mechanics of Composites*, North-Holland (1985).
9.12. Kanninen, M. F. and Popelar, C. H., *Advanced Fracture Mechanics*, Chapter 6, Oxford Univ. Press, pp. 392–436 (1985).
9.13. Jayatilaka, A. de S., *Fracture of Engineering Brittle Materials*, Chapter 7, Academic Science Publ., pp. 216–280 (1979).
9.14. Cherepanov, G. P., *Mechanics of Brittle Fracture*, Chapter 9 (translated from Russian by R. deWit and W. C. Cooley), McGraw-Hill, pp. 616–731 (1979).
9.15. Gdoutos, E. E., Photoelastic analysis of composite materials with stress concentrations, in *Photoelasticity in Engineering Practice* (eds S. A. Paipetis and G. S. Holister), Elsevier Applied Science Publ., pp. 157–179 (1985).
9.16. Sih, G. C., Paris, P. C. and Irwin, G. R., On cracks in rectilinearly anisotropic bodies, *International Journal of Fracture* 1, 189–203 (1965).
9.17. Lekhnitskii, S. G., *Anisotropic Plates* (translated from Russian by S. W. Tsai and T. Cheron), Gordon and Breach Science Publ. (1968).
9.18. Sih, G. C. and Chen, E. P., Fracture analysis of unidirectional composites, *Journal of Composite Materials* 7, 230–244 (1973).
9.19. Sih, G. C., Chen, E. P., Huang, S. L. and McQuillen, E. J., Material characterization on the fracture of filament-reinforced composites, *Journal of Composite Materials* 9, 167–186 (1975).
9.20. Waddoups, M. E., Eisenmann, J. E. and Kaminski, B. E., Macroscopic fracture mechanics of advanced composite materials, *Journal of Composite Materials* 5, 446–454 (1971).
9.21. Whitney, J. M. and Nuismer, R. J., Stress fracture criteria for laminated composites containing stress concentrations, *Journal of Composite Materials* 8, 253–265 (1974).
9.22. Morris, D. H. and Hahn, H. T., Mixed-mode fracture of graphite/epoxy laminates: Fracture strength, *Journal of Composite Materials* 11, 124–138 (1977).
9.23. Wu, E. M., Application of fracture mechanics to anisotropic plates, *Journal of Applied Mechanics, Trans. ASME* 34, 967–974 (1967).
9.24. Wright, M. A. and Iannuzzi, F. A., The application of the principles of linear elastic fracture mechanics to unidirectional fiber reinforced composite materials, *Journal of Composite Materials* 7, 430–447 (1973).
9.25. Zhen, S., The D criterion in notched composite materials, *Journal of Reinforced Plastics and Composites* 2, 98–110 (1983).
9.26. Sih, G. C., *Handbook of Stress-Intensity Factors*, Chapter 5, Institute of Fracture and Solid Mechanics, Lehigh Univ. (1973).
9.27. Lifshitz, J. M., Nonlinear matrix failure criteria for fiber-reinforced composite materials, *Composite Technology Review* 4, 78–83 (1982).
9.28. Dharani, L. R., Jones, W. F. and Gorce, J. G., Mathematical modeling of damage in unidirectional composites, *Engineering Fracture Mechanics* 17, 555–573 (1983).
9.29. Kanninen, M. F., Rybicki, E. F. and Griffith, W. I., Preliminary development of a fundamental analysis model for crack growth in a fiber reinforced composite material, *Composite Materials: Testings and Design (Fourth Conference)* ASTM STP 617, American Society for Testing and Materials, Philadelphia, pp. 53–69 (1977).
9.30. Ouyang, C. and Lu, M. Z., On a micromechanical fracture model for cracked reinforced composites, *International Journal of Non-Linear Mechanics* 18, 71–77 (1983).
9.31. Meyers, B. L., Slate, F. O. and Winter, G., Relationship between time dependent deformation and microcracking of plain concrete, *Journal of the American Concrete Institute Proceedings* 66, 60–68 (1969).
9.32. Testa, R. B., and Stubbs, N., Bond failure and inelastic response of concrete, *ASCE*

Journal of the Engineering Mechanics Division **103**, 295–310 (1977).
9.33. Gdoutos, E. E., Kattis, M. A., Kourounis, C. G. and Zacharopoulos, D. A., A mathematical model for the interpretation of the non-linear behavior of concrete, *Proceedings of the Eighth National Greek Conference on Concrete* **1**, 340–350 (1987).
9.34. Ditommaso, A., Evaluation of concrete fracture, in *Fracture Mechanics of Concrete* (eds A. Carpinteri and A. R. Ingraffea), Martinus Nijhoff Publ., pp. 31–65 (1984).
9.35. Slate, F. O. and Hover, K. C., Microcracking of concrete, in *Fracture Mechanics of Concrete* (eds A. Carpinteri and A. R. Ingraffea), Martinus Nijhoff Publ., pp. 137–159 (1984).
9.36. Kaplan, M. F., Crack propagation and the fracture of concrete, *Journal of the American Concrete Institute* **58**, 591–610 (1961).
9.37. Glucklish, J., Fracture of plain concrete, *ASCE Journal of the Engineering Mechanics Division* **89**, 127–138 (1983).
9.38. Mindess, S., Fracture toughness testing of cement and concrete, in *Fracture Mechanics of Concrete* (eds A. Carpinteri and R. Ingraffea), Martinus Nijhoff Publ., pp. 67–100 (1984).
9.39. Carpinteri, A. and Ingraffea, R. (eds), *Fracture Mechanics of Concrete*, Martinus Nijhoff Publ. (1984).
9.40. Sih, G. C. and Ditommaso, A., *Fracture Mechanics of Concrete*, Martinus Nijhoff Publ. (1985).
9.41. Shah, S. P. (ed.), *Application of Fracture Mechanics to Cementitious Composites*, Martinus Nijhoff Publ. (1985).
9.42. Hillerborg, A., Modeer, M. and Paterson, P. E., Analysis of crack formation and crack growth in concrete by means of fracture mechanics and finite elements, *Cement and Concrete Research* **6**, 773–782 (1976).
9.43. Wecharatana, M. and Shah, S. P., Predictions of nonlinear fracture process zone in concrete, *ASCE Journal of Engineering Mechanics* **109**, 1231–1246 (1983).
9.44. Ballarini, R., Shah, S. P. and Keer, L. M., Nonlinear analysis for mixed mode fracture, in *Application of Fracture Mechanics to Cementitious Composites* (ed. S. P. Shah), Martinus Nijhoff Publ., pp. 51–83 (1985).
9.45. Ingraffea, A. R. and Walter, H. G., Nonlinear fracture models for discrete crack propagation, in *Application of Fracture Mechanics to Cementitious Composites* (ed. S. P. Shah), Martinus Nijhoff Publ., pp. 247–285 (1985).
9.46. Sih, G. C., Mechanics of material damage in concrete, in *Fracture Mechanics of Concrete* (eds A. Carpinteri and A. R. Ingraffea), Martinus Nijhoff Publ., pp. 1–29 (1984).
9.47. Sih, G. C., Non-linear response of concrete: Interaction of size, loading step and material property, in *Application of Fracture Mechanics to Cementitious Composites* (ed. S. P. Shah), Martinus Nijhoff Publ., pp. 3–23 (1985).
9.48. Carpinteri, A. and Sih, G. C., Damage accumulation and crack growth in bilinear materials with softening: Application of strain energy density theory, *Theoretical and Applied Fracture Mechanics* **1**, 145–159 (1984).
9.49. McGonnagle, W. J., Nondestructive testing, in *Fracture – An Advanced Treatise*, Vol. III, *Engineering Fundamentals and Environment Effects* (ed. H. Liebowitz), Academic Press, pp. 371–430 (1971).
9.50. Coffey, J. M. and Whittle, M. J., Non-destructive testing: its relation to fracture mechanics and component design, *Philosophical Transactions of the Royal Society of London* **A299**, 93–110 (1981).
9.51. Coffey, J. M., Ultrasonic measurement of crack dimensions in laboratory specimens, in *The Measurement of Crack Length and Shape During Fracture and Fatigue* (ed. C. J. Beevers), EMAS, pp. 345–385 (1980).
9.52. Catou, M. M., Flavenot, J. F., Flambard, C., Madelaine, A. and Anton, F., Automatic measurement of crack length during fatigue. Testing using ultrasonic surface waves, in *The Measurement of Crack Length and Shape During Fracture and Fatigue* (ed. C. J. Beevers), EMAS, pp. 387–392 (1980).
9.53. Richard, C. E., Some guidelines to the selection of techniques, in *The Measurement of Crack Length and Shape During Fracture and Fatigue* (ed. C. J. Beevers), EMAS, pp. 461–468 (1980).
9.54. Krautkramer, J. and Krautkramer, H., *Ultrasonic Testing of Materials* (2nd ed), Springer-Verlag (1977).

Author index

Achenbach, J. D., 231, 252
Adachi, J., 5, 14
Adams, N. J., 152, 160, 178, 192
Afanasev, E. F., 234, 253
Ahmad, J., 183, 185, 193
Aifantis, E. C., 274, 275, 277
Alavi, M. J., 48, 62, 66, 73
Andersson, H., 247, 253
Andrews, W. R., 183, 192
Andrianopoulos, N., 225, 229
Anton, F., 302, 305
Aoki, S., 185, 193
Atanmo, P., 260, 276
Atkins, A. G., 146, 147, 160
Atkinson, C., 234, 253
Au, N. N., 285, 304

Baker, B. R., 234, 237, 252
Ballarini, R., 298, 305
Bamford, W. H., 183, 192
Barenblatt, G. I., 99, 110
Barnes, C. R., 185, 193
Barsom, J. M., 142, 160, 263, 273, 276
Barsoum, R. S., 53, 62, 66, 73, 75, 106, 110
Barthelemy, B. M., 267, 270, 276
Batson, R. G. C., 6,14
Begley, J. A., 170, 175, 192, 261, 276
Beinert, J., 251, 254
Benzley, S. E., 106, 110
Berry, J. P., 233, 252
Bieniawski, Z. T., 221, 227
Bilby, B. A., 49, 73, 99, 109, 134
Bishop, T., 4, 14
Bluhm, J. I., 250, 254
Bowie, O. L., 40, 72
Boyd, G. M., 4, 14
Boyle, R. W., 133, 136, 149, 152, 159
Brace, W. F., 233, 252
Bradley, W. B., 251, 254
Brinson, H. F., 99, 109
Broberg, K. B., 234, 237, 252
Brown, W. F., Jr, 46, 72, 133, 142, 149, 152, 159, 160
Bucci, R. J., 171, 192
Budiansky, B., 185, 193
Bueckner, H. F., 42, 44, 45, 72, 122, 159
Burdekin, F. M., 99, 110, 187, 193
Bush, A. J., 183, 192
Caddell, R. M., 146, 160

Carlson, A. J., 185, 193
Carpinteri, A., 220, 221, 228, 297, 298, 301, 305
Carpinteri, Andrea, 221, 228
Cartwright, D. J., 45, 72
Catou, M. M., 302, 305
Cha, B. C. K., 216, 227
Chan, C. F., 244, 254
Chang, K. J., 157, 161
Chao, C. K., 221, 226, 228, 229, 272, 276
Chen, C., 220, 221, 223, 228
Chen, E. P., 234, 242, 244, 252, 253, 291, 294, 295, 304
Cherepy, R. B., 46, 73
Cherepanov, G. P., 122, 159, 162, 183, 191, 193, 234, 253, 273, 276, 291, 304
Cheung, Y. K., 50, 73
Chow, C. L., 146, 160, 225, 229
Clark, A. B. J., 247, 253
Clark, H. M., 260, 276
Clarke, G. A., 178, 192
Clausing, D. P., 146, 160
Coffey, J. M., 302, 305
Cook, T. S., 49, 73
Copley, L. G., 71, 75
Corten, H. T., 178, 192
Cotterell, B., 247, 253
Cottrell, A. H., 99, 109, 134, 186, 193
Coulomb, C. A., 2, 13
Craggs, J. W., 234, 252
Czoboly, E., 197, 227

Dai, P. K., 81, 99, 109
Dally, J. W., 58, 74, 249, 251, 253, 254
Dawes, M. G., 188, 193
Dean, R. H., 189, 194
De Koning, A. V., 265, 276
DeLorenzi, H. H., 183, 192
Dharan, C. K. H., 291, 304
Dharani, L. R., 295, 304
Ditommaso, A., 296, 297, 305
Dobreff, P. S., 71, 75
Docherty, J. G., 6, 14
Donahue, R. J., 260, 276
Dowling, N. E., 261, 276
Drakos, G., 225, 229
Drucker, D. C., 122, 159
Drugan, W. J., 189, 193
Duffy, A. R., 5, 14, 234, 252

Dugdale, D. S., 96, 98, 99, 109
Dulaney, E. N., 233, 252

Eftis, J., 25, 72, 152, 153, 160
Eiber, R. J., 5, 14, 234, 252
Eisenmann, J. E., 294, 304
Elber, W., 265, 276
Elle, D., 231, 252
Ellinger, G. A., 4, 14
Embley, G. T., 99, 110, 244, 253
Emery, A. F., 48, 62, 66, 73
Engle, R. M., 260, 276
Erdogan, F., 40, 41, 45, 49, 71, 72, 73, 75, 154, 160, 231, 252, 260, 261, 276
Ernst, H. A., 181, 183, 192, 193
Eshelby, J. D., 49, 73, 162, 191, 234, 253
Etheridge, J. M., 57, 74, 249, 251, 253
Evans, W. T., 76, 109
Ewing, P. D., 155, 160

Faria, L., 279, 303
Fawkes, A. J., 108, 111, 231, 234, 251, 252, 253, 254
Fenner, R. T., 62, 66, 75
Field, F. A., 99, 109
Flambard, C., 302, 305
Flavenot, J. F., 302, 305
Folias, E. S., 71, 75
Forman, R. G., 260, 276
Fourney, W. F., 249, 251, 253
Freund, L. B., 231, 234, 251, 252, 253, 254

Gales, R. D. R., 44, 99
Gallagher, P. H., 55, 73
Gamble, R. M., 183, 192
Garwood, S. J., 191, 194
Gdoutos, E. E., 30, 55, 57, 58, 72, 74, 80, 99, 108, 109, 110, 111, 197, 205, 211, 220, 223, 225, 227, 228, 229, 275, 277, 280, 285, 291, 296, 304, 305
German, M. D., 173, 192
Gillemot, F., 197, 227
Glucklish, J., 296, 305
Goldman, N. L., 105, 110
Goodier, J. N., 99, 109
Gorce, J. G., 295, 304
Grandt, A. F., 45, 72
Green, A. E., 62, 74
Griffith, A. A., 3, 5, 7, 8, 9, 10, 11, 12, 13, 113, 116, 120, 121, 122, 154
Griffith, W. I., 295, 304
Gross, B., 46, 72
Günther, W., 162, 191
Gupta, G. D., 49, 73
Gurney, C., 125, 146, 147, 159, 160

Hahn, G. T., 82, 99, 109
Hahn, H. T., 294, 304
Haigh, B. P., 198, 227
Harrison, N. L., 158, 161
Harrop, L. P., 99, 110

Hartranft, R. J., 45, 48, 62, 64, 65, 69, 72, 74, 75, 122, 137, 159, 216, 227, 244, 253
Hartzitrifon, N., 223, 228
He, M. Y., 105, 110
Head, A. K., 259, 276
Hellen, T. K., 55, 73
Hertzberg, R. W., 273, 277
Heyer, R. H., 152, 160
Hillerborg, A., 198, 305
Hilton, P. D., 53, 55, 73, 77, 105, 106, 109, 110
Hinton, E., 105, 110
Hoek, E., 221, 227
Hopkinson, B., 231, 252
Hopkinson, J., 231, 252
Hover, K. C., 296, 305
Hsu, T. M., 45, 72
Hsu, T. R., 108, 111
Huang, G.-H., 191, 194
Huang, S. L., 294, 295, 304
Hudson, C. M., 142, 160
Hult, J. A. H., 82, 109
Hunt, J., 125, 146, 147
Hussain, M. A., 53, 73, 158, 161
Hutchinson, J. W., 53, 73, 78, 100, 102, 103, 104, 105, 109, 110, 173, 181, 189, 192, 194

Iannuzzi, F. A., 294, 304
Inglis, C. E., 7, 14, 15, 71
Ingraffea, R., 297, 298, 305
Irwin, G. R., 5, 6, 11, 14, 15, 16, 28, 31, 45, 48, 62, 71, 73, 74, 84, 93, 95, 109, 133, 147, 152, 159, 160, 247, 249, 251, 253, 254, 292, 304
Isida, M., 62, 66, 75

Jayatilaka, A., de S., 225, 229, 291, 304
Jenkins, I. J., 225, 229
Jones, D. L., 152, 153, 160
Jones, D. P., 108, 111
Jones, W. F., 295, 304
Joyce, J. A., 183, 193

Kaiser, S., 178, 192
Kalthoff, J. F., 234, 251, 254
Kaminski, B. E., 294, 304
Kanninen, M. F., 99, 108, 110, 111, 122, 159, 183, 185, 192, 193, 234, 252, 273, 277, 291, 295, 304
Kantorovich, L. V., 47, 73
Kaplan, M. F., 296, 305
Karmarsch, I., 9, 14
Kassir, M. K., 62, 72, 74, 120, 159
Katsamanis, F., 251, 254
Kattis, M. A., 296, 305
Kaufman, J. G., 142, 160
Kearney, V. E., 260, 276
Keer, L. M., 99, 110, 298, 305
Kerkhof, F., 231, 252
Kfouri, A. P., 108, 111, 122, 159
Kiefer, B. V., 108, 111, 220, 223, 228

Kies, J. A., 133, 147, 159, 160
Kim, Y. J., 108, 111
Kinsel, W. C., 46, 73
Kipp, M. E., 225, 228
Kishimoto, K., 185, 193
Klesnil, M., 261, 276
Knauss, W. G., 158, 161
Knott, J. F., 142, 159
Knowles, J. K., 69, 75, 162, 191
Kobayashi, A. S., 46, 48, 61, 62, 66, 73, 74, 75, 244, 251, 254
Kobayashi, T., 57, 74, 249, 251, 253, 254
Kommers, J. B., 7, 14
Koskinen, M. F., 91, 109
Kostrov, B. V., 234, 235, 253
Kourounis, C. G., 296, 305
Krafft, J. M., 133, 136, 149, 152, 159
Krautkramer, H., 302, 305
Krautkramer, J., 302, 305
Kröner, E., 162, 192
Krylov, V. I., 47, 73
Kuhn, P., 5, 14
Kumble, R., 260, 276

Lachenbruch, A. H., 48, 73
Lam, P. M., 146, 160
Landes, J. D., 170, 171, 175, 178, 183, 191, 192, 193, 194
Lau, K. J., 146, 160
Lee, J. D., 108, 111
Lekhnitskii, S. G., 292, 304
Levy, N., 108, 111
Liebowitz, H., 16, 25, 72, 108, 111, 118, 122, 152, 153, 158, 160, 279, 291, 303, 304
Lifshitz, J. M., 295, 304
Liu, A. F., 157, 161
Loeber, J. F., 244, 253
Lu, M. Z., 295, 304
Lucas, P., 261, 276
Luxmoore, A. R., 76, 109

Macdonald, B., 207, 221, 227
MacGregor, C. W., 25, 72
Madelaine, A., 302, 305
Madenci, E., 108, 110, 220, 228
Mai, Y. W., 146, 147, 160
Maiti, S. K., 157, 160
Mall, S., 251, 254
Manogg, P., 58, 74
Marcal, P. V., 108, 111
Maxey, W. A., 5, 14, 234, 252
McCabe, D. E., 152, 160, 183, 191, 193, 194
McClintock, F. A., 11, 84, 109
McClure, G. M., 5, 14, 234, 252
McEvily, A. J., 260, 276
McGonnagle, W. J., 302, 305
McGowan, J. J., 57, 63, 74
McMeeking, R. M., 45, 72, 108, 111, 171, 172, 173, 178, 192
McQuillen, E. J., 294, 295, 304
Merkle, J. G., 176, 178, 192

Meyers, B. L., 296, 304
Milios, J., 251, 254
Miller, K. J., 108, 111
Mills, N. J., 44, 99
Mindess, S., 297, 305
Mises, R. von, 2, 13
Modeer, M., 290, 305
Mohr, O., 2, 13
Morris, D. H., 294, 304
Mott, N. F., 231, 252
Moyer, E. T., Jr, 108, 111, 225, 229, 271, 276
Munro, H. G., 178, 192
Mura, T., 99, 110
Muskhelishvili, N. I., 16, 17, 36, 37, 38, 39, 41, 49, 72

Nadai, A., 2, 13, 198, 227
Neuman, J. C., Jr, 62, 75, 99, 110
Ngan, K. M., 146, 160
Nillson, F., 234, 252
Noguchi, H., 62, 66, 75
Nuismer, R. J., 158, 161, 294, 304

Obreimoff, J. W., 122, 143, 159
Orowan, E., 152, 160
Osias, J. R., 108, 111
Ostergren, W. J., 108, 111
Owen, D. R. J., 105, 108, 110, 111
Owen, R. C., 146, 160
Ouyang, C., 295, 304

Palaniswamy, K., 158, 161
Pan, J., 105, 110, 183, 193
Papadopoulos, G., 251, 254
Papakaliatakis, G., 108, 111, 220, 228
Paris, P. C., 16, 40, 41, 43, 45, 72, 171, 176, 181, 183, 192, 193, 260, 261, 276, 292, 304
Parker, E. R., 4, 14
Parks, D. M., 108, 111, 172, 173, 192
Paterson, P. E., 298, 305
Pazis, D., 251, 254
Pearson, K., 6, 14
Petroski, H. J., 45, 72, 99, 110
Pook, L. P., 157, 161
Popelar, C. H., 183, 192, 193, 273, 277, 291, 304
Post, D., 231, 251, 252, 254
Poulose, P. K., 108, 111
Prasad, S. V., 225, 229
Pu, S. L., 53, 73, 158, 161

Raftopoulos, D., 251, 254
Raju, I. S., 62, 66, 75
Ratwani, M., 261, 276
Ravera, R. S., 244, 253
Reissner, E., 67, 69, 75
Rice, J. R., 44, 55, 72, 73, 76, 78, 89, 92, 100, 108, 109, 110, 111, 122, 159, 162, 168, 170, 171, 172, 176, 185, 189, 191, 192, 193
Richards, C. E., 302, 305

Roberts, D. K., 233, 234, 252
Rolfe, S. T., 142, 160, 273, 276
Rosakis, A. J., 249, 251, 253, 254
Rose, L. R. F., 237, 253
Rosenfield, A. R., 82, 109
Rosengren, G. F., 78, 100, 110
Rossmanith, H. P., 249, 251, 253, 254
Rubio, A., 5, 34
Rudd, J. L., 45, 72
Rybicki, E. F., 295, 304

Sack, R. A., 62, 74
Sadowsky, M. A., 62, 74
Sakata, M., 185, 193
Sanders, J. L., Jr, 71, 75, 162, 191
Sanford, R. J., 58, 74
Schardin, H., 231, 247, 252, 253
Schijve, J., 265, 276
Schroedl, M. A., 57, 63, 74
Segedin, C. M., 62, 74
Seward, S. K., 142, 160
Shah, R. C., 48, 66, 73, 75, 158, 161
Shah, S. P., 297, 298, 305
Sham, T.-L., 189, 193
Shank, M. E., 4, 14
Shih, C. F., 105, 110, 173, 183, 185, 192, 193
Sih, G. C., 10, 12, 14, 16, 24, 25, 40, 41, 43,
  45, 48, 55, 62, 64, 65, 68, 69, 70, 71, 72,
  73, 74, 99, 108, 110, 111, 118, 120, 122,
  137, 154, 155, 159, 160, 197, 207, 209, 216,
  218, 220, 221, 223, 225, 226, 227, 228, 229,
  231, 234, 241, 242, 244, 248, 252, 253, 267,
  271, 272, 276, 278, 281, 282, 291, 292, 293,
  294, 295, 297, 298, 300, 301, 303, 304, 305
Skudra, A. M., 291, 304
Slate, F. O., 296, 304, 305
Smith, C. W., 57, 58, 63, 74, 75
Smith, D. G., 58, 74
Smith, E., 99, 109
Smith, F. W., 48, 62, 66, 73
Smith, H. L., 133, 159
Smith, R. A., 157, 161
Sneddon, I. N., 15, 48, 49, 62, 71, 73, 74
Sönnerlind, H., 178, 192
Sorensen, E. P., 108, 111, 172, 189, 192, 193
Spencer, A. J. M., 122, 159
Srawley, J. E., 133, 149, 152, 159
Stanton, T. E., 6, 14
Sternberg, E., 62, 74, 162, 191
Stone, D. E. W., 99, 110, 187, 193
Struth, W., 231, 252
Stubbs, N., 296, 304
Sullivan, A. M., 133, 136, 149, 152, 159
Swedlow, J. L., 105, 108, 110, 111
Swinden, K. H., 99, 109, 134

Tada, H., 45, 72, 181, 183, 192
Tamuzs, V. P., 291, 304
Tan, C. L., 62, 66, 75
Testa, R. B., 296, 304

Testa, R. B., 296, 304
Theocaris, P. S., 30, 55, 58, 72, 74, 76, 99,
  109, 110, 225, 229, 251, 254
Timo, D. P., 5, 14
Timoshenko, S. P., 5, 14
Todhunter, I., 6, 14
Tracey, D. M., 53, 55, 62, 66, 73, 75, 172,
  192
Traube, I., 8, 14
Tresca, M. H., 2, 13
Tsai, Y. M., 235, 253
Turner, C. E., 189, 194
Tzou, D. Y., 221, 225, 226, 228, 229

Underwood, J., 158, 161
Unger, D. J., 274, 277

Vassilaros, M. G., 183, 193

Waddoups, M. E., 294, 304
Walker, H., 178, 192
Walter, H. G., 298, 305
Wang, N. M., 69, 70, 75
Webb, D., 234, 253
Wecharatana, M., 298, 305
Wells, A. A., 186, 193, 231, 233, 234, 251,
  252, 254
Wessel, E. T., 142, 160
Westergaard, H. M., 15, 23, 28, 71
Wheeler, O. E., 264, 276
Whitney, J. M., 294, 304
Williams, J. G., 155, 160, 273, 276
Williams, M. L., 4, 14, 15, 18, 46, 67, 71, 75,
  105, 110
Wilson, R. K., 274, 276
Wilson, W. K., 53, 73, 108, 111
Winter, G., 296, 304
Wittle, M. J., 302, 305
Wright, M. A., 294, 304
Wu, E. M., 294, 304
Wu, H. C., 157, 161
Wu, Sh.-X., 191, 194
Wu, X. R., 62, 75

Xiao, Y.-G., 191, 194
Xu, Jilin, 225, 227

Yang, W. H., 105, 110
Yao, R. F., 157, 161
Yipp, M. C., 157, 161
Yi-Zhou, Chen, 99, 110
Yoffe, E. H., 231, 234, 238, 247, 252
Yoshida, T., 62, 66, 75
Yukawa, S., 5, 14

Zacharopoulos, D. A., 296, 305
Zahoor, A., 181, 192
Zehnder, A. T., 249, 253
Zhen, S., 294, 304
Zienkiewicz, O. C., 50, 54, 73, 105, 110
Zorski, H., 162, 192

Subject index

Acoustic emission, 303
Alternating method, 47
Anisotropic materials, 291
Antiplane (tearing) mode, 17, 21, 34, 80, 84, 92
Arrest toughness, 249
ASTM standard
   E 399-81, 82, 138
   E 561-81, 151
   E 813-81, 178

Bend specimen, 138, 176
Boundary collocation method, 46
British standard BS 5762, 190
Brittle failures, 3
Bueckner's principle, 42

Caustics, 58
Chevron notch, 140
Compact tension specimen, 139
Compliance, 129
Complex potentials method, 36
COD design curve, 187
Continuous dislocations method, 49
Conventional failure criteria, 1
Concrete, 295
   cohesive model, 297
   damage model, 298
   fracture mechanics, 296
Coulomb–Mohr yield criterion, 2
Crack
   branching, 247
   central, 26, 29, 35
   edge, 45
   elliptical, 65, 212
   in anisotropic bodies, 291
   periodic array, 27
   semi-infinite, 18
   three-dimensional, 16
   two-dimensional, 61
   antiplane tearing mode, 17, 21, 34, 80, 84, 92
   opening mode, 16, 20, 24, 26, 29, 78
   sliding mode, 16, 21, 25, 28, 33, 79
Crack arrest, 249
Crack branching, 247
Crack detection methods, 301
   acoustic emission, 303
   dye penetrant, 302
   eddy currents, 302
   magnetic particles, 302
   radiography, 302
   ultrasonics, 303
Crack driving force, 117
Crack growth
   stable, 142, 180, 189, 221
   environment-assisted, 273
Crack growth resistance curve method, 147
Crack opening displacement criterion
   design curve, 187
   outline, 185
   stable crack growth, 189
   standard test, 190
Crack speed, 233
Crack stability, 142
Crack tip plastic zone
   approximate determination, 78
   numerical solutions, 105
Crack tip stress field
   dynamic, 230
   elastic, 15
   plastic, 76
Cylindrical shell, 285

Da Vinci experiments, 5
"Dead-load" loading, 124, 130
Double cantilever beam specimen, 130
Dugdale's model, 96
Dye penetrant method, 302
Dynamic crack propagation, 230
Dynamic fracture, 230
Dynamic fracture toughness, 241, 249

Eddy current method, 302
Effective crack length, 95, 96
Eigenfunction expansion method, 18
Elastic-plastic crack problems, 76
   antiplane (tearing) mode, 84, 92
   numerical solutions, 105
   opening mode, 100
   work hardening materials, 100
Elliptical crack, 120
Elliptical hole, 118
Energy balance, 112, 113

Energy release rate
  criterion, 157
  critical, 153
  dynamic, 239
  elastic, 117
  elastic-plastic, 153
Engineering applications, 278
Environment-assisted fracture, 255, 273
Experimental methods, 55

Failure criteria, 1
Fatigue crack propagation laws, 257
Fatigue fracture, 255
  mixed mode, 266
  nonlinear analysis, 270
  variable amplitude load, 262
Fatigue life, 261
Fiber-reinforced composites, 290
  anisotropic model, 291
  discrete models, 294
Finite element method, 50
"Fixed grips" loading, 123, 129
Fracture
  slant, 135
  square, 135
Fracture criterion
  crack opening displacement, 185
  stress intensity factor, 132
  J-integral, 170
  strain energy density, 195
  strain energy release rate, 132
Fracture mechanics design philosophy, 279
Fracture toughness
  definition, 12
  dynamic, 241, 249
  plane strain, 135
  plane stress, 133
  thickness dependence, 133

Griffith theory, 5, 116
Green's function method, 42

HRR solution, 100
Hilbert–Riemann problem, 38

Impact loads, 241
Irwin's model, 93
Irwin–Orowan theory, 152
Integral transforms method, 48

J-integral, 162
  definition, 164
  experimental determination, 173
  for notches and cracks, 165
  fracture criterion, 170
  mixed-mode crack growth, 183
  multiple-specimen method, 175
  relation between J and G, 168
  single specimen method, 176
  stable crack growth, 180

K, see stress intensity factor
Kirchhoff plate theory, 67

Linear elastic stress field in cracked bodies, 15

Magnetic particles method, 302
Maximum stress criterion, 154
Mises yield criterion, 2
Mixed mode crack growth, 154, 183
Mott's model, 231

Nondestructive evaluation, 302, 303
Numerical methods, 41, 105

Opening mode, 16, 20, 24, 26, 29, 78

Paris law, 260
Path independent integrals, 163
Photoelasticity, 55
Plate (bending) crack problems, 66
Propagating crack
  stress field, 234

Radiography, 302
Reissner plate theory, 69
Resistance curve, 147
Rotating disk, 281

Shell crack problems, 71
Singular stress fields
  antiplane mode, 34
  opening mode, 29
  sliding mode, 33
Slanted crack in thickness direction, 282
Sliding mode, 16, 21, 25, 28, 33, 79
Small scale yielding approximation, 76, 84
Strain energy density theory, 195
  basic hypothesis, 201
  bending of cracked plates, 216
  bodies without pre-existing cracks, 223
  criteria based on energy density, 225
  development of crack profiles, 222
  ductile fracture, 219
  energy dissipation, 220
  fatigue nonlinear analysis, 270
  inclined crack under compression, 210
  inclined crack under tension, 209
  opening-mode, 206
  resistance curves, 221
  sliding-mode, 207
  three-dimensional crack problems, 212
  two-dimensional crack problems, 203
  uniaxial extension of an inclined crack, 205
Standards
  ASTM E399-81, 82, 138
  ASTM E561-81, 151
  ASTM E813-81, 178
  British BS 5762, 190

Subject index

Stress corrosion cracking, 273
Stress intensity factor
  antiplane mode, 23, 34
  bending plates, 68, 70
  compact tension specimen, 139
  critical value, 132, 137
  definition, 21
  double cantilever beam specimen, 130
  dynamic, 251
  experimental determination, 55
  fracture criterion, 132
  numerical determination, 41
  opening mode, 20, 29
  periodic arrays of cracks, 28
  sliding mode, 21, 33
  tearing mode, 23, 34
  three-dimensional cracks, 65
  three-point bend specimen, 138

Tearing modulus, 181
Thickness effect, 81, 133
Three-point bend specimen, 138
Toughness, see fracture toughness
Tresca yield criterion, 2
Three-dimensional crack problems, 61, 212

Ultrasonics, 303

Velocity
  crack propagation, 232
Volume strain energy density, 197

Westergaard method, 23
Weight function method, 42
Waves interacting with cracks, 244

ENGINEERING APPLICATION OF FRACTURE MECHANICS

Editor-in-Chief: George C. Sih

---

1. G.C. Sih and L. Faria (eds.): *Fracture Mechanics Methodology*. Evaluation of Structural Components Integrity. 1984 ISBN 90-247-2941-6
2. E.E. Gdoutos: *Problems of Mixed Mode Crack Propagation*. 1984
ISBN 90-247-3055-4
3. A Carpinteri and A.R. Ingraffea (eds.): *Fracture Mechanics of Concrete*. Material Characterization and Testing. 1984 ISBN 90-247-2959-9
4. G.C. Sih and A. DiTommaso (eds): *Fracture Mechanics of Concrete*. Structural Application and Numerical Calculation. 1985
ISBN 90-247-2960-2
5. A. Carpinteri: *Mechanical Damage and Crack Growth in Concrete*. Plastic Collapse to Brittle Fracture. 1986 ISBN 90-247-3233-6
6. J.W. Provan (ed.): *Probabilistic Fracture Mechanics and Reliability*. 1987
ISBN 90-247-3334-0
7. A.A. Baker and R. Jones (eds.): *Bonded Repair of Aircraft Structures*. 1987
ISBN 90-247-3606-4
8. J.T. Pindera and M.-J. Pindera: *Isodyne Stress Analysis*. 1989
ISBN 0-7923-0269-9
9. G.C. Sih and E.E. Gdoutos (eds.): *Mechanics and Physics of Energy Density*. Characterization of Material – Structive Behaviour with and without Damage. (forthcoming) ISBN 0-7923-0604-X
10. E.E. Gdoutos: *Fracture Mechanics Criteria and Applictions*. 1990
ISBN 0-7923-0605-8
11. G.C. Sih: *Mechanics of Fracture Initiation and Propagation*. Surface and Volume Energy Density Applied as Failure Criterion. (forthcoming)
ISBN 0-7923-0877-8

KLUWER ACADEMIC PUBLISHERS – DORDRECHT / BOSTON / LONDON